PRIMATE SOCIAL SYSTEMS

PRIMATE SOCIAL SYSTEMS

ROBIN I.M. DUNBAR

Department of Zoology
University of Liverpool

Comstock Publishing Associates
a division of Cornell University Press
Ithaca, New York

© 1988 Robin Dunbar

All rights reserved. Except for brief quotations in a review, this book, or parts thereof, must not be reproduced in any form without permission in writing from the publisher. For information address Cornell University Press, 124 Roberts Place, Ithaca, New York 14851.

First published 1988 by Cornell University Press

Library of Congress Cataloging-in-Publication Data

Dunbar, R.I.M. (Robin Ian MacDonald), 1947-
    Primate social systems.

    Bibliography: p.
    Includes indexes.
    1. Primates — Behavior. 2. Social behavior in animals.
3. Mammals — Behavior. I. Title.
QL737.P9D76 1987   599.8′0451   87-47595
ISBN 0-8014-2087-3
ISBN 0-8014-9412-5 (pbk.)

Printed in Great Britain

Contents

Preface

This book grew from small beginnings as I began to find unexpected patterns emerging from the data in the literature. The more I thought about the way in which primate social systems worked, the more interesting things turned out to be. I am conscious that, at times, this has introduced a certain amount of complexity into the text. I make no apologies for that: what we are dealing with is a complex subject, the product of evolutionary forces interacting with very sophisticated minds. None the less, I have done my best to explain everything as clearly as I can in order to make the book accessible to as wide an audience as possible.

I have laid a heavy emphasis in this book on the use of simple graphical and mathematical models. Their sophistication, however, is not great and does not assume more than a knowledge of elementary probability theory. Since their role will inevitably be misunderstood, I take this opportunity to stress that their function is essentially heuristic rather than explanatory: they are designed to focus our attention on the key issues so as to point out the directions for further research. A model is only as good as the questions it prompts us to ask. For those whose natural inclination is to dismiss modelling out of hand, I can only point to the precision that their use can offer us in terms of hypothesis-testing.

Inevitably with a book like this, its final form owes a great deal to many different people, most of whom contributed to it inadvertently. The manuscript was produced while I was supported by a University Research Fellowship from the University of Liverpool, and I am grateful to the Department of Zoology for providing me with the resources necessary for undertaking a project like this. The text itself benefited from discussions with both students and colleagues, in particular Geoff Parker. I am especially indebted to John Lazarus and Sandy Harcourt for their detailed comments on the manuscript, as well as to Robert Hinde and Saroj Datta for their comments on specific chapters. Many other individuals generously provided me with access to unpublished data or drew my attention to papers that I would otherwise not have seen: in particular, I thank Jeanne Altmann, Dick Byrne, Caroline Harcourt, Julie Johnson, Bill McGrew, Thelma Rowell, Robert Seyfarth, Joan Silk and Andy Whiten. I am also very grateful to those who provided me with unpublished data on their baboon study populations, in particular Connie Anderson, Tony Collins, Juliet Oliver, Dennis Rasmussen, Hans Sigg, Alex Stolba and, of course, Martin Sharman who worked with me on the analysis of these data. Finally, this book owes a great deal to the influence of John Crook, Hans Kummer and Thelma Rowell: in many ways, it stands as a tribute to their influence on the field.

Robin I.M. Dunbar

Chapter 1
Primates and their Societies

Primates are among the most intensely social of all animals. This sociality forms an integral part of each individual's attempts to survive and reproduce successfully in a world that is not always conducive to successful survival and reproduction. So it is that when we observe one monkey approach and groom another, we do not see an isolated event occurring in a social vacuum. Rather, it is the end-product of a long series of interactions that can be traced back through those two individuals' past histories. That particular interaction is just one of a sequence of instances in which grooming is solicited and given, of requests that are granted or occasionally declined.

Prompted by such knowledge, the thoughtful observer will want to know what it is about grooming that makes it so important and widespread a component of primate social life. Yet, even in asking such an innocuous question, we raise issues that probe deeply into the very heart of primate biology. For time spent grooming is time that the animal cannot devote to biologically important activities like feeding and caring for dependent young. Noting that time spent grooming with one individual is time that cannot be spent grooming with another, we are prompted to probe further and ask why an animal chooses the particular social partners that it does. These, in turn, raise questions about the animals' ecological adaptations, about the structure of their social groups and the role that these play in the animals' lives. As we continue to ponder these questions, we find ourselves moving outwards beyond the particular animals we happen to be watching to consider other species, to ask why grooming should differ in both its frequency and its uses in different species, and to wonder how these differences came to be during the course of evolutionary history.

What seemed at first sight to be an isolated event, a simple interaction between two animals, spawns questions that oblige us to look far beyond the immediate context in which two animals are interacting to explore not only the intricacies of the social system in which they live, but also the ecological context in which they are embedded. It is this complexity that makes social behaviour so challenging a topic to study, and nowhere is this more true than in the primates, the most overtly social of all animal species.

1

This book is about the social behaviour of primates. But my intention is not simply to give a succinct summary of what we know about primate social behaviour. I want to go beyond that to show why, in trying to understand their behaviour, we need to broaden our horizons to include many other aspects of their biology. For in such species, social behaviour is not simply a matter of evoking a response to gain some immediate reward in the way that a rat presses the lever in a Skinner box. Rather, social interactions are part of a continuous process of developing relationships that animals negotiate and service from the day they are born to the day they die. To explain what is happening in any given interaction, it is not enough just to describe the events that occur; nor is it enough to identify the physiological mechanisms that underlie them. We also need to see those interactions in the context of the animals' personal histories, to ask why it is that they behave in the way they do, to identify the social goals they are trying to achieve and to determine how and why such behaviour has evolved.

There are several different senses in which we can be said to explain behaviour, and a failure to keep these clearly separated has often generated confusion in the behavioural sciences. I shall begin, then, by saying a little more about the kinds of explanation we can offer for the behaviour we see. Once this has been clarified, I shall devote the rest of this chapter to introducing the primates and to establishing a general framework for studying their societies. I shall go on in the second chapter to establish the essential theoretical background that we shall need in order to be able to explore the functional biology of social systems. By functional here, I mean not only evolutionary function but also the nature of the interactive processes that integrate the various components of a species' biology.

Chapters 3 to 7 deal with the system constraints that an animal has to cope with in organising its day-to-day social relationships, namely the problems of survival posed by the environment and the effects that these factors have on both demographic processes (notably birth and death rates) and on the way an animal allocates its time to different activities. These are fundamental biological problems that an animal has to solve in order to be able to survive well enough to have any kind of social life. In the last of this group of chapters, I consider the factors that determine grouping patterns in primates.

The next group of chapters (Chapters 8 to 11) are concerned with the social behaviour of individual animals that are constrained to live within a particular socio-ecological system. The emphasis here will be on the ways in which individuals create and exploit options in the form of alternative mating, rearing and social strategies. Finally, in the last two chapters, I try to pull all these disparate strands together again by examining the evolution of social structure in more detail. I shall argue that each species' social system is the product of a set of general principles working themselves out in a particular ecological context from the standpoint of a unique evolutionary history. To suppose that we can understand a species' behaviour by reference to those general principles alone is to ignore the most salient feature of an animal's existence, namely the fact that it is deeply embedded into a biological system that itself creates many of the problems that the animals seek to solve by what they do. My approach here will be to concentrate on a few well studied species in order to illustrate how we can use the ideas developed in this book to help us understand how and why primate social evolution has taken the course it has.

Asking the Right Questions

In a seminal paper on the nature of ethology, Tinbergen (1963) pointed out that as simple a question as asking why an animal behaves in a certain way in fact conceals at least four radically different, though related, kinds of answer. In asking why one monkey grooms another, we might be asking for an explanation in terms of: (1) the motivations or other physiological or behavioural factors that prompted it to groom another individual (a question about *proximate causes*); (2) the experiences it has had during its lifetime that prompt it to groom in a given way or to groom only certain individuals (a question about ontogeny or *development*); (3) the purpose being served by its grooming another individual (a question about *function*); and (4) the sequence of changes in behaviour that led to the evolution of grooming in that species (a question about *evolutionary history*).

At a certain level, we can legitimately deal with each of these types of question in isolation. Our answer to any one of them is not necessarily dependent on our answers to any of the others. From a functional point of view, for example, it may not matter what motivations prompt a monkey to groom so long as it does so. Nor need it matter what functions grooming serves if we only want to explain how grooming develops in the individual as it matures. None the less, there is a sense in which to answer only one of these questions while ignoring the others is to provide only a partial answer. In particular, when we offer an explanation of, say, grooming in terms of its proximate causation, development or evolutionary history, we unavoidably beg the question as to why the animals bother to perform the behaviour at all. In other words, hidden within any question about the origins or causes of behaviour (or any other morphological structure or process) is an implicit question about function.

Put simply, this is only to say that a physiological process does not exist in a vacuum. It is part of an organism, and it must serve some function within that organism's biology. That knowledge alone should prompt us to ask questions about the function of behaviour. Questions about function inevitably lead us back to evolutionary history, since all biological systems are a consequence of evolution. Although the study of function and evolutionary history is often very much more difficult than the study of proximate mechanisms, we should not be deterred from making the effort to grapple with these issues, for doing so will force us to ask deeper questions about the animals' biology.

There is, in addition, a more important reason for considering function when we come to examine social behaviour. Behaviour in general is primarily concerned with an animal's attempts to organise its world in a way that makes its continued survival as certain as possible. By the same token, social behaviour is geared to organising fellow members of the species so that they will behave in a way that most effectively allows that animal to survive and reproduce. Communication, in fact, is designed to manipulate other individuals by persuading them to change their behaviour (see Altmann 1967, Krebs and Dawkins 1984). Thus, function is a particularly prominent consideration when we come to discuss social behaviour. Indeed, we cannot explain social behaviour without recourse to functional considerations, for what makes social behaviour interesting is the *context* in which it occurs — in other words, the function it serves in an animal's overall life strategy. We cannot explain an animal's behaviour in this sense merely by describing the way in which a stimulus elicits a response or by detailing the physiological processes that underlie it. We can only understand what is happening when we *interpret* the animal's behaviour in macroscopic terms against the background formed by the network of relationships of which it is a part.

From a purely methodological point of view, Tinbergen's four senses of 'Why' in fact provide a valuable heuristic device. If we insist that an explanation in terms of any one of the questions has to be accompanied by explanations in terms of all the others, we greatly strengthen our interpretation of what is happening at that level simply by virtue of the coherence that such interlinked accounts confer. This is particularly important for functional and evolutionary explanations. If we can establish the existence of a physiological process that could produce a presumed functional consequence, then it necessarily increases the plausibility of that functional relationship. For example, the behaviour of animals when they are being groomed is such as to suggest that they find it relaxing: in some species, animals actually go to sleep while being groomed. This might suggest that animals solicit grooming in order to reduce the stresses and tensions that develop during the day. This functional explanation would obviously receive considerable support if we could also show that (a) animals did show a tendency for signs of stress (e.g. levels of adrenaline in the blood) to increase with time, and (b) these were reduced following grooming. Note once again how incomplete a purely physiological explanation would be: it begs the question as to why animals should need to reduce tension in this way and why this tension should build up in the first place. By ensuring that our answers hang together, we make it less likely that we will be led astray by a plausible (but quite incorrect) answer to any one of the four questions.

The fact that these levels of explanation can and do interact with each other raises another important consideration to which allusion has already been made, namely the fact that an animal is an integrated system. Its elements have to function effectively together or it will not survive to reproduce. There are two lessons to be learned from this.

One is that in order to understand an organism's behaviour we need to understand how that behaviour affects the rest of the animal's biology. That biological system, note, includes not just its physiological machinery, but also those aspects of its ecology on which survival depends *and* the other individuals with whom it interacts. Everything an animal does, it does within an ecological and a demographic context, and that will have important implications for many other aspects of its own behaviour. We cannot, in other words, study particular kinds of behaviour in isolation because if we do we may not be aware of the limitations and costs imposed on an animal's freedom of movement by other features of the biological system within which it has to operate. If an animal begins to groom another, for example, we must ask not only what it has given up in order to do so, but also why it should consider grooming that individual to be more important than these other options. Although in practice we cannot, of course, collect data on every relevant aspect of an animal's biology, we do need to be aware that concentrating on one issue to the exclusion of all others is to risk doing no more than tinkering with Nature. It is a bit like trying to understand how a car works simply by fiddling with the carburettor.

The second point is that studying whole systems in this way is inevitably a demanding business and we cannot expect to come away with simple answers. In any system, activity in one component inevitably ramifies throughout the system, which in turn will feed back on to the component in question. In many such cases, we need to resort to mathematical modelling in order to be able to work through all the complexities involved. We generally find it difficult to envisage more than two dimensions at once, so that complex cause–effect relationships that involve many variables are difficult to conceptualise. Only by converting them into models where equations or computers do the work for us are we likely to be able to explore these complex relationships without making mistakes.

The Primate Heritage

The primates emerge as a distinct group from the mists of time about 70 million years ago during the Cretaceous era. Their origins lie within the Insectivores, a family of small, relatively specialised, terrestrial and largely nocturnal animals that include species like shrews, tenrecs and the hedgehog. From these humble origins, the ancestral primates eventually diversified into a wide range of new ecological niches that include leaf and fruit eaters as well as insectivores, and arboreal as well as terrestrial ways of life. Primates, today, are found throughout the tropical zones of South America, Africa and

Table 1.1. The main primate families and their characteristics

Family	Living species	Common names of some member species	Weight[a] (kg)	Habitus[b]	Group size	Group type[c]
A. *Strepsirhini*						
Lemuridae	9	lemur	0.6–2.5	A,D	3–12	multimale
Cheirogaleidae	6	dwarf lemur, mouse lemur	0.1	A,N	1–2	solitary
Indriidae	4	indri, sifaka	3.5–12.5	A,D	2–4	monogamous
Daubentoniidae	1	aye-aye	0.1	A,N	1	solitary
Lorisidae: Lorisinae	4	loris, potto	0.2–1.2	A,N	1	solitary
Galaginae	5	galago	0.1–1.0	A,N	1	solitary
B. *Haplorhini*						
Tarsiidae	3	tarsier	0.1	A,N	1	solitary
Cebidae: Cebinae	6	capuchin, squirrel monkey	0.6–2.3	A,D	7–33	multimale
Callitrichinae	17	marmoset, tamarin	0.3–0.6	A,D	3–5	monogamous
Atelidae: Atelinae	13	howler, spider monkey	5.7–6.8	A,D	15–25	multimale
Pithecinae	11	titi, uakari, saki, night monkey	0.7–1.4	A,D	1–5	various
Cercopithecidae:						
Cercopithecinae	44	macaque, baboon, gelada, guenon	1.1–17.0	T/A,D	4–150	various
Colobinae	34	colobus, langur	3.6–10.5	A/T,D	5–50	various
Hominidae: Hylobatinae	7	gibbon, siamang	5.0–10.5	A,D	3–5	monogamous
Ponginae	4	orang utan, chimpanzee, gorilla	33.0–90.0	A/T,D	1–30	various

Sources: Taxonomic classification follows Szalay and Delson (1979); behavioural ecology based on Clutton-Brock and Harvey (1977).

[a] Range in mean female body weight for individual species.

[b] A, arboreal; T, terrestrial; N, nocturnal; D, diurnal; some species are exceptional within their families.

[c] Some species are exceptional to the general pattern typical of their family.

Asia. In previous times, they also occurred in Europe, but aside from ourselves no species of primates now survives there in the wild. Within those continental areas where they do occur, primates occupy all types of habitat, from climax rainforest and moorland on high mountain ranges to open savannah and desert habitats.

Primates are characterised, as a group, by a number of anatomical features that distinguish them from other groups of mammals. The most important of these are (1) a shortened snout with a corresponding reduction in the sense of smell, (2) an increased dependence on sight and, in particular, on binocular vision, (3) unspecialised hands and feet that retain the primitive five-digit pattern, (4) nails instead of claws on all digits, (5) a thumb that is more or less opposable, so enabling the animals to grip and manipulate small objects, and (6) relatively large brains for their body size. While we can argue about the relative importance of the first five characters and their ecological significance, it is the sixth that is of crucial significance in the present context, for it is the greatly improved cognitive capacity provided by so large a brain that makes it possible for primates to engage in the kinds of sophisticated social interactions that form the core of this book.

The primates divide fairly neatly into two major groups, the prosimians (the lemuroids and lorisoids) and the anthropoid primates (monkeys and apes). Each of these in turn consists of a number of distinct families, themselves subdivided further into finer groupings. Table 1.1 lists the main subdivisions and gives examples of representative species, together with an indication of these species' ecological strategies and social systems. Following the conventional view, I have categorised social systems simply in terms of their demographic structure and temporal stability. One of the themes I shall want to draw out in the chapters that follow, however, is that this classification grossly underestimates the complexity that really exists in the societies of individual species of primates. I shall have more to say on this in the next section, but for present purposes this classification suffices to provide us with a broad picture of what we are trying to explain.

The prosimians, or more strictly the Strepsirhine primates, are less a group of primitive primates than a diverse group of highly specialised species that have undergone considerable evolution from the original ancestral primates of the Cretaceous. Indeed, if we want to gain some idea of what these ancestral primates were like, the tarsiers (members of the so-called 'higher' primates) are probably the closest we can get. The prosimians consist of two main groups, the lorisoids and lemuroids. The latter represent the end-product of the dramatic flowering of the primitive primate stock that occurred on the island of Madagascar when this became separated from the African mainland some 60 million years ago. The lorisoids, on the other hand, are the descendants of the original primate stock that remained on the African mainland.

The ring-tailed lemur is, in many ways, typical of the larger members of the lemuroid group. It is about 2.5 kg in body weight and inhabits a variety of habitats from gallery forest along watercourses to dry scrub in southern Madagascar. It is diurnal and, although it commonly travels on the ground, it feeds predominantly in trees on a variety of leaves and fruits. It typically lives in groups of 12–25 individuals, with each group occupying a territory of about 6 ha. Groups generally maintain control over their territories by engaging in 'scent fights' by using their bushy tails to waft scent at each other. These lemurs stand in marked contrast to the smaller members of the prosimian group which tend to be nocturnal, insectivorous and semi-solitary. Examples include lemuroids like the diminutive aye-aye and the mouse lemur, and lorisoids like the galagos (or bushbabies) and pottos. These tend to weigh about 1 kg and occupy individual territories that are actively defended against members of the same sex. Males tend to occupy larger territories than females do, and each male's territory will usually include the territories of several females.

Among the higher primates, three main groups can be distinguished: the New World monkeys, the Old World monkeys and the apes. These differ markedly in a number of key anatomical characteristics, including the detailed structure of the skull and teeth. In addition, the New World monkeys are confined to South and Central America, whereas the Old World monkeys and apes are distributed widely throughout Asia, Africa and (at least in pre-historic times) Europe. Unlike the prosimians where nocturnal habits are common, the higher primates are, with the exception of only two species, exclusively diurnal.

The New World monkeys, which separated from the main primate stock in the Old World during the Eocene 35–40 million years ago, consist of two main groups, the callitrichids (marmosets and tamarins) and the larger cebids (of which capuchin and howler monkeys are typical examples). None of the callitrichids weighs more than 0.6 kg and many weigh considerably less. The common marmoset, a typical member of this group, is arboreal, and tends to live in small family groups containing a single male and a single female and their dependent young; it feeds predominantly on a variety of small high-energy food sources (fruits, gum, nectar and insects) in territories of about 1 ha in size. Capuchins, on the other hand, are larger (around 3 kg), more strictly frugivorous and live in larger groups (typically 10–20 individuals) in very much larger ranges (80–150 ha). The cebids are more variable than the callitrichids in both their size and their social systems. Some, like the titis, are monogamous, while others, like the spider monkeys, live in loosely structured communities which rarely travel in groups of more than two individuals at a time.

The Old World monkeys also consist of two main groupings, the leaf-eating colobines (the African colobus monkeys and the Asian langurs) and the cer-copithecine monkeys (the macaques of Asia and the baboons and guenons of Africa). These vary considerably both in ecological adaptation and in their social systems, but by and large the forest-based arboreal colobines and guenons live in small groups (typically 10–20 individuals, but groups of up to 100 occur in some species) in territories that are normally around 15 ha in size. The baboons and macaques, on the other hand, live in larger groups (usually 30–80 in size) that often contain many adult males. Among the more open-country species, ranging areas can be up to 15 km² in area, often with considerable overlap in the ranges of neighbouring groups. Such large ranges are rarely defended, unlike the smaller territories of the forest monkeys, but baboon and macaque groups may none the less be antagonistic towards each other when they do meet.

The apes consist of just a handful of species that differ from the monkeys in lacking an external tail and in their larger size and greater intelligence. The lesser apes or gibbons are relatively small by ape standards (about 5 kg in weight), and live in monogamous pairs in large defended territories (typically around 20–40 ha in size). In contrast, the great apes (orang utan, gorilla and two species of chimpanzee) are a much less homogeneous group other than in the fact that they are the largest of all living primates (with body weights varying from 30–50 kg in the chimpanzee to 130 kg in the gorilla) and the fact that they are all essentially frugivorous rather than leaf-eating. Their social systems vary from the semi-solitary (the orang) through the loosely associated multimale communities of the chimpanzees to the small but cohesive one-male groups of the gorilla.

Such vast differences in size inevitably have profound implications at an ecological level (Schmidt-Nielsen 1984), which in turn have significant consequences for those demographic variables that influence the context in which social behaviour takes place. In the chapters that follow, we shall explore some of these consequences in some detail. But for present purposes it is important to note that the primates consist of a diverse group of animals whose ecological niches and social systems cover almost the full array of possibilities. Such diversity will not easily by explained by simplistic theories, and part of our task will be to try to account for such variability and its evolution.

Primate Social Systems

The conventional view has been that primates live in groups that can be described in terms of a few key structural dimensions. Of these, mating system and spatio-temporal cohesiveness have been particularly important (Gray 1985). Attention has tended to focus on just four different kinds of grouping pattern: semi-solitary (typified by the galagos and the orang utan), monogamous pairs (as in marmosets and gibbons), one-male groups (e.g. colobus monkeys and many forest guenons) and multimale groups (as in baboons, macaques and howler monkeys).

The difficulty with this classification is that considerable distortion is often necessary to make every species conform to it. Gelada and hamadryas baboons, for example, live in one-male units that associate together in stable bands of 60–100 individuals (Kawai *et al.* 1983, Sigg *et al.* 1982): their social systems can be classified equally well as one-male-grouped or multimale-grouped according to inclination. Other species such as chimpanzees and spider monkeys live in stable communities that consist of a number of mature males and females, but these are rarely seen in groups of more than two or three individuals (Klein and Klein 1977, Wrangham 1977, Izawa *et al.* 1979): are their social systems multimale or semi-solitary?

Over and above these apparently anomalous species, however, the traditional classification seriously underestimates the flexibility and variation found within the social systems of those species that ostensibly fit neatly into it. Species like the Guinea baboon and the pigtail and longtailed macaques live in multimale groups, but these continually subdivide into smaller units that forage independently (Byrne 1981, Sharman 1981, van Schaik *et al.* 1983b, Caldecott 1986a). Similar dispersal into subgroups has even been noted in some populations of common baboons, a species group whose multimale grouping pattern has generally been regarded as archetypal (e.g. Aldrich-Blake *et al.* 1971, Anderson 1981). In other cases, populations of the same species may exhibit a mixture of group types (e.g. both one-male and multimale groups in the same langur and colobus populations: Yoshiba 1968, Dunbar and Dunbar 1976).

The classical view was based on the premiss that groups have formal boundaries across which interactions are rare and within which relationships are fairly homogenous. This assumption has been undermined by recent research demonstrating quite unequivocally that not only are group boundaries often very much more permeable than had been supposed (e.g. colobus: Oates 1977b; guenons: Cords 1984, Cords *et al.* 1986; patas: Chism and Rowell 1986), but that the relationships between the various groups in a local population are far from homogenous. Not only may individuals know the identities of animals living in the groups around them (see Cheney and Seyfarth 1983), but particular groups are also more likely to exchange members with each other than they are with other groups (vervets: Henzi and Lucas 1980, Cheney and Seyfarth 1983; baboons: D.R. Rasmussen 1981a; macaques: Colvin 1983a). In other cases, particular groups are more tolerant of each other's proximity and are more prepared to share sleeping sites (e.g. drills: Gartlan 1970; mandrills: Jouventin 1975, Hoshino *et al.* 1984; hamadryas baboons: Kummer 1968, Sigg and Stolba 1981). Moreover, the groups themselves are far from being homogenous in terms of the relationships between their members. The larger groups, in particular, often consist of subgroups or cliques of individuals who interact with each other and support each other far more than they do with other group members; indeed, their relationships with these other group members may be more regularly antagonistic than their relationships with members of other neighbouring groups. Such structuring of relationships within groups has been documented in macaques (Sade 1972, Grewal 1980), baboons (Seyfarth 1976), gelada (Dunbar 1979a) and vervets (Seyfarth and Cheney 1984).

Part of the problem is that the traditional conception of societies has treated sociality as being synonymous with group-living. Indeed, there has been a covert assumption underlying most discussions of animal social systems to the effect that social complexity is roughly commensurate with group size: the large multimale groups of baboons are presumed to be more complex than the monogamous groups of marmosets, which in turn are more complex than the semi-solitary social systems of galagos. Such species as the orang utan are then viewed as socially degenerate (Mackinnon 1974).

But this is a very simplistic conception of what is involved in a social exist-ence, at least in mammals as advanced as primates. In these species, groups exist through time not because animals are arbitrarily forced to associate, as, for example, dungflies do on a cow dropping, but because of the relationships that they have with each other. Those relationships exist independently of whether or not the animals actually live in physical proximity in the same group. Just because animals live semi-solitary lives, it does not follow that their relationships with each other are not as complex as those in species that live in formal groups. A case in point is provided by the chimpanzee. At one stage in the continuing saga of our understanding of their complex social life, it was thought that they lacked any coherent relationships other than those between a mother and her offspring. More intensive study over a longer period has revealed that even though chimps typically travel in groups of only two or three animals, the relationships within a community are in fact highly structured with a well defined dominance hierarchy among the males and a social system of very great dynamic complexity (see, for example, Simpson 1973, de Waal 1982).

The idea that group-living is all there is to sociality is ultimately part of the intellectual baggage we inherited from behaviourism. The behaviourists argued (and still argue) that we can only legitimately study observable phenomena (i.e. things like behaviour and grouping patterns). Since the rela-tionships that animals have with each other are essentially cognitive (or mental) phenomena, they are not directly susceptible to observation by con-ventional scientific means, so they ought not to form a legitimate part of any science of behaviour. Underlying this view is an implicit assumption that animals are in fact incapable of maintaining knowledge about relationships through time and that they can only express what relationships they have by remaining in each other's physical proximity. Not only is this a very naive view of animals' cognitive capacities (see Humphrey 1976, Griffin 1981, 1984), but it also seriously underestimates the complexity that the social inter-actions of primates can often aspire to (see Kummer 1982). We need, in effect, to develop new ways of describing primate societies that break the old moulds within which our thinking has been constrained.

The past decade has seen the gradual emergence of a new approach, partly prompted by the shift of emphasis from groups to individuals brought about by the rise of sociobiology. Its beginnings can perhaps be traced back to Hinde (1976) who pointed out that it is more appropriate to view social systems in terms of their constituent relationships. He argued that social systems are no more than the patterning and quality of relationships between individuals, and that it is essentially the observer, and not the animals, who creates the groups by abstracting an overall picture from the network of relationships that are the main concern of the individual animals. These relationships in turn are no more than the patterning and quality of the interactions involving those individuals.

We can use Hinde's scheme as a basis for suggesting that a local population consists of a large set of animals who have relationships of varying intensity with each other. Certain individuals will associate and interact frequently, others only rarely, and in yet other cases the relationships may be purely antagonistic. Some of these relationships may be exploitative, others mutualistic, yet others neutral. Where the sets of relationships of several individuals coincide and are reciprocated, a formal group emerges as a stable cohesive unit. Such groupings can be created at many different levels within the same population at the same time. Figure 1.1 illustrates this for the gelada: the patterns of association between individual animals reveal a hierarchically organised series of clusters at five levels. These clusterings are sufficiently distinctive in their cohesion in time and space that we can define them as levels of grouping (listed on the right-hand side of the diagram). Not all these grouping levels are necessarily obvious in the field to an observer, but the use of more sophisticated analyses of appropriate kinds of observational data will usually reveal them where they exist.

The gelada pattern is not, of course, likely to be universally valid, since, as I shall argue in the next chapter, its particular structure is the outcome of the ecological and social problems that individual gelada need to solve in order to survive and reproduce successfully. None the less, the gelada are by no means unique in having so complexly structured a social system. An analogous series of hierarchically structured grouping patterns has been described for the hamadryas baboon (Kummer 1968, Sigg et al. 1982). In this case, definable groupings occur in the form of one-male units, clans, bands and sleeping troops. These groupings are similar to those of the gelada only in respect of the functional significance of one-male units and bands. Otherwise, they differ significantly from those of the gelada in size, composition, structure, dynamics and function (Dunbar 1983b). Moreover, each hierarchy contains groupings that do not exist even in analogous form in the other: hamadryas one-male units are not structured into matrilineally related coalitions like those of the gelada, and gelada bands lack all trace of the patrilineally organised clans so characteristic of the hamadryas.

Comparable complexity can be seen in other species besides the hamadryas

Figure 1.1. Notional association patterns among individual animals in a gelada population, illustrating the tendency for relationships to cluster at certain definable levels. Terms used to refer to these levels are listed on the right-hand side.

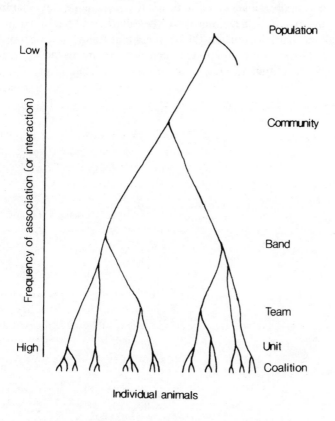

and the gelada. Relationships within and between groups of both common baboons and vervet monkeys also form series of clusters at levels that have been referred to as coalitions, troops and communities (see D.R. Rasmussen 1981a, Cheney and Seyfarth 1983). Similarly, Garber *et al.* (1984) discuss the social system of moustached tamarins in terms of a two-tier structure involving unstable breeding pairs formed temporarily within an essentially multimale community. Nor, in fact, are such hierarchically structured social systems unique to the primates. They have been shown to exist in prairie dogs (King 1955), elephants (Moss and Poole 1983) and at least some birds (e.g. bee-eaters: Hegner *et al.* 1982).

This emerging consensus emphasises the fact that the social systems of primates are very much more complex than we have previously supposed. This complexity itself not only raises interesting questions about evolutionary function, but also encourages us to ask questions about both the proximate factors that underlie it and the social and cognitive processes that animals require to generate and maintain such a diversity of relationships. I shall argue in the next chapter that the relationships that give rise to these structures are essentially alliances of a political nature aimed at enabling the animals concerned to achieve more effective solutions to particular problems of survival and reproduction. In the chapters that follow, our task will be to elucidate and explore the interactions between proximate factors and functional consequences that give rise to this diversity and complexity.

Chapter 2
Theory of Reproductive Strategies

Progress in science (by which I mean the steady development in our ability to explain and predict what happens in nature) depends on our having theories. Without a theory, we can only describe what we see; we cannot explain why things are as they are because facts only acquire significance when they are interpreted within a particular theoretical structure. (Some philosophers even doubt whether it is possible to describe events in the world without a theory, but there is no need to go this far in the present context: for lucid introductions to the philosophy of science, see Chalmers 1978 and Gale 1979.) A theoretical framework allows us to formulate questions about the phenomena we see, thus allowing us to interrogate nature directly by formulating and testing hypotheses about the causes and consequences of these events. Obviously, the more precise the questions we ask, the more likely we are to find the right answer, even if the theory itself is in fact wrong: precise questions allow us to rule out incorrect hypotheses more quickly, so forcing us to examine the biology involved more closely.

In this chapter, I shall begin by setting out a general theoretical perspective on the evolution of behaviour that provides us with just such a powerful quantitative theoretical structure. I shall then outline a more specific framework within which we can study social systems more effectively. My theme will be that, in the final analysis, social behaviour is concerned with the business of reproduction. Successful reproduction, however, requires that an animal solve at least three primary sets of problems: day-to-day survival, mating and the rearing of offspring to maturity. These problems impose conflicting demands on an animal's time and energy budgets, and it is in the resolution of these conflicts of interest that social strategies come to the fore.

An Evolutionary Perspective

During the 1960s and 1970s, functional approaches to the evolution of social systems were dominated by what was then known as socio-ecology. This was founded on the twin premises that social systems could be described in terms of the demographic structure of the population (i.e. grouping patterns) and that this structure could be understood as a consequence of the way in which environmental factors determined the dispersion of animals in a habitat. Partly because of its focus on groups and partly because of the intellectual climate of the time, socio-ecological explanations tended towards an inherent group selectionism (i.e. a tendency to argue that evolution occurs for 'the good of the species'). Since the conditions under which group selection can occur are so stringent as to make the process of limited interest in promoting evolutionary change (Maynard Smith 1976), it was perhaps inevitable that socio-ecology should have suffered an eclipse following the rise of sociobiology in the later 1970s.

Sociobiology introduced an important new perspective on behavioural evolution by showing how the rigorous mathematics of population genetics could be applied to something as nebulous as behaviour. It shifted the focus of attention very firmly from the group on to the individual and the individual's personal response to the problems of survival and reproduction. This was undoubtedly both desirable and necessary, but it had the inadvertent effect of distracting attention away from the fact that what an animal can do is closely constrained by the ecological and demographic context in which it lives. In other words, socio-ecology provides the context in which sociobiological rules are worked out.

The two perspectives are, then, complementary rather than mutually exclusive, and, now that we have had time to absorb the lessons of sociobiology, it is essential that a concerted effort be made to integrate them more effectively (see also Emlen 1980). Unless we can achieve a satisfactory articulation of the two bodies of theory, progress is likely to be piecemeal and erratic precisely because in applying sociobiological principles we tend to overlook the fact that their predictions depend crucially on the context in which they are applied. Only in a few cases will the principles be so context-free that we can make a direct test without having to take other aspects of the species' biology into account.

The most important point to emerge from the new perspective offered by sociobiology is a purely methodological one. Evolution occurs when genes are propagated at different rates, thereby altering the genetic composition of future generations. The force of this is such that organisms will sooner or later evolve in such a way as to seek to maximise their contributions to their species' gene pool. Dawkins' (1976a) emphasis on 'selfish genes' provides us with a sharp reminder that, in calculating the costs and benefits of a given action, it is genes that we must use as the basis of our accounting since it is only genes that are passed on from one generation to the next to produce evolutionary change.

It is this assumption that provides us with a fine-edged tool with which to explore the workings of nature. It provides an unequivocal criterion by which we can assess the functional value of behavioural or morphological traits, so making it possible to make very precise predictions about what ought to happen under particular conditions. Such precise predictions make it more difficult for inadequate hypotheses to slip through the net. In addition, the selfish-gene paradigm provides a common currency into which we can translate the costs and benefits of any action or morphological trait in order to be able to compare their respective values directly. In this way, we can now compare phenomena which previously seemed to belong to completely different dimensions of an animal's life. This, in turn, has allowed us to see how different aspects of an animal's biology can impinge on or constrain its behavioural decisions.

Conventionally, an individual contributes to its species' gene pool by reproducing. In sexual species, this obviously means finding a mate of the opposite sex and then ensuring that the offspring so produced are successfully reared to maturity in order to make certain that they, in their turn, will be able to reproduce. In evolutionary terms, offspring that do not breed cannot influence their parents' contribution to future generations and are thus of no significant interest.

The success with which a particular gene is propagated in future generations is termed its *fitness*. Note that, strictly speaking, fitness is a property of a character (or gene), not of an individual (for a more detailed discussion of this, see Dunbar 1982a). But in studying social behaviour we are rarely concerned with the evolution of particular traits as such and are more interested in the performance of individual animals. I shall often use the term in a looser sense to refer to the fitness of an individual, but by this I will usually mean the fitness of an individual *who behaves in a certain way*: in other words, I shall be referring to the fitness of a given character or strategy.

Hamilton (1964) pointed out that, in social species, conventional measures of fitness such as lifetime reproductive success actually pool together two quite different sources of genetic replication, namely offspring born as a result of the individual's own personal efforts and offspring which it produces *only* as a result of help received from its relatives. This second component can in fact be interpreted as a contribution to the relative's fitness since it helps to propagate genes that they share by virtue of descent from a common ancestor. Hamilton then went on to show that an alternative way of calculating the fitness of a given gene was to consider an individual's impact on its relative's reproduction. He defined this new measure, which he called *inclusive fitness*, as the individual's personal fitness *stripped of all the effects* (both positive and negative) *due to the behaviour of its relatives* plus the additional genes contributed by all those relatives *as a direct result of the individual's intervention in their reproductive activities*. (For a lucid summary of Hamilton's argument, see Grafen 1984.)

Unfortunately, calculating inclusive fitness turns out to be considerably more difficult than is commonly supposed. Indeed, most textbooks (e.g.

Wilson 1975, Barash 1982) define it incorrectly, usually by assuming that all of an individual's own reproductive output counts towards its inclusive fitness. In fact, Hamilton's definition rests on the assumption that, under most conditions (see below), reproductive success and inclusive fitness are exactly equivalent because reproductive success will usually include the effects of all relatives' behaviour on the individual's reproduction. Hence, by including all of the individual's own reproductive output, we count the gains from helping behaviour twice over.

Because it is often difficult to determine exactly what proportion of a relative's reproductive output is due to the assistance of the individual concerned, Grafen (1982, 1984) has suggested that we will usually be better off simply measuring reproductive success. Where helping behaviour is strictly reciprocal (i.e. each act of assistance is directly repaid by a comparable act), this is a valuable solution. But when helping is not reciprocal, it does not solve the problem. Grafen (1982) has suggested that we can compare the reproductive success of individuals that perform the activity in question with that of individuals that do not. There remain, however, two conditions where we will not be able to do even this. One is when all members of the species perform the activity of interest, but we still want to know whether it has evolved because of the advantages it confers in terms of individual survival and reproduction (i.e. *personal fitness*) or because of the advantages conferred through kin selection on the survival and reproduction of relatives (i.e. what we might call *collateral fitness*). The other circumstance where it may be difficult to implement Grafen's suggestion is with long-lived species like primates: the sheer logistic difficulty of measuring lifetime reproductive success in a species that may live for 20 years or more leaves us with little alternative but to cost out the gains and losses that would result if an animal behaved in a given way.

In this respect, it is particularly important to assess the costs incurred when an animal pursues a given course of action. These costs come in two forms. There are the direct costs incurred by the action (e.g. injuries received) and there are indirect costs incurred by having to forgo doing something else. The second of these is often known as the *opportunity cost* or *règret*. It is important to take this into consideration when evaluating the evolutionary benefit of a given course of action because sometimes the short-term benefits of acting in a given way can be offset by longer-term costs, so that an animal might be better off in the long run by not pursuing such a course of action (see also Grafen 1980).

If we cost these out, we find, as Hamilton (1964) showed, that a gene for any altruistic behaviour (i.e. behaviour that depresses the reproductive output of the actor while benefiting that of the recipient) will spread only providing:

$$rB > C \qquad (2.1)$$

where r is the coefficient of relationship between the two individuals concerned, B is the benefit to the recipient in terms of the number of *extra* offspring gained through the altruist's help, and C is the cost to the altruist in terms of the number of offspring lost. This is known as *Hamilton's Rule*. Essentially, it says that a gene for altruism will spread only so long as the number of copies of the gene that an altruist loses from its own reproductive output by behaving in this way towards a given relative is less than the number of copies of that same gene gained through the recipient producing more offspring than it would otherwise have done.

Several points need to be noted about equation 2.1. First, it is often misunderstood as including *all* the offspring of the beneficiary (including those born before the altruistic act is performed). The term B refers *only* to those offspring born as a direct result of the altruist's intervention: those offspring that the beneficiary would have produced anyway, irrespective of whether or not the altruist had come to its aid, obviously are not part of the *net* gain from behaving in that way. Secondly, intervening on another individual's behalf necessarily incurs a cost: it might be as little as the time wasted in doing so or the energy used in fighting, or it might be as severe as injuries received or risks of death incurred. This cost, however small, can be translated into lost reproductive opportunities for the intervening animal, and these have to be set against the extra offspring gained by the recipient of the intervention. These are losses of reproduction that the intervening animal would not have incurred had it not intervened.

The third point to note is that a relative's value to an individual depends on the likelihood that the two of them share the gene in question. In diploid species (all vertebrates are diploid), each individual has two sets of chromosomes, only one of which is passed on to its offspring at each reproductive event. Consequently, the likelihood of two individuals sharing a given gene is a simple function of their relatedness because each reproductive event that separates them in the pedigree to which they belong reduces by exactly one-half the probability that a copy of a particular gene will be passed on to both of them by the same ancestor. Hence, the number of extra offspring produced by a relative as a result of an individual's intervention must be devalued by the probability that they share the gene in question, in order to determine exactly how many copies of that gene are transmitted to the next generation by such altruistic behaviour. This probability is known as their *coefficient of relationship* (usually symbolised as r).

We can calculate the coefficient of relationship between any two individuals by counting up the number of separate reproductive events that separate them in the pedigree to which they belong. Figure 2.1 shows a pedigree for a particular individual, SELF, to illustrate various kinds of relatives that it might have (siblings, parents, grandparents, aunts, cousins, etc.). The values of r between SELF and some of these relatives are given in Table 2.1. (It should be noted that this method of calulating r is in fact only approximate,

Figure 2.1. A notional pedigree illustrating the main genetic relationships that an individual (SELF) has to other members of its extended family

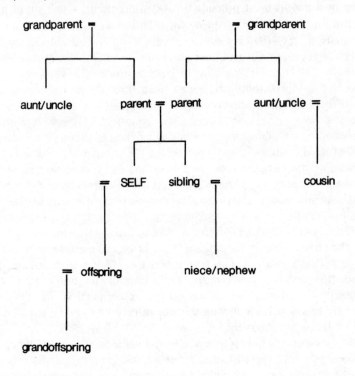

Table 2.1. Coefficients of relationship between various relatives in relation to whether or not siblings have the same father

Relationship	Coefficient of relationship (r)	
	Same father	Different father
Parent–offspring	0.5	0.5
Brothers, sisters	0.5	0.25
Aunt/uncle to niece/nephew	0.25	0.125
Cousins	0.125	0.063
Grandparents–grandchildren	0.25	0.25
Identical twins	1.0	—

though it will produce distorted results only when the levels of selection are relatively high: see Grafen 1984.)

One consequence of the structure of equation 2.1 is that the conditions for the spread of an altruistic gene by kin selection alone are quite stringent. Because the coefficient of relationship devalues the gain to the altruist by a considerable amount, the corresponding gain to the beneficiary must be substantial if the loss of personal reproduction by the altruist is to be offset. Thus, for every offspring that the altruist loses by helping the beneficiary, a full sibling or a parent must produce two extra offspring, a half-sib or a grand-parent must produce four, a cousin eight, and so on.

In general, personal reproduction is always more valuable than helping a relative to reproduce more successfully. Hence, altruistic behaviour will usually be worthwhile only when the cost to the altruist is small or the altruist has little chance of reproducing for itself (Dunbar 1983a; see also Rubenstein and Wrangham 1980). It is, perhaps, worth pointing out here that Hamilton's Rule cuts both ways. A particular course of action may be disadvantageous despite a significant increase in the performer's personal fitness if it reduces the fitnesses of its relatives to such an extent that its inclusive fitness is less than it would otherwise have been.

Genetic reproduction thus consists of two major components: personal reproduction and gains via the increased reproduction of collateral relatives generated by the individual's assistance. An animal will maximise its contri-bution to future generations by allocating its available resources (including time and energy) to these two components in whatever proportion yields the highest net returns in terms of genes contributed to the species' future gene pool. In some circumstances, this may well mean forgoing one option in order to concentrate exclusively on the other. However, in general, animals benefit most by compromising in order to gain the best of both worlds.

Most attempts to apply kin selection theory to real organisms have failed to calculate out the components of inclusive fitness correctly (see Grafen 1980, 1982), so let me summarise very briefly by way of conclusion the most common errors. These have been: (1) not calculating *lifetime* reproductive success; (2) not determining the costs to the altruist in terms of lost personal reproduction; (3) not determining the opportunity cost that an animal incurs by failing to pursue alternative strategies; (4) counting the whole of the altru-ist's reproductive output without discounting the proportion that was due to help from relatives; and (5) counting the whole of the recipient's reproductive output.

Optimal Strategy Sets

In this section, I want to focus on the individual's personal fitness in order to specify a little more precisely just what is involved in the process of contri-buting genes (i.e. offspring) to the next generation. The burden of my argu-ment will be that reproduction is not a simple event, but necessarily involves all aspects of an organism's biology. In fact, personal fitness is a consequence of the way in which three key biological processes interact over the course of an individual's lifetime. These processes are the ability to survive from one day to the next, the ability to generate offspring, and the ability to rear these offspring to maturity (see Goss-Custard *et al.* 1972, Crook *et al.* 1976).

The problems of personal survival and reproduction form a natural hierarchy. In order to contribute genes to the next generation, an animal has to produce offspring that will in their turn breed; to do this, it needs to invest time and energy in rearing its offspring; and in order to do this, it must mate with a member of the opposite sex, which in turn requires that it find and then successfully court such an individual; but in order to be able to do even this, it must ensure that it survives long enough to perform all these activities and this in turn means that it must find sufficient food of the right kinds while at the same time it must avoid falling prey to a predator.

This sequence of events corresponds to a series of decisions that the animal has, in some sense, to make. The sequence can be viewed as a series of increasingly close approximations to genetic fitness. Thus, we can often use, say, the successful solution of a foraging problem as an approximate index of fitness (at least for problems that are concerned, in this case, with foraging). In doing so, we assume that the relationship between ultimate genetic fitness and a successful solution to a lower level problem (such as foraging) is at least monotonic (i.e. that they are correlated at least in terms of rank order). Krebs and McCleery (1984) discuss this issue in more detail.

There are, however, occasions when this assumption does not hold true. This will usually be because an animal has to solve all its problems of survival and reproduction within a fairly limited period of time: it cannot always solve them piecemeal, one at a time. Thus, even though it might like to spend all day mating (and might produce large numbers of offspring if it did), its physical survival (and so ability to mate in the future) depends on its devoting at least some time each day to feeding. This means that the costs and benefits of different activities have to be traded against each other, and the animal is, thus, obliged to find whatever mix of these activities will generate the largest possible number of descendants (i.e. the highest fitness).

So, although it might in principle be possible to solve a particular problem in an ideal way, the perfect solution will not always be a possible option because it conflicts too sharply with the demands placed on the animal's time and energy budgets by the need to solve the other problems of survival and reproduction. An animal's solution to a given problem will often have to be a compromise forced on it by the need to solve several different problems more or less at the same time. One important implication of this is that, when a particular problem is examined in isolation, an animal may sometimes seem to be behaving in a way that is sub-optimal. But this may only be because the perspective in which the problem is being viewed is too narrow. To be certain that an animal really is behaving sub-optimally, we need to be sure that other aspects of the biological system within which it is embedded are not having a significant impact on its freedom of choice.

Individuals may, of course, differ in the weighting they give to the various components of reproduction. Among mammals, for example, females can be expected to place a greater emphasis on the rearing component because their reproductive biology commits them to bearing the relatively heavy costs of gestation and lactation. (Why they should be obliged to do so is a problem that is beyond the scope of this book: it has to do with the evolution of mammals and as far as primates are concerned is simply a constraint that they have to live with.) From a female mammal's point of view, the genetic variance in male quality may be trivial by comparison with the potential variance in rearing conditions. In contrast, male mammals are unable to contribute significantly to rearing in most cases, so they can be expected to place a greater emphasis on maximising the number of females they mate with. Note that this is not true of birds, for example, where the male is just as capable as the female of hatching the eggs and feeding the nestlings: this inevitably has profound consequences for the kinds of mating system that have evolved within the birds (see Ralls 1977).

Social mammals like primates face an additional problem. Successful reproduction invariably hinges on the animal being able to coordinate its own behaviour with that of a number of other individuals in order to be able to benefit from associating with them in a group. This creates a number of problems for an animal, which I will discuss in more detail in Chapter 6. Here, I simply draw attention to the fact that an animal's freedom of choice in its behavioural strategies will depend closely on what other individuals choose to do. Animals that want to live in groups have to coordinate their behaviour in order to maintain the group's cohesion through time, otherwise the group will disintegrate and all the potential benefits will be lost.

Similar problems arise in competitive contexts where the resources for which the animals compete are limited. As the number of animals competing for a given resource increases, so the average expectation of gain for each individual decreases because there is less to go round. If dominance comes into play, some individuals may even be excluded altogether from access to the resource. Frequency-dependent effects of this kind play an important role both in population genetics and in behaviour at the individual level (Ayala and Campbell 1974, Maynard Smith 1982; for an introduction to this area, see Parker 1984). One familiar example from every-day life is the problem of deciding which check-out queue to join at a supermarket: it obviously depends on the length of each queue, and that is partly a consequence of the choices made by previous individuals arriving at the check-outs.

The most important consequence of the frequency dependence of costs and benefits is that, if animals fail to gain access to an essential resource, they may seek alternative means of doing so. Classic examples of this occur in relation to mate acquisition (for a recent review, see Dunbar 1982b). In many species of vertebrates, for example, males compete for territories that give them access to oestrous females. Where there are too few territories for all males to have one, those males that are excluded may attempt to displace those that have them, so that residency times (and hence reproductive success) will decline. In addition, as the level of competition rises, so the fighting will become more ferocious: males may be seriously injured or even killed before having the opportunity to mate. At this point, those males whose chances of defeating an incumbent male are poor may seek to gain access to mates in other ways. This may involve becoming a satellite on another male's territory (so as to be able to mate with any surplus females that may come on to the territory) or it may involve moving rapidly from one group to another over a wide area (in order to capitalise on local variations in the sex ratio that are more favourable to it).

In order to be able to evaluate the relative value of the different behavioural options open to an individual animal, we need to be able to integrate the various components of the biological system into a unified whole. Doing so will allow us to see how costs and benefits can be traded between different dimensions of the system, thereby giving us a much better idea both of the options that an individual has and of the limitations that are placed on these options. We can only do this if we understand how the various components relate to each other and if we have a basis for converting the costs and benefits that arise in the different component sub-systems into a common form where their effects can be added together. That common currency, as we have seen, is provided by the 'selfish gene' perspective. The systemic nature of primate biology is the subject matter of this book.

A Question of Ontogeny

Throughout the preceding discussion, I have spoken rather loosely of animals choosing between various options. Before we can proceed any further, it is essential that I clarify just what I mean by this kind of language. Conventionally in sociobiological analyses, features of an animal's biology are analysed as though a specific gene was making the decision. This usage is, of course, strictly metaphorical (see Dunbar 1982a), and it conveys no implications about the biological status of the phenomenon in question. A decision format works equally well irrespective of whether the decision is completely under genetic control or is one that the animal itself makes in a genuinely cognitive sense.

In practice, not all of an animal's responses are infinitely flexible. While some behavioural responses can be altered from moment to moment, others are less flexible because they depend on underlying morphological features that cannot be changed so quickly. In fact, there are at least three distinct degrees of flexibility that need to be distinguished:

(1) Some features are characteristic of the species as a whole: these are strategic decisions made in evolutionary time by genes. Since these are under close genetic control, individual animals cannot influence them

directly but have to accept them as constraints. Examples include features such as reproductive and digestive physiology, and gross aspects of the dietary niche such as gut morphology or limb structure. These usually involve specialised biological machinery that can only be created at the genetic level on an evolutionary time scale.

(2) At the other extreme, certain decisions are obviously made by individual animals on a moment-by-moment basis. The decision on whether or not to interact with a particular individual at a given moment is a decision made by the animal itself in the light of the circumstances it finds itself in. It is clearly not a response that is coded into its genes. Moreover, it is a decision that is often made at a cognitive level where the animal can meaningfully be said to assess its options and choose between them on the basis of appropriate cues.

(3) In between these two levels, there is a third level where behavioural or morphological strategies are not, strictly speaking, genetically determined, yet are not under the direct conscious control of the animal itself. These include features such as body weight which, while subject to a degree of genetic influence, are also heavily influenced by environmental variables. Animals grow faster and become larger under ecologically favourable conditions than they do when food is in short supply. In most cases, this will be a consequence of such environmental variables as climate and habitat productivity, but the mother's ability to provide enough food for her offspring may also be important considerations, as will be serious setbacks such as illnesses at crucial stages of rapid growth. Because the conditions experienced early in life determine an animal's lifetime growth trajectory (see Graham 1968, Guzman 1968), early experience can radically alter an animal's competitive ability later in life independently of its genetic inheritance.

What we are dealing with, then, is an interaction between three general kinds of biological determinism. An animal's decision about how to behave in a given instance is made within the context of (and is constrained by) (1) the genetically determined aspects of its species' biological strategies and by those aspects of its particular genotype that are inherited from its parents; (2) the environmentally determined responses of some of these genetic strategies during its own development; and (3) the demographic and ecological circumstances that limit its behavioural responses in a particular context. To see this, consider an animal's decision on whether or not to fight a rival. Suppose the decision depends simply on its own size relative to that of its opponent. This decision is a flexible response to the position it has been placed in by a combination of the general body-size characteristics that it inherited from its parents and the response of these to the particular circumstances encountered during the formative period of its immaturity. Its decision whether or not to risk a fight despite these constraints will then depend on the extent to which it can spare the time and energy to do so given the benefits that are likely to accrue to it if it wins. The animal's decision is thus a tactical response within the general strategic trajectory on which its biological inheritance has set it.

Sociobiological explanations have, none the less, become synonymous with genetic determinism, largely, I suspect, because people have been seduced by the use of the language of population genetics. While it is no doubt true that certain kinds of behaviour or psychological predispositions do have a significant genetic basis, the question of ontogeny is not of itself directly relevant to sociobiological explanations, for these are explanations of function and not of ontogeny. The same logic applies whether an animal's behaviour is the product of its biology or a consequence of a rational cognitive decision. Doubtless the question of the ontogeny of behaviour is an interesting one, but it is one we are still far from being able to answer and, more importantly in the present context, it is one we do not particularly need to answer in order to undertake functional analyses in a sociobiological framework. This much was made clear by Tinbergen (1963) two-and-a-half decades ago. The fact that natural selection provides our criterion for assessing the value of behavioural strategies in no way commits us to supposing that behaviour is genetically determined. To suppose so is to have a surprisingly naive view of the evolutionary process. That organisms are programmed to maximise their fitness need only mean that they have a genetically inherited goal state at which to aim and some machinery for allowing them to recognise when a mismatch occurs between this goal state and their actual current condition and decide how that mismatch could be rectified. (For a further discussion of the state-space approach to behaviour, see McFarland and Houston 1981.) Moreover, it is important to remember that evolutionary processes like natural selection do not act directly on fitness: fitness is a consequence of selection which is acting directly on more proximate aspects of the organism's biology. It is these proximate goals that organisms themselves seek to maximise, though the goals have been selected during evolutionary time precisely because they happen to correlate well with genetic fitness. Thus, an animal has an evolved interest in sexual activity because in most circumstances effective execution of this behaviour results in offspring being contributed to the next generation and, therefore, in the organism having high fitness.

Finally, a comment is necessary on the meaning of the terms *strategy* and *tactic*. Their use in sociobiological arguments has often been criticised on the grounds that either their definition is so wide as to include everything an animal is or does or that they imply rational conscious problem-solving by animals. The answer on both counts is that we need a term that is sufficiently general not to commit us to any assumptions about the ontogenetic origins of the organism's behaviour. With particular regard to the rationality implied by these terms, it is important to appreciate that rationality is precisely the criterion that is presumed to lie at the heart of evolution. Hence, the same logic is involved irrespective of whether evolutionary forces make the 'decision' at a genetic level or whether it is done by an individual organism at a cognitive level. It must also be understood that the assumption of rationality here is a heuristic device: it is an assumption that allows us to make very fine-grained predictions which can then be subjected to more precise testing. Whether or not this assumption turns out to be valid is less important in the long run than whether or not making this assumption speeds up the progress of science. And as a general rule, progress in science is invariably faster when powerful theories are available to promote the generation of very precise hypotheses (see Platt 1964).

This much having been said, I do want to draw an important distinction between a strategy and a tactic. One view that has emerged recently (see, for example, Dominey 1984) is to reserve the term 'strategy' for genotypic responses made by members of a particular species as a whole, with tactics being the essentially phenotypic options open to individuals within this. Unfortunately, this distinction raises the spectre of the old nature/nurture controversy, for it seems to be based on an implicit assumption that we can clearly differentiate biological phenomena into those that are largely genetically determined and those that are not. Oyama (1986) has pointed out that despite the nominal resolution of this hoary chestnut, most people still think in terms of simple dichotomies of this kind. Aside from the fact that it is rarely possible to partition effects so neatly into genetic and environmental components, so simple a dichotomy fails to recognise the hierarchy of constraints operating on many behavioural decisions. A male's decision about mating, for example, is not a single simple decision: rather, it is a series of nested decisions, with a decision at each step opening up a new set of options for the male to choose among. Thus, a male might be faced by the choice between defending individual females as and when they come into oestrus or defending a whole group throughout the year in order to have priority of access when any do come into oestrus. Once a decision on this has been made (say, to monopolise a group), he is faced with a choice of options on how to achieve this strategic goal: should he challenge an incumbent harem-holder or should he join a group as a submissive follower in the hope of taking the group over when the incumbent male dies? Having made a choice on this (say, to opt for a takeover), he faces a new set of options aimed at bringing this about: should he go for a small group or a large one, should he seek to challenge the owner directly, or will he do better by soliciting desertion by the females?

It is crucially important in analysing the functional status of behaviour to keep these layers of options clearly separated: failure to do so may lead to a mistaken belief that options pursued at different levels are of equivalent logical and evolutionary status, and we may then be perplexed to find that they do not yield similar expectations of genetic fitness. A nice example of this is provided by tree frogs. Perrill et al. (1978) found that while some males acquired small territories around the edge of the breeding pond from which they called to attract females with whom to mate, other males became silent satellites on these male's territories. We might be tempted to regard these two as alternative mating strategies: calling is the normal strategy, but some males attempt to capitalise on the calling prowess of territorial males to steal females that are attracted to them. But satellite males very seldom actually acquire females to mate, so that it would seem to be a very ineffective mating strategy. In fact, it turns out not to be a mating strategy at all, but rather a strategy for acquiring control of a calling site. It is part of a set of tactical options available to males for acquiring calling sites, one of which is presumably to wander around in search of a vacant site.

For this reason, I prefer to adopt a more conventional definition which views tactics as a set of choices specific to a particular strategic option. This leaves us free to use the terms in a strictly relative sense so that what at one level of analysis is a tactic within a particular strategy itself becomes a strategy with its own array of tactical options at a lower level of analysis. Once again, this avoids any need to commit ourselves prematurely to making assumptions about the ontogenetic status of the behaviour in question. I shall maintain this distinction throughout this book so it is important to remember that, when I use these terms, it is strictly with respect to a particular level of analysis in the chain of decisions with which we happen to be concerned.

Structure and Function in Primate Society

In Chapter 1, I suggested that societies should be viewed as sets of relationships. This has two important implications. One is that it stresses the dynamic nature of social systems. The other is that it invites questions about the functional significance of different grouping patterns and, thus, of social relationships in general. Why do animals form the particular relationships that they do?

Although the size and composition of a group imposes certain constraints on the behaviour of its members (see Chapter 6), a group is, in the final analysis, only a collection of individuals that find it convenient to remain together through time and space. Thus, the emphasis on individual relationships immediately gives us a more dynamic perspective on social systems (see Hinde and Stevenson-Hinde 1976). Because an animal's willingness to form relationships with different individuals is likely to change over time as its own circumstances alter, so the structure and composition of the social groupings to which it belongs can also be expected to change. I shall discuss this in more detail in later chapters (notably Chapter 11).

This structural flexibility raises a further important distinction between the 'deep structure' (the rules that guide an individual's decisions in a social context) and the 'surface structure' (the actual pattern of interactions created when those rules are given expression in a particular demographic context). The same deep-structure rule (e.g. 'Form an alliance that maximises your

dominance rank') may find expression in a variety of actual surface structures: for one individual, this may be achieved by allying with a close relative,
for another by forming a coalition with the most dominant female in the
group, for a third by an alliance with a male. (Specific examples of this will be
discussed in more detail in Chapter 10.) The distinction between deep and
surface structure is crucial because differences in surface structure do not
necessarily signal differences in deep-structure rules. It may simply indicate a
difference in the context in which those particular rules are being applied by
different individuals.

Short-term fluidity is a familiar component of temporary coalitions, but
there is no reason why it should not also occur at higher levels of social
organisation. Thus, hamadryas sleeping troops form where suitable refuges
are scarce, but do so only rarely where cliffs are plentiful. Moreover, these
large aggregations disperse rapidly into their constituent bands once the
animals leave the sleeping cliffs on the day's foraging since such large-scale
groupings are then a hindrance rather than a benefit.

We cannot sensibly discuss the formation of such groupings without asking
why they form, and we are, of course, used to thinking of certain types of
grouping in terms of their function. Thus, sleeping troops are interpreted as a
response to a chronic shortage of suitable refuges, and intra-group coalitions
are seen as being a functional response to the disadvantages incurred from
low dominance rank within a group. Kummer (1978) has stressed the essentially strategic nature of primate social relationships. He pointed out that an
animal invests time and energy in developing relationships with other individuals in order to be able to exploit those individuals for its own advantage.
Such exploitation is possible only as a result of careful assessment of both the
ways in which another animal might be of value and of that animal's
behaviour so as to be able to manipulate it more effectively. Kummer (1971)
has also pointed out that primate societies are essentially group solutions to
ecological problems. In other words, a social group is basically a coalition of
individuals that find it convenient to cooperate in solving a given set of problems (see also Nagel 1979). To an observer, they may seem to constitute a
formal group, but in reality that group only has a degree of temporal stability
and spatial cohesion imposed on it by the nature and persistence of the prob-
lem(s) it is designed to solve.

This is clearly illustrated by the gelada, for here there is a separation of
functions into discrete grouping levels (see Table 2.2). In a very real sense, all
these groupings can be interpreted as coalitions of one kind or another
(Dunbar 1986). Communities function as genetic demes within which genetic
material can be exchanged on a larger scale and individuals have more choice
of mating opportunity. Bands function as ecological units, since each band
has its own ranging area within which it has preferential access to ecological
resources. Reproductive units function from a male's point of view in terms of
priority of access to reproductive females, whereas from the females' point of

Table 2.2. Levels of grouping in gelada society and their functions

Grouping level	Function or context
Coalitions	Minimising harassment and reproductive suppression
One-male unit	Breeding, rearing, minimises stress in large herds
Team	Unknown
Band	Ecological unit, genetic grouping
Community	Ranging flexibility, source of alternative one-male units for takeover by males

view they function as buffers against the harassment incurred by living in large herds. Finally, because the reproductive units are themselves moderately large and become compressed within large herds, females form small coalitions among themselves that help to buffer them against the stresses incurred by the close proximity of other individuals.

Some of these problems are reproductive and others demographic. But the nature of the coalitions formed to solve them are essentially the same and are underwritten by similar kinds of interaction. The frequency and persistence of those interactions depend largely on the biological prominence of the problems in a given environmental context. Thus, the extent to which an observer is struck by the permanence and cohesion of a particular grouping pattern depends on the functional importance of the problem that particular coalition is designed to solve in that environment.

So far, the networks of coalitions that we have discussed have been essentially hierarchical, as exemplified by Figure 1.1. Each higher level of clustering consists of all the units at the next lower level. Hierarchical structures of this kind have a certain organisational simplicity (Dawkins 1976b). They allow large groups to be built up out of smaller units and then to be dispersed again as circumstances require. Since, in many cases, the functional significance of coalitions lies in their size rather than their composition, this allows coalitions of different sizes to be built up very rapidly in species where animals do not normally associate at random.

In ontogenetic terms, the most obvious basis on which to build such a hierarchical system is the female kin group or family unit. This is simply because members of the same family group inevitably have most opportunity to become familiar with each other and to build up the relationships on which coalitions are necessarily founded. With the family unit as a building block, it is easy to construct larger and larger groupings by pooling together families on the basis of their genetic relatedness. This is certainly the case with the gelada, for example, whose five-tier social system is based on genetic relatedness (Shotake 1980, Dunbar 1984a). Studies of a number of species have indicated that grouping patterns tend to follow genetic lines in these cases too (e.g. galagos: Nash 1986; macaques: Chepko-Sade and Sade 1979, Grewal 1980).

But it is by no means necessary that all such groupings should be strictly hierarchical in a genetic sense. There is no intrinsic reason, for example, why animals should not also form coalitions that run across the genetically based hierarchical format. Two examples will serve to illustrate the point. Gelada reproductive units form an additional type of grouping aside from the ones listed in Table 2.2: they form loose associations termed herds that have no temporal stability and no formal structure (Kawai et al. 1983). These are formed as a defence against predators by whatever units happen to be in a particular vicinity at a given time (Dunbar 1986). Consequently, a herd may consist of some or all of the units of one or more bands, so that a given unit may sometimes be found in a herd that contains no units with which it is particularly closely related. This level of grouping thus lies on a separate dimension at right angles to the 'natural' grouping based on kinship. A second example is provided by vervets. Although kinship is an important basis on which coalitions between different individuals are founded, Cheney (1983) has shown that individuals may sometimes form alliances with unrelated individuals when it is to their advantage to do so.

I shall have more to say about this in Chapter 10, but it is important to note here that even though kin-based coalitions may be both advantageous and in some sense 'natural', there may often be circumstances in which animals find it to their advantage to form coalitions with unrelated individuals. An animal has, in effect, to weigh up the relative merits of different potential allies and then choose the best. Which turns out to be the best will depend on the constellation of ecological and demographic conditions that prevail in any given individual's case. In summary, then, non-hierarchical alliance patterns can and do occur and may sometimes even be integrated within a strictly hierarchical system of alliances built up on kinship.

These considerations lead us to distinguish between two quite separate questions: (1) why do primates live in groups? and (2) given that they decide to live in groups, why do they choose to form groups with the particular individuals that they do? With a few notable exceptions (e.g. van Schaik and van Hooff 1983), these two questions have seldom been clearly distinguished in discussions of the evolution of primate social systems. Instead, there has been a tendency to search for environmental factors that uniquely determine either specific forms of social organisation (e.g. Crook and Gartlan 1966, Eisenberg et al. 1972, Wrangham 1980) or specific forms of social behaviour (e.g. Clutton-Brock and Harvey 1976). The result has been a tendency to confuse different grouping levels and the selective advantages that they confer on their members. Attempts to erect unitary explanations for the grouping patterns of all primates are doomed at the outset because primate societies, like those of all higher vertebrates, are not unitary phenomena. There is no single all-pervading functional problem that is so dominant that it dictates the form of all social systems. Rather, social systems are multifaceted attempts to solve a number of often radically different problems. What we see is invariably a compromise thrown up by the animals' attempts to respond to the conflicting demands of these various problems.

Additional confusion has been generated by a tendency to lump together as functionally identical grouping levels that in different species are designed to solve different problems. In many cases, the only real similarity between the groupings concerned is that they are the most easily observed because they are the most stable. But that in itself carries no implications whatsoever about their function. Many authors have assumed, for example, that the social systems of the gelada and the hamadryas baboon are identical not only in structure and dynamics, but also in function and evolutionary origin (see, for example, Crook and Gartlan 1966, C.J. Jolly 1970, Denham 1971). In fact, there are almost no points of similarity between them other than the fact that the reproductive units have a single breeding male: aside from this, they differ significantly in both size and dynamic organisation (Dunbar 1983b). The point I want to emphasise here is that we should not make any unwarranted assumptions about the nature of individual species' social systems. The first task must always be to examine a species' biology closely as an integrated whole before attempting to interpret the functional significance of its behaviour.

One important implication of this view of an animal's social biology is that it calls into question any attempt to search for behavioural universals. This is not to say that there are no principles that are universally true, either for primates in general or for individual species. Rather, the point is that what an animal does is not necessarily a simple, direct reflection of the underlying principle that is guiding its behaviour. In species as behaviourally flexible as primates, much of what an individual does consists of attempts to apply universal deep-structure rules (which might well be genetically programmed) in specific contexts. Since the optimal solutions to those rules depend on particular contextual features (and especially the demographic context), attempts to search for behaviour patterns that occur in all animals in all contexts are likely to prove disappointing. Being reminiscent of the emphasis on species-specific forms of behaviour that dominated classical ethology more than three decades ago, such views would seem to be a retrogressive step.

For primates, then, sociality is one of the chief means that individuals use to reproduce successfully in a world that is not always conducive to successful reproduction. But while group-living may solve one problem, the processes on which social life depend (coordinating behaviour with others, forming coalitions, etc.) themselves create new problems along other dimensions. It is in the give and take of these countervailing benefits and costs that the driving force behind the complexity of primate social relationships is to be found. With this providing us with a framework, we can now turn to real primates and begin to explore this complexity in more detail.

Chapter 3
Survival Strategies

In order to survive from day to day, an animal has to solve two major sets of problems simultaneously. One is to obtain enough food to balance its energy budget; the other is to avoid being caught by a predator while doing so. The difficulty from the animal's point of view is that these two requirements are often incompatible. Obtaining food invariably obliges an animal to move around, for a given food patch can support an animal for only a limited period of time. But movement of any distance inevitably forces an animal to enter areas where it is more vulnerable to predators. An animal's problem, then, is to optimise the time it devotes to foraging in the light of the predation risk it is likely to suffer.

Nutritional Requirements

Successful survival from day to day depends on having a diet that provides an adequate balance of all essential nutrients. Briefly, an animal has to acquire enough energy, protein, carbohydrate, fat, fibre, vitamins and trace elements to fuel its biological processes, while at the same time minimising its intake of secondary compounds and other toxins that either inhibit digestion or poison its system. (For a review of this area, see Richard 1985.)

An animal's energy demand is determined by four key processes: basal metabolism, active metabolism, growth and reproduction. Basal metabolic rate (or BMR) is the rate of energy consumption required to maintain life when the animal is at complete rest. For mammals in a thermally neutral environment,

$$\text{BMR} = 70 \ W^{0.75} \ \text{kcal/kg/day},$$

where W is the animal's body weight in kilograms. Note that BMR scales at the three-quarters power of body weight. This means that the energy requirement per kilogram of body weight for maintaining basic metabolic processes is greater for small animals than for large ones. This does *not* mean that large animals need to eat less than small ones, but only that they need less per unit body weight. Because of this, it is essential that we take body size into consideration when undertaking comparisons of most ecological variables between species. The value of $W^{0.75}$ (sometimes known as the *metabolic body weight* of an animal) is often used as a basis for comparison in this respect.

Among homeotherms that maintain a constant body temperature, BMR is also partly determined by the costs of thermoregulation. In most eutherian mammals, body temperature is around 37–39°C (Mount 1979). When the ambient temperature drops substantially below this, energy has to be expended to maintain body temperature at its normal value. Energy consumption appears to be a linear function of the difference between the critical body temperature and the ambient temperature (Kleiber 1961, Tokura *et al.* 1975, Mount 1979). However, small animals gain and lose heat more rapidly than large ones, so that small species face more severe problems in this respect. The energy expended in maintaining body temperature has to be obtained by consuming more food (or foods of a higher energy value). Iwamoto and Dunbar (1983) found that the time devoted to feeding by wild gelada increased with declining air temperatures in exactly the way predicted by the energy consumption equations when differences in the nutritional content of the forage were taken into account. Fa (1986) also found that time spent feeding by Barbary macaques increased significantly during winter months when ambient temperatures were at their lowest. Increases in time spent feeding as ambient temperatures fall have been documented in moose (Belovsky and Jordan 1978) and klipspringer antelope (Dunbar 1979b).

The active metabolic requirement reflects the fact that any activity over and above resting requires additional energy to fuel it. The quantity of additional energy depends on the amount of work being done. Energy consumption increases linearly with both the speed of movement and the load being carried (including body weight) (Taylor *et al.* 1982). Estimates of the activity-specific energy requirements for medium-sized monkeys are given in Table 3.1. Coelho *et al.* (1976) calculated that the amount of extra energy required for activity is about 15–20% of BMR in wild howler monkeys and a little less in spider monkeys.

Resting and active metabolic requirements are essentially geared to maintaining the *status quo* in an animal's life: they permit an animal only to survive. Two other important components are necessary for an animal to be able to fulfil its primary biological function of propagating its genes: these are growth and reproduction itself. Growth (or, in later life, the repair of damaged tissue) is a major drain on energy and is the main reason why immature animals eat a great deal more than one would expect on the basis of body weight alone.

The energy costs of reproduction are most conspicuous in the case of females, for these have to be able to achieve a minimum body condition in

Table 3.1. Activity-specific energy consumption for a 5-kg primate

Activity	kcal/min
BMR	0.1625
Grooming	0.3949
Playing	0.4625
Threatening	0.2888
Moving	0.2683[a]
Feeding	0.2190
Inactive	0.2066

Source: data given by Coelho (1974).
[a] Calculated assuming 2.309 kcal/km for a day journey of 6 km with 23% of the day spent moving (based on data given for baboons by Altmann and Altmann 1970).

order to be able to ovulate. I shall discuss this in more detail in the next chapter: here, we need only note that these demands are present. After fertilisation has occurred, gestation imposes an increased energy demand on the female, with an energy requirement that is about 25% above normal during the second half of pregnancy; lactation adds to this burden, requiring an energy intake approximately 50% higher than normal (Abrams 1968, Portman 1970, Hanwell and Peaker 1977). Presumably the same considerations apply to other nutrients, though little is known about this aspect of primate nutrition. Gautier-Hion (1980), however, did note that there was a significant shift in the diets of female guenons to foods with a high protein content (young leaves, insects) during that part of the year when females were pregnant or lactating. Riopelle *et al.* (1976) found no significant difference in either birth weight or skeletal maturity in infant rhesus monkeys born to females fed diets of differing protein content, and they suggested that the mothers are able to monitor protein intake so as to ensure that the fetus receives adequate quantities (even if this means going into protein deficit themselves). Although male primates do not have to bear these costs, they do have to compete with each other for access to reproductive females, and this inevitably requires a heavy expenditure of energy not only to fuel any fighting that may occur but also for the growth and tissue repair required to give them maximum advantage in contests.

Portman (1970) has suggested that, taking all these requirements into account, the total energy requirement of an animal the size of a baboon or macaque is about double its BMR. For humans, 1.2 times BMR is considered to be the minimum required to sustain life and about 1.8 times BMR that required by a socially and physically active individual.

Energy alone is not the only nutritional factor that animals need to be concerned about. They must also ensure an adequate intake of other elements. Deficiencies in any of these can have pathological consequences (see reviews by Portman 1970, Kerr 1972). Table 3.2 gives data on the dietary intakes of two groups of captive baboons, one fed a commercial synthetic diet and the other a synthetic diet with a large vegetable supplement. Also shown are estimates of the minimum requirements for each nutrient needed to maintain a healthy animal. Note that both groups of animals obtained rather similar quantities of each nutrient, and that these values compare favourably with the minimum requirements.

Table 3.2. Approximate daily nutritional intakes of adult baboons fed on two different diets, together with estimated minimum daily requirements

Nutrient	Units per kg body weight		
	Purina monkey chow[a]	Brookfield Zoo (Chicago)[b]	Minimum requirement[c]
Energy (kcal)	172.4	114.9	100.0
Protein (g)	10.6	4.2	2.5
Fat (g)	2.04	1.07	—
Carbohydrate (g)	15.2	19.8	—
Fibre (g)	1.08	0.92	—
Ash (g)	2.44	1.42	—
Calcium (g)	0.384	0.139	0.150
Phosphorus (g)	0.220	0.099	—
Iron (g)	0.012	0.011	0.019
Sodium (g)	0.140	0.071	0.044
Potassium (g)	0.412	0.339	0.152
Vitamins: A (IU)	1200	2387	400
thiamin (g)	0.0005	0.0025	0.0003
riboflavin (g)	0.0004	0.0001	0.001
niacin (g)	0.0036	0.0017	0.002
B_{12} (g)	—	—	0.0001
C (g)	0.03	0.0163	0.0006
D (IU)	264	59	25

[a] Data given by Hummer (1970) assuming a ration of 40 g chow per kg body weight.
[b] Analyses kindly provided by J. Roberts for a diet based on a mixture of two commercial preparations plus fresh vegetables and fruit.
[c] Based on recommendations given by Portman (1970), Greenberg (1970) and Kerr (1972). Except for thiamin, riboflavin and vitamin B_{12}, all these requirements derive from studies of rhesus macaques. Data for these three vitamins derive from baboons.

A word of caution is necessary concerning all data on nutritional requirements. First, most of the data that are available derive from studies of baboons or macaques. Their relevance to other species of primates remains uncertain. Secondly, all data derive from studies of captive animals. While it seems likely that free-living animals will have similar requirements, this really needs to be checked. The standard nutritional definition of an adequate diet for humans has recently been called into question by evidence that human populations may be able to adjust to diets that are officially considered to be nutritionally inadequate. Finally, the criteria used to define adequate intake levels are not necessarily those relevant to evolutionary considerations. Most studies have used weight gain as their criterion for macro-nutrients (i.e. energy, protein, etc.) and the alleviation of specific clinical symptoms for micro-nutrients (i.e. vitamins, trace elements, etc.). Studies of human nutrition have often used a long and vigorous life as their criterion, and this, as Portman (1970) points out, is not always the criterion that is maximised by natural selection.

Two more general points about primate nutrition can be mentioned. One is that all primates require water, though actual consumption rates will depend on the water content of the plants eaten, the humidity of the environment and the ambient temperature. The other point is that bulk (i.e. fibre) is an important dietary element. Macaques fed a bulk-free semi-purified diet suffered erosion of the caecal wall due to the fact that hair (ingested during grooming) formed boluses that obstructed the passage of faecal material (Greenberg 1970).

Young primates require proportionately more nutrients than adults because of the heavy demands of growth. Glassman *et al.* (1984) found that growth rates varied considerably over the course of an animal's life. For males, growth rates averaged 0.01 kg/day during the first year, fell to around 0.008 kg/day during the second and third years and then rose to a peak of 0.017 kg/day during the fifth year. Females followed a similar pattern but at slightly lower growth rates. Other studies of captive baboons have yielded rather similar growth rates (Buss and Reed 1970, Coelho 1985). A study of free-ranging baboons by Altmann and Alberts (1987) obtained growth rates of 0.005 kg/day which were more similar to those obtained by Buss and Reed (1970) for captive animals fed a low protein diet. Small and Smith (1986) report a growth rate of about 0.005 kg/day for captive rhesus macaques (a species that weighs about half the body weight of baboons). Macaques too undergo distinct periods of rapid growth alternating with periods of much slower growth (van Wagenen and Catchpole 1956). Portman (1970) estimated that newborn macaques require double the energy requirement per unit body weight of an adult in order to maintain steady growth, with protein requirements that are probably of about the same order.

The energy and protein required to fuel this growth has to be provided exclusively by the mother through lactation until the infant is old enough to fend for itself. Since baboon breast milk provides only about 0.76 cal/ml of energy and 0.016 g/ml of crude protein (Buss 1968), a 1-kg infant requires some 250–300 ml of milk a day to meet its energy and protein requirements by milk alone. Lactation, however, is a relatively inefficient use of energy: Blackburn and Calloway (1976) calculated that in humans the conversion ratio for lactation (i.e. the efficiency with which energy ingested by the mother is converted into energy in milk) is only about 80%. Brody (1945) gives a value of 60% for mammals in general. This means that the mother has to ingest about 25–66% more energy than her infant actually needs in order to ensure that it gets what it wants.

Dietary Strategies

Although most primates are omnivorous and eat a wide variety of food types, the majority fall into one of three classes: those that concentrate on fruits (frugivores), those that concentrate on leaves (folivores) and those that concentrate on insects (insectivores).

Fruits generally provide a rich source of energy, but suffer from the disadvantage that their distribution is often very patchy both in time and space. Hladik (1977), for example, found that during the non-fruiting season frugivorous hanuman langurs are obliged to rely on leaves: as a result, they suffer a net energy deficit. Caldecott (1986a) likewise noted that the relative paucity of suitable fruiting trees in the dipterocarp forests of south-east Asia forces the groups of large-bodied monkeys like the pigtail macaque to disperse while foraging.

Folivores circumvent this problem because leaves are more densely and evenly distributed. One consequence of this is that folivores can usually subsist on fewer species of plants than frugivores. Clutton-Brock (1974) found that just two species of trees accounted for more than 90% of the feeding time of the more folivorous black-and-white colobus, whereas the five most commonly eaten species accounted for only 50% of the total feeding time of the more frugivorous red colobus. Similarly, Hladik (1977) found that just three species accounted for 70% of the dietary intake of folivorous grey langurs compared with 10 species for the more frugivorous hanuman langurs. Comparable differences between folivorous and frugivorous species have been reported for gibbons (Raemakers 1979) and lemurs (Susman 1977).

The dietary advantages of leaves are offset, however, by the fact that most of the nutritional value is locked up in the structural material of the cell walls. These consist largely of cellulose and lignin that are indigestible in their natural state (Sullivan 1966, van Soest 1977). Herbivores have evolved three general strategies for dealing with this problem.

The best known of these is that adopted by the ruminants (ungulates that chew the cud). Ruminants have large sacculated stomachs in which the ingesta are fermented by specialised bacteria that live there (McQueen and van Soest 1975). To assist this process, the food is passed back up to the mouth from the stomach to be re-masticated after it has been partially digested. A ruminant-like strategy has evolved in one group of Old World monkeys (the colobines: Bauchop and Martucci 1968, Kay et al. 1976) and one group of New World monkeys (the howler monkeys: Milton and McBee 1982). Although these do not chew the cud, ingested foliage is broken down by microbial fermentation in the digestive tract. Experimental studies of howler monkeys have suggested that these animals acquire 25–35% of their daily energy requirements from the by-products of fermentation (Nagy and Milton 1979, Milton et al. 1980).

The alternative strategy evolved by the non-ruminant ungulates is bulk feeding. The animal accepts a low rate of nutrient extraction and compensates by increasing proportionally the amount of food ingested (Bauchop 1978, van Soest 1980). This strategy is particularly associated with the horses and their allies (Janis 1976), but it has also evolved among geese, one of the few groups of birds that habitually eat foliage (Sibley 1981). Among primates, bulk feeding seems to have evolved in only two taxa, the gelada (Dunbar 1984a) and the gorilla (Fossey and Harcourt 1977, Goodall 1977).

The last and perhaps least common strategy for handling structural cellulose is coprophagy (the eating of faeces). In this case, bacterial fermentation in the large intestine breaks down the structural cellulose, but does so at a point in the digestive system where the products cannot easily be extracted. The nutritionally rich contents can only be recovered if the excreted digesta are re-ingested so as to permit a second pass through the system. This strategy is most closely associated with the lagomorphs (rabbits and their allies). Charles-Dominique and Hladik (1971) have suggested that the sportive lemur might pursue this strategy because the amount of energy assimilated during feeding was estimated to be only half that required to maintain basal metabolism. Their suggestion remains unsubstantiated, however.

Animal matter provides a significant proportion of the diet for only a small number of species, most of whom are insectivorous. Although insects constitute a high-energy food source, individual items are generally small and seldom occur in large concentrations. This makes them difficult to locate and limits the rate at which they can be ingested. As a result, insectivorous primates are invariably small and usually nocturnal (Clutton-Brock and Harvey 1977, Sailer et al. 1985). Examples include tarsiers, galagos and the smaller lemurs (Charles-Dominique 1977, Hladik 1979). Cords (1986) has shown that not only was the smaller of two species of forest guenons more insectivorous and less folivorous than the larger species, but that also, within species, smaller individuals (e.g. adult females and juveniles) eat more insects and fewer leaves than larger individuals (e.g. adult males). Insects that habitually occur in concentrations large enough to sustain animals of significant size are species like the colonial bees and termites whose colonies are well protected either by sting-bearing defenders or by the concrete-like walls of their mounds. These require special anatomical adaptations (thick hide, powerful claws) such as are found in the large insectivores (e.g. aardvarks, anteaters, echidnas), most of which are probably obliged to be solitary in order to avoid excessive competition over such concentrated food sources.

Although it usually accounts for only a small proportion of the diet, meat-eating by certain Old World monkeys and apes has attracted considerable attention (e.g. Teleki 1973, Harding 1975, Hausfater 1976, Strum 1981). Perhaps the most important point to note is that the hunting of vertebrate prey such as reptiles, birds and small mammals is both energetically expensive and time consuming, so its value as a foraging strategy tends to be marginal if more sedentary plant foods are readily available. Because of this, hunting seems to be mainly a reserve strategy for most primates, something they can resort to when vegetable food sources are in short supply. Among baboons, for example, hunting is more common in poorer quality habitats, and, within habitats, it is more common during the dry season when food supplies are most limited (Figure 3.1). Similar findings have been reported for vervet monkeys: Kavanagh (1978) found that invertebrates formed a significant proportion of the diet only during the dry season when vegetable foods were scarce, while both Skinner and Skinner (1974) and Galat and Galat-Luong

Figure 3.1. Frequency of predation attempts in four populations of Papio *baboons, ranked in relation to the quality of the habitats they live in (with forested habitats being the richest). Data from two populations also indicate that the lower availability of vegetable foods during the dry season favours meat-eating*

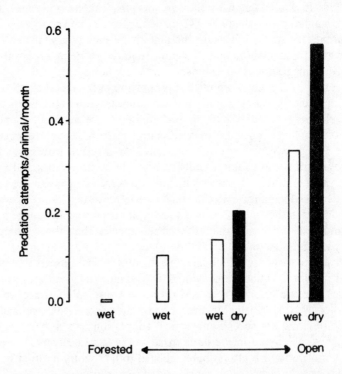

Source: Data for populations at, from left to right, Bole and Awash, Ethiopia, Gilgil and Amboseli, Kenya, summarised by Dunbar 1984b.

(1977) reported a marked increase in predation on vertebrates (mainly birds and rats) during periods of severe food shortage.

In many cases, the successful exploitation of food resources depends on the evolution of specialised anatomical features. Thus, the ruminant-like fermentation of foliage requires a large stomach to handle the quantities of food that have to be ingested. Other dietary strategies impose comparable requirements. Some South American callitrichids, for example, rely on nectar during the dry season when fruits and insects are in short supply, but they can only rely on this resource because they are so small (Terborgh 1983). Similarly, capuchin monkeys exploit energy-rich palm nuts during the dry season (Terborgh 1983): they are the only animals capable of doing so because they have evolved the large molars and massive jaw musculature needed to crack the tough shells of these nuts (Kinzey 1974). These are, of course, extreme examples, but they illustrate the point that major shifts of diet are not necessarily free of cost. Specialised structures may both be expensive to evolve and impose major limitations on the animals' options in other respects.

Optimal Foraging

Optimal foraging theory (see Schoener 1971, Pyke *et al.* 1977) suggests that animals should select the most nutritionally valuable items (where value is defined in terms of the rate of nutrient extraction per unit weight of plant ingested). Nutritional value, however, cannot be seen in isolation from the factors that prevent animals from extracting all the nutrients available in a plant. These include (1) structural elements that make plants relatively indigestible (discussed above), (2) secondary compounds like tannins and phenolics that bind the nutrients into indigestible compounds within the cell, and (3) toxins that actively poison consumers.

Although studies of the nutritional ecology of free-living primates face formidable problems, not least because the animals often forage at considerable heights above ground, it has been possible to demonstrate for a number of species that animals' preferences for particular plants are related to their nutritional content. Hamilton *et al.* (1978), for example, determined the nutritional contents of three species of plants commonly eaten by chacma baboons. Those parts that were eaten had higher crude protein levels, lower crude fibre levels and higher gross energy contents than those parts of the same plants that the baboons discarded. Waterman and Choo (1981) found positive correlations between the digestibility of the mature leaves of different plant species and the frequency with which they were eaten by populations of African and Asian colobines. They noted, however, that the degree of correlation differed markedly between habitats. This they attributed to differences in the production of digestion-inhibiting secondary compounds that were probably related to local differences in soil conditions. They suggested that forested habitats may differ significantly in their suitability for a folivorous strategy.

In another study, Wrangham and Waterman (1981) showed that vervet monkeys select among the parts of two species of acacia in a way that is related to the total levels of phenolics and condensed tannins, preferring the gum of one and the seeds of the other. A similar tendency to eat plant parts that have high sugar, protein and fat contents and reject those that do not has been reported for both wild baboons (Hausfater and Bearce 1976) and guenons (Sourd and Gautier-Hion 1986).

Milton (1980) analysed the nutritional contents of the most common items in the diet of free-ranging mantled howler monkeys and concluded that the protein/fibre ratio was the most important factor determining the monkeys' choice of seasonal foods. Phenolic content seemed to be a significant factor only in flower selection. Glander (1981) also found that red howler monkeys tended to select leaves that had high protein and low fibre and secondary compound contents. Data obtained during a succession of seasons indicated that the monkeys maintained a balance of nutrients in their diets by continuously adjusting the rate at which they ate the various plant parts as these changed in abundance and nutritional content from one season to the next.

On a more general level, many studies have been able to show that primates feed selectively, choosing particular species or parts of plants at rates that are much greater than their relative abundance in the habitat (e.g. mangabeys: Waser 1977a; colobus: Oates 1977a). Gaulin and Gaulin (1982) found that howler monkeys consistently selected the least tough leaves and tended to pick the largest and softest fruits available to them in a given tree. Marked seasonal shifts in the diets of primates in response to the seasonal availability of preferred foods (especially fruits, young leaves and insects) have been described in forest guenons (Gautier-Hion 1980, Cords 1986), gelada (Dunbar 1977, Iwamoto and Dunbar 1983) and various South American cebids and callitrichids (Terborgh 1983).

The amount of energy that animals can extract from their environment has important implications for their growth patterns. Evidence in support of this comes from a number of studies reporting correlations between habitat quality and body weight. Froehlich and Thorington (1982), for example, found a significant correlation between the quality of individual ranges and the body weights of the animals living in them in the Barro Colorado (Panama) howler monkey population. Gautier-Hion and Gautier (1976) found that body weights for captive animals living under optimal nutritional conditions were significantly greater than those for wild individuals in a number of Old World monkey species. Similarly, A. Mori (1979a) and Sugiyama and Ohsawa (1982a) found that mean body weight declined following the withdrawal of artificial feeding in the two populations of Japanese macaques.

An animal's ability to extract nutrients from the food it eats is a key factor determining how much food it will actually have to consume in order to meet its nutritional requirements. Goodall (1977), for example, found that the energy intake of wild gorillas was three times greater than that required to maintain minimum metabolic activity (BMR as determined by body weight alone). Similarly, Iwamoto (1979) reported that the mean energy intake of adult male gelada was three to four times the expected, even when active metabolic requirements were taken into account. Fa (1986) reports rather similar results for a provisioned population of Barbary macaques. Even allowing for a total energy requirement that is double BMR, this still represents an energy intake that far exceeds the animals' apparent needs. This excess is largely due to the fact that animals cannot utilise all the energy they ingest, particularly that proportion that is locked up in plant cell walls. Goodall (1977) estimated that the proportion of a plant's energy content that a gorilla can assimilate varied between 11% and 77% for the 40-odd species of plants that were regularly eaten by them.

Hladik (1977) found a significant difference in the ability to extract nutrients from the environment between two species of langurs. The more frugivorous hanuman langurs exhibited greater fluctuations in both protein and lipid (fat) intake over the course of a year than did the more folivorous grey langurs. This was mainly because fruits are highly seasonal in their availability, and a frugivore cannot fall back on leaves with the kind of digestive efficiency that is necessary for a folivorous diet. In addition, hanuman langurs had to expend much more energy in harvesting fruits because these were so much more patchy in their distribution.

Similar problems are encountered by Japanese macaques, another predominantly frugivorous species. During winter, when few fruits are available, they are forced to subsist mainly on the leaves of evergreen trees and shrubs. Because leaves have a high fibre content, the animal's stomach fills with food before the animal can digest it. Consequently, both energy and protein intakes drop below the monkeys' daily requirements during winter months (Iwamoto 1982). A. Mori (1979b) was able to demonstrate experimentally that the macaques lost weight when fed a diet consisting entirely of leaves. When they were subsequently switched to a diet of wheat, not only did the amount of time spent feeding each day decline dramatically but energy intakes increased and the animals rapidly regained their pre-experimental body weights.

These features of primate dietary strategies have important implications for a number of behavioural variables, in particular the distance that the animals have to cover each day while foraging (day journey length) and the amounts of time that have to be devoted both to moving and to feeding. In poor habitats or during dry seasons, animals may have to travel further and spend more time moving in order to find the food they need each day; when they do find it, the amount that has to be eaten will tend to be higher due to the problems of digestibility, and this in turn will mean that a higher proportion of the day has to be spent feeding. These dietary problems will thus

put pressure on the amount of time the animals have available for social interaction. I shall deal with day journey length and other ecological variables in
the following section. Time budgets and their implications will be dealt with in
Chapter 6.

Exploiting the Habitat

With the exception of a small number of specialist feeders, most vertebrates
cannot obtain their food without moving about their habitat. Herbivores, for
example, would soon exhaust their food supplies if they remained within a
confined area for any length of time. At the same time, most species benefit
by remaining within a circumscribed area rather than being completely
nomadic. Doing so allows the animals to become familiar with the distribution
and phenological cycles of the food plants, the location of safe refuges and
waterholes and the shortest routes between resource patches. By minimising
the time taken to locate key resources, animals can both minimise the time for
which they are exposed to the risk of predation and waste less time and energy
in travel. Sigg and Stolba (1981), for example, found that hamadryas baboon
bands made more effective use of their own core ranging areas than did other
visiting bands because they were more familiar with the location of peripheral
and ephemeral resource patches off the main foraging routes (see also Ripley
1970).

    Although knowledge of the local environment is an important consider-
ation promoting the regular use of a specific area as a range, groups may
sometimes migrate between seasonal ranges that are some distance apart (e.g.
hamadryas baboons: Kummer 1968, Dunbar and Dunbar 1974a; chacma
baboons: Whiten *et al.* in press; macaques: Wada and Ichiki 1980; squirrel
monkeys: Terborgh 1983). In all these cases, the habitats occupied by the animals are seasonally impoverished and the animals apparently attempt to
reduce competition by making use of areas outside their normal range.

    The size of area needed to ensure a steady supply of food depends on at least
five factors: the type of diet, the nutritional quality of the food in a particular
habitat, its distribution in time and space, the number of animals that forage
together and the species' normal body weight. These factors influence a
number of interrelated variables, in particular the overall density at which the
animals live, the size of the ranging area and the length of the day journey.
Essentially, the better the quality of the food, the more densely it is distributed and the smaller the animals that have to live on it, the higher can be the
density, the larger the foraging group size, the smaller the ranging area and
the shorter the day journey that is required to ensure that every group
member obtains all the food it needs.

    Generally speaking, frugivores require larger ranging areas for a given
group size than folivores do in the same habitat (see Mace and Harvey 1983),

mainly because fruits are more patchily distributed than leaves. This is reflected in higher population densities for folivores as a general rule. Grazing gelada baboons, for instance, can maintain densities that are three times greater than those of frugivorous olive baboons in the same habitat (Dunbar and Dunbar 1974b). They also have shorter day journeys and can manage to live in very much larger groups (Table 3.3). Hladik (1977) found that frugivorous hanuman langurs lived in much larger ranges than the more folivorous grey langurs, and similar results have been reported for different species of lemurs (Sussman 1977), colobus monkeys (Clutton-Brock 1974) and gibbons (Raemakers 1979).

Table 3.3. Comparative behavioural ecology of sympatric populations of olive baboons and gelada in the Bole valley, Ethiopia

	Baboons	Gelada
Group size (mean)	20.3	60.0
Population density (n/km^2)	26.0	82.0
Day journey (m)	1210	630
Diet: grasses, herbs (%)	39.3	97.6
bushes, trees (%)	38.9	1.5

Source: Dunbar and Dunbar (1974b).

Within dietary classes, habitat quality is probably the most important factor influencing population density and range size. Two aspects of habitat quality can usefully be distinguished: gross primary productivity (the total amount of vegetation produced each year) and the nutritional quality of the forage. Both are determined by fundamental features of the climate, though other factors such as the presence of groundwater and the chemical nature of the soils play an important role. Annual rainfall, however, has been shown to be a reliable index of a habitat's annual primary production (Rosenzweig 1968, Coe *et al.* 1976). A better estimate is actually given by evapotranspiration (a composite index of rainfall, temperature and humidity), but since this involves the measurement of more variables it is often less easy to determine for a given site. The fact that rainfall alone is a good index of productivity, however, does mean that we can use it as a basis for comparison between habitats, especially when we have a number of habitats that differ widely in rainfall. Comparisons between habitats that differ only slightly in rainfall and comparisons between micro-habitats within the same general area will, of course, require finer-grained analyses, usually in terms of the actual productivity of the habitat (see, for example, Iwamoto 1979) or the number of food trees available or an index of these weighted by their size and/or productivity (see, for example, Struhsaker 1974, Caldecott 1986a, Dunbar in press).
Comparisons based on analyses of these kinds have demonstrated that

population density declines as habitats become increasingly impoverished (gibbons: Caldecott 1980; macaques: Ménard *et al.* 1985, Caldecott 1986a; baboons: Anderson 1981). Conversely, groups of similar size have been shown to require larger ranges in poorer quality habitats than in good quality ones (macaques: Takasaki 1981, Caldecott 1986a; vervets: Struhsaker 1967a, DeMoor and Steffens 1972; chimpanzees: Suzuki 1979; orang utans: Mackinnon 1974). Data from baboons show that the area required to support one individual (the *per capita* range size) increases as habitat quality declines (Figure 3.2). A particularly clear demonstration of this is given by changes that occurred in the Amboseli baboon population: in 1963, the main study group of 40 animals used a range of approximately 17 km^2 (Altmann and Altmann 1970), but by 1975, when the annual rainfall had fallen by a third and the habitat quality had deteriorated markedly as a result of other macro-climatic changes, the group (now only marginally larger at 48 animals) used a ranging area of 40 km^2 (Post 1978).

The actual size of a group's ranging area is also determined by the group's total energy requirements (itself mainly a function of the group's total biomass). A number of studies have demonstrated that, within the Order Primates as a whole, range size is correlated either with group biomass (Milton

Figure 3.2. Median per capita *ranging area for populations of* Papio *baboons living in habitats of different quality (sample sizes are 12, 10, 3 and 2 populations respectively). Vertical bars indicate interquartile ranges*

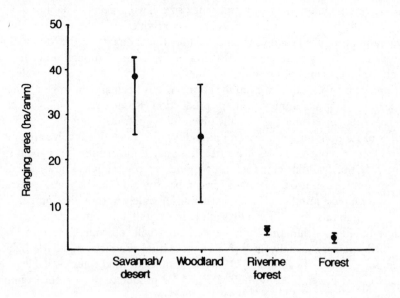

Source: as listed in Figure 7.8.

and May 1976, Clutton-Brock and Harvey 1977) or with the group's 'metabolic weight' (Harvey and Clutton-Brock 1981). Within species, metabolic requirements are held constant, so range size should be a simple function of group size. This has been confirmed for populations of howler monkeys (Froehlich and Thorington 1982), colobus monkeys (Marler 1972, Suzuki 1979), langurs (Sugiyama 1967), baboons (Davidge 1978), macaques (Makwana 1978, Teas et al. 1980, Takasaki 1981) and gelada (Iwamoto and Dunbar 1983). Unintended natural experiments have confirmed that this is a flexible response made by the animals: Ikeda (1982) found that when the size of a troop of Japanese macaques was reduced from 120 to 40 animals for control reasons, the size of the ranging area fell from 4.7 km^2 to 2.7 km^2.

Not all studies have found such a relationship between group size and range size (e.g. howler monkeys: Neville 1972). However, exceptions to this rule can often be shown to be due to the influence of other factors. Iwamoto and Dunbar (1983), for example, found that although range size was related to band size in gelada, the relationship was not linear: most of the deviation from expected values could be attributed to the energetic load placed on the animals by altitudinal variations in ambient temperature. Figure 3.3 shows very clearly how habitat quality can interact with group size to determine the size of ranging area that a group requires.

Range size may also be influenced by the local population density, especially in areas where animals cannot easily emigrate to find new territories. In such circumstances, ranges may behave like elastic discs and become compressed by the pressure of the groups around them (Huxley 1934). Where population density is low, ranges may be able to expand and contract in response to changes in the size of the group, but in high-density populations they may become compressed down to the minimum size that can support a group. Compression of ranges in this way is most likely to occur in small forest blocks where the animals have to cross large tracts of open country to reach new areas of forest. The costs of migration (e.g. risk of predation) may then be so great that the animals prefer to remain where they are and tolerate crowded conditions. Among black-and-white colobus monkeys, large (typically 15–20 ha) territories whose areas correlate closely with group size have been reported from a number of large forest blocks where densities are low (e.g. Marler 1969, Oates 1977b, Suzuki 1979), whereas small (typically 2 ha) territories whose areas tend to remain fixed despite changes in group size have been reported from a number of small forest blocks where densities are high (Schenkel and Schenkel-Hulliger 1967, Dunbar and Dunbar 1974c, Dunbar in press). Similar results have been reported for red colobus: overlapping ranges of 60–110 ha in large forest blocks (Struhsaker 1974, Clutton-Brock 1975), non-overlapping territories of 5–20 ha in small patchy forests (Gatinot 1975, Marsh 1981a). Freeland (1979) also found an inverse relationship between population density and the extent to which territories overlapped in mangabeys.

Figure 3.3. Territory sizes for individual black-and-white colobus monkey groups living in two different types of forest in the Budongo Forest, Uganda, plotted against group size

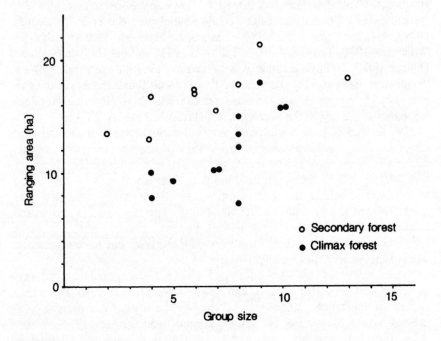

Source: Suzuki 1979; note that the points on this graph differ from those in Suzuki's original Figure 2 since they are based on the data given in his Table 2.

While range size determines the amount of food and other resources available to animals during the course of a year, the animals do not use their entire ranging area each day. The area used by a group each day (its day range) or its linear dimension (day journey length) are both influenced in similar ways by the same set of variables. Day journey length, for example, is positively correlated with foraging group biomass in frugivores, but not in folivores (Clutton-Brock and Harvey 1977). This is because fruits are more patchy in their distribution than leaves. Once a patch of fruit has been exhausted, a group may have to move a considerable distance to find another; in contrast, folivores could in principle spend days at a time in the same group of trees. Within species, day journey is an approximately linear function of group size in capuchin monkeys (de Ruiter 1986), mangabeys (Quris 1975, Waser 1977a), baboons (Sharman and Dunbar 1982), gelada (Iwamoto and Dunbar 1983), macaques (Ikeda 1982, van Schaik *et al.* 1983a) and chimpanzees (Wrangham and Smuts 1980).

Day journeys are also longer in poorer quality habitats than in rich ones in baboons (Figure 3.4; see also Anderson 1981), vervet monkeys (Struhsaker 1967a) and sifaka (Richard 1978). In some species, groups also range further during the dry season (when food is in short supply) than during the wet season (colobus: Clutton-Brock 1975, McKey and Waterman 1982; baboons: Anderson 1981), though some populations exhibit either no significant difference between seasons (baboons: Altmann and Altmann 1970) or the reverse trend (sifaka: Richard 1978). Marsh (1981a) reported that red colobus occupying marginal habitats in eastern Kenya ranged more widely during months when the availability of young leaves (the animals' principal food supply) was low than when it was high, but Struhsaker (1974) found no such correlation for a population inhabiting climax forest in Uganda.

Figure 3.4. Median per capita *day journey length for populations of* Papio *baboons living in habitats of different quality (sample sizes are 10, 6 and 4 populations, respectively). Vertical bars are interquartile ranges*

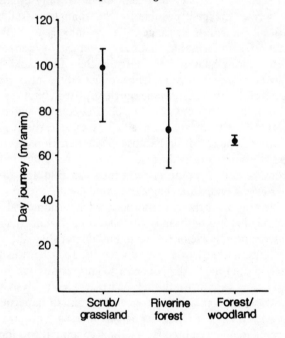

Source: from sources listed in Figure 7.8.

When conflicting results such as these are found, it is important with observational (as opposed to experimental) data to avoid being over-hasty in rejecting the original hypothesis. In most such cases, it will turn out that the original prediction has overlooked the effects of other variables that are only prominent in certain habitats. A more useful approach, therefore, is to look carefully at individual cases to see whether the hypothesis has been phrased in an appropriate way. In the present case, for example, the hypothesis assumes that the dry season is the nutritionally most stressful, and hence predicts that *?da* day journeys will be longest at this time of year because the animals will be forced to travel further to find enough food. However, not only may the dry season be a period of relative abundance in some habitats for species with the appropriate dietary preferences (notably granivores), but some species may opt for alternative solutions to the problem of food shortage (e.g. trying to ride it out by remaining for long periods in the same area so as to reduce travel costs). Raemakers (1980), for example, argued that when food was in short supply, gibbons preferred to cut their losses by minimising the distance travelled in order to make do with what food was available in the immediate vicinity.

Raemakers (1980) pointed out that gibbons can afford to reduce activity during bad times because the greater diversity of south-east Asian plant communities and their marked lack of synchrony in fruiting cycles both within and between species makes it possible for the animals to find at least some food in almost all parts of their ranges. This is unlikely to be the case in the more seasonal African savannah habitats occupied by baboons, however. Moreover, other forest habitats that differ in their floral and climatic characteristics may be more seasonal in their productivity. McKey and Waterman (1982), for example, argue that, in the nutrient-poor habitats of West African forests, the normally folivorous black colobus adopts a frugivorous strategy out of necessity. When fruits are not available, however, they cannot fall back on leaves as other populations do because of the nutritionally poor quality of these plant parts in such forests. Instead, they are forced both to broaden their diet so as to make use of whatever food they can find and to move significantly further over a wider area in order to find enough to eat.

A further illustration of the interplay between environmental factors is provided by the gelada. Iwamoto and Dunbar (1983) found that day journey length increased with the size of the band, but did not show the corresponding increase with altitude that was expected from the fact that the animals spent proportionately more time feeding in order to compensate for the higher costs of thermoregulation at higher altitudes. More detailed analysis revealed that, in the particular sample of habitats concerned, rainfall happened to increase with altitude, with the result that grass density increased proportionately. The increase in grass biomass required to maintain an animal was fortuitously offset by this increase in grass density, such that the area each animal needed to graze each day remained roughly constant.

Economics of Territoriality

In theory, animals can maximise their access to resources if they keep competitors away by defending a ranging area as a formal territory. Whether it is worth an individual (or group) defending a territory depends on how easily competitors can be kept out and on whether any of the resources it contains are in limited supply. If the resource in question is abundant, there is little point in trying to defend an area for exclusive use since territorial defence requires time and energy that is wasted if competitors can find all they need nearby. Where resources are worth defending, an animal's ability to defend a territory is determined by the ease with which it can detect and evict intruders. This will depend partly on the frequency of invasions and partly on the size of area that is required to sustain the animal(s) that have to live in it.

Generally speaking, the defendability of a territory declines as the area to be defended increases (solid curve, Figure 3.5). Since a group usually forages as a coordinated unit rather than dispersing throughout its ranging area, only a limited area can be patrolled effectively: consequently, only a limited number of intruders can be evicted at any one time. On the other hand, the value of a territory to an animal in terms of the resources that can be extracted from it initially increases as territory size increases, but soon reaches an asymptote at a point where all the animals' nutritional requirements are met (broken curve, Figure 3.5). The point at which these two curves intersect marks a point of diminishing returns. Ranges that lie to the left of this point are worth defending; those that lie to the right are not because they cost more to defend than the animals can expect to gain from them.

Figure 3.5. Relationship between the size of a ranging area and (a) the benefit gained through exclusive access to a resource (broken line), and (b) its defendability (a composite measure of both the physical costs of defence and the success with which intruders can be excluded: solid line), where both of these are measured in terms of their effect on the lifetime reproductive output of an individual animal. The point at which the two graphs cross gives the size of territory, T_{opt}, which yields the greatest gains in relation to the ease with which the territory can be defended

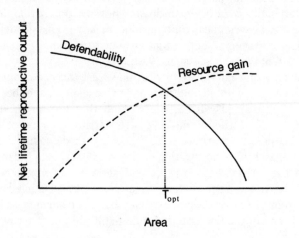

On the assumption that defendability is primarily related to the frequency with which a group can patrol its territorial boundary, Mitani and Rodman (1979) argued that the ease with which a territory could be defended would be a function of the relationship between its area and the length of the day journey. From simple geometric considerations, they suggested that defendability, D, was related to the size of the range area (A km^2) and the mean length of the day journey (d km) in the following way:

$$D = d/(4A/\pi)^{0.5} \tag{3.1}$$

They found that all species which actively defended territories had values of $D \geq 1$. This implies that territories are defended only when the animals can easily traverse their territory during the course of a normal day's ranging. Mitani and Rodman also noted that, while species with $D < 1$ never defended territories, those with $D > 1$ did not always do so. Having a relatively small ranging area makes it *possible* to defend a territory *if* other factors make it worth the animals' while to do so.

Defendability also depends on the density of animals in the population as a whole, since this determines the frequency of invasions. When the density is low, there are likely to be few intruders on to a territory and little is gained by patrolling the area regularly in order to evict them. Overlapping ranges that are only weakly defended should thus result. As population density rises, territory sizes will be compressed and considerable pressure will be exerted on territorial boundaries both by neighbouring groups attempting to expand their ranges and by other individuals looking for vacant territories to settle on. Smaller actively defended territories should then result. Shifts from undefended overlapping ranges at low densities to non-overlapping vigorously defended territories at higher densities have been documented in colobus (Dunbar in press), langurs (Yoshiba 1968), vervets (Gartlan and Brain 1968, Kavanagh 1981) and tamarins (Dawson 1979). Even species that do not normally defend territories may begin to do so when the population density is high, providing the geographical configuration makes it easy to defend the area and the range contains resources that are worth defending (e.g. baboons: Hamilton *et al.* 1976; macaques: Lauer 1980).

As population density continues to rise, however, the defendability of a resource will reach an asymptote and start to decline again due to the fact that the frequency of intrusions will become so high that territory-holders are unable to prevent all intruders entering their territory and exploiting its resources. At this point, a territory is no longer worth defending; territory-holders are then better off concentrating on defending access to the individual resource patch they happen to be using and allowing other individuals or groups to use the other parts of the range. Examples of such behaviour include the sharing of scarce sleeping cliffs by bands of hamadryas baboons (Kummer 1968) and of the few waterholes that still contain water at the end of the dry season by patas monkeys (Struhsaker and Gartlan 1970). Similar shifts from territorial defence to resource sharing as the frequency of intruders rises have been documented in many other taxa (reptiles: Brattstrom 1974; birds: Davis 1959; mammals: Davis 1958, LeBoeuf 1974).

The Problem of Predation

The problem with having to move any distance in search of food is that it exposes animals to the risk of being caught by predators. The safest place for most primates to be is in trees, but, even among arboreal species, feeding may oblige animals to spend time in the canopy where they are more exposed to the risk of capture by birds of prey.

The nature of predation risk depends on the type of habitat and on the size of the animals, as well as on the density of predators. To some extent, it also depends on whether there are other species present that predators prefer to hunt. Generally speaking, species that live in open country are exposed to a higher risk of predation than those that live in forested habitats. This is partly because animals in open habitats have fewer trees in which to escape predators. Small species are also vulnerable to more predators than are larger species, simply by virtue of their size. Thus, Struhsaker (1967a) pointed out that 15 to 20 kg baboons in the Amboseli area of Kenya are preyed on by only four species of predator, whereas 5 kg vervets are prey for up to 16 different predator species (one of which is the baboon).

Despite the seriousness of predation risk, deaths due to predation have not been widely documented in primates. A number of studies have reported occasional losses to predators: baboons (Bert *et al.* 1967, Anderson 1981, D.R. Rasmussen 1983); mandrills (Jouventin 1975); langurs (Schaller 1967); colobus (Badrian and Badrian 1984a); talapoin monkeys (Gautier-Hion 1973); guenons (Gartlan and Struhsaker 1972). In addition, Terborgh (1986) found that 39% of the prey items brought back to the nest by New World harpy eagles were the remains of monkeys. But only two studies have suggested that predation constitutes a major cause of mortality. Busse (1977) recorded 39 kills by chimpanzees in a community of red colobus over a two-year period, a figure estimated to represent 8–13% of the total population. In the second study, Cheney *et al.* (1986a) found that only 19 of the 68 deaths in their vervet monkey population could be attributed to disease, and the rest could be attributed to predation (12 known and 37 probable cases).

Because mortality due to predation appears to be negligible, there has been a tendency to conclude that predation is of little consequence for the animals and, therefore, that it has had only a negligible impact on either their behaviour or the evolution of their social systems (see, for example, Wrangham 1983, Cheney and Wrangham 1987). Unfortunately, this confuses a crucial distinction between *actual* predation (as a cause of mortality) and *risk* of predation (as a cause of evasive behaviour). The fact that predation occurs only rarely does not mean that predation risk does not influence the behavioural strategies of the animals concerned (see also Busse 1977, Rowell and Chism 1986a).

In fact, the risk of predation experienced by animals does seem to be significant even in those cases where actual losses to predators are rare. Terborgh (1983), for example, observed attacks by seven species of large raptorial birds on his five species of South American monkeys. Though no observed attack was successful, the animals clearly treated them as a very serious threat. Precipitate flight reactions or nervousness in response to the appearance of both terrestrial and aerial predators have been reported for terrestrial species (e.g. baboons: Rowell 1966a, Aldrich-Blake *et al.* 1971) and arboreal species (e.g. forest guenons and colobines: Struhsaker and Gartlan 1970, Gautier-Hion 1973, Dunbar and Dunbar 1974b; capuchins: de Ruiter 1986). Other studies have reported that animals become nervous and move rapidly in close formation (a) when crossing exposed ground where they are far from the safety of trees (Byrne 1981), (b) when in dense vegetation where poor visibility makes them vulnerable to surprise attack (D.R. Rasmussen 1983), or (c) when passing places where predators have recently attacked them (Altmann and Altmann 1970). Finally, the fact that all species have (and regularly use as well as respond to) predator alarm calls suggests that primates are sensitive to the risk of predation. Vervet monkeys, for example, possess three acoustically distinct alarm calls, each given exclusively to a different category of predator (terrestrial, aerial and snake: see Struhsaker 1967b, Seyfarth *et al.* 1980). Most primates are also very attentive to their surroundings, continually glancing up while feeding on the ground or in exposed positions.

Taken together, these observations indicate that predation is not regarded as a negligible risk by the animals themselves, even if their losses to predators may not be especially significant at the end of the day. The problem is that the absence of predation can just as easily be explained by the claim that the animals' anti-predator strategies are very effective as it can by the claim that they suffer no predation risk. To distinguish between these two possibilities, we need to look closely at the behaviour of the animals themselves. Doing so clearly indicates that the first explanation is the more likely to be correct. This should caution us against the naive assumption that behaviour is ever directly determined by a single environmental variable. In practice, what we observe is the outcome of the interaction between a number of environmental factors and the animals' own biological characteristics. Consequently, we need to be very careful about dismissing as unimportant factors that leave no observable trace at the more conspicuous phenomenal level (see p. 50 for another example discussed by Iwamoto and Dunbar 1983).

Chapter 4
Demographic Processes: (1) Lifehistory Variables

Demography deals with two different, but related, aspects of an animal's biology. *Lifehistory processes* are concerned with the rates at which animals give birth, die and migrate; *demographic structure* is concerned with the composition of a population in terms of various age and sex classes and the way that the animals are distributed among social groups. Lifehistory processes are the more fundamental of the two because it is rates of birth, death and migration that are largely responsible for determining the actual composition and size of social groups.

Both aspects, however, have significant implications for the behaviour of animals, albeit for different reasons. The demographic structure of the population is important because an individual's choice of social partners is ultimately limited by their availability. Because the group provides the context within which each animal has to make its social and reproductive decisions, a knowledge of the demographic structure of groups (and perhaps of the population in general) may help to clarify why animals behave in the way they do. In addition, an understanding of the relevant lifehistory processes is crucial for any analysis of reproductive strategies. Reproductive success at any given time tells us very little about the evolutionary consequences of particular behavioural strategies. What matters in evolutionary terms is not the numbers of offspring produced on any one day or in any one year, but the number generated over a lifetime. Producing many offspring now may be counterproductive if the effort of doing so results in premature death. Hence, in order to assess the evolutionary value of a given strategy, we need to be able to map its consequences on to the animal's lifehistory trajectory. In other words, we need to determine the pattern of mortality and fecundity over a typical animal's lifetime.

The basic demographic structure of a particular population of primates is determined by the species-typical factors that dictate its general grouping patterns and its reproductive capacities. Within this framework, however, demographic structure can vary widely under the influence of a number of environmental and social variables that directly affect processes such as birth and death rates. Both group size and population age–sex structure are also affected by the processes of group fission (whereby groups divide into two or more new independent units) and migration (whereby individuals or groups leave their ranging area either to join a neighbouring group or to set up a new ranging area elsewhere). These particular processes, together with other population phenomena like sex ratio and population dynamics, will be discussed in the following chapter.

Lifehistory Variables

Lifehistory variables are usually expressed in what is known as a *life table*: this gives an at-a-glance summary of the probability of dying or reproducing during successive time intervals throughout an animal's life. Life tables are normally determined by considering a single cohort (all the animals born during one year) and determining the proportion that die or give birth in each successive time period (usually a year) until all the members of the cohort have died. Separate life tables are normally prepared for each sex, since the two sexes often differ in terms of lifehistory characteristics.

Table 4.1 gives an example of a life table, in this case for female rhesus macaques from Cayo Santiago in the Caribbean, based on data provided by Sade *et al.* (1976). The two left-hand columns give values for the two fundamental variables, mortality rate, q_x, and fecundity, m_x, where the subscript x denotes a specific interval of time. Note the distinction between a female's

Table 4.1. Life table for female rhesus macaques on Cayo Santiago (Caribbean)

Age (years)	Mortality rate (q_x)	Fecundity (m_x)	Survivorship (l_x)	Life expectancy (e_x)	Reproductive value (v_x)
0–2	0.196	0.000	1.000	7.5	1.06
2–4	0.165	0.064	0.804	10.5	1.53
4–6	0.146	0.633	0.671	12.6	2.05
6–8	0.129	0.800	0.573	13.5	1.95
8–10	0.188	0.645	0.499	15.9	1.47
10–12	0.153	0.712	0.405	16.4	1.20
12–14	0.166	0.587	0.343	17.9	0.83
14–16	0.336	0.000	0.286	18.5	0.37
16–18	0.247	0.500	0.190	19.2	0.60
18–20	0.503	0.000	0.143	19.5	0.20
20–22	1.000	0.500	0.071	22.0	0.45

Source: Sade *et al.* (1976).
Note: Owing to small sample sizes, values for the last few age classes are more variable than would be expected.

birth rate, b_x (the number of infants born to her during each interval), and her *fecundity,* m_x (the number of *female* infants born during each interval). In primates, the ratio of male to female infants at birth usually approximates 1:1 (see Chapter 5), so that $b_x = 2m_x$ for most practical purposes. Lifehistory analyses generally find it more convenient to use m_x (fecundity) because this makes it easy to calculate various derivative indices of demographic and evolutionary significance. The remaining columns in Table 4.1 give a number of other important indices that can be derived from mortality and fecundity rates, all of which are age-specific. These include survivorship, l_x (the proportion of individuals that survive from birth to age x), life expectancy, e_x (the mean age at death of animals that survive to age x), and reproductive value, v_x (the future reproductive potential of an individual that reaches age x). Details of how to calculate these and other indices are given by Eisenberg *et al.* (1981, pp. 135–75) and Dunbar (1987). Caughley (1977) provides a lucid account of the theoretical background to these calculations (see also Wilson and Bossert 1971) and explains some of the common pitfalls in their use.

Survivorship, l_x, can be calculated in one of two ways: either directly by observing a cohort born in the same year and determining the proportion of individuals that survive to each successive time interval (the longitudinal method), or indirectly from the observed age-specific mortality rates, d_x (the cross-sectional method). The second method can be used to estimate survivorship when data are available only for a limited period (e.g. a one-year field study). Survivorship can also be estimated from the age structure of the population (sometimes known as the *standing crop*) by assuming that each successive age class results from mortality acting to reduce the number of animals in the immediately preceding age class. However, this method is valid *only if* (1) there is no emigration or immigration into the population, and (2) the population itself is demographically stationary (i.e. deaths exactly balance births: see below, p. 76). Since this is rarely likely to be the case in primates, the method should not be used.

From survivorship and fecundity, we can calculate a composite index known as the *net reproductive rate,* R_0. This is defined as:

$$R_0 = \sum_x l_m m_x \qquad (4.1)$$

This index can be interpreted as the average number of female offspring that a female can expect to produce during her lifetime. Since it therefore estimates the female's success at replacing herself, it can also be used as an index of a population's demographic health. If $R_0 = 1$, each female manages to produce only just enough daughters to replace herself when she dies, so the population would obviously remain stable in size. If $R_0 < 1$ the population will go into decline, and if $R_0 > 1$ it will increase in size. It will be seen that $R_0 = 1.9$ for the rhesus population in Table 4.1.

Note that this does not mean that the Cayo Santiago population is doubling its size almost every year. R_0 is purely a measure of the female's ability to replace herself during her lifetime, and takes no account of whether or not those offspring can themselves survive to reproduce. The growth rate of a population is given by r, the intrinsic rate of increase (sometimes also known as r_m, the Malthusian parameter). (It should not be confused with the coefficient of relationship for which, confusingly, the same symbol is used!) Although r can be derived from R_0, the relationship between them is not straightforward, and r usually has to be estimated by trial and error (see Wilson and Bossert 1971, Caughley 1977). It is usually more convenient to estimate r directly from the observed difference in population size in two successive time periods. Since the size of the population at time t is related to its size at time 0 by:

$$N_t = N_0 e^{rt}$$

(where N is the size of the population at a given time, e is the base of natural logarithms, namely 2.718, and t is the time interval), it is easy to see that:

$$r = t^{-1} \ln (N_t / N_0)$$

where t^{-1} signifies the reciprocal of t (i.e. $1/t$) and ln signifies the natural logarithm. When the population is stationary (i.e. $R_0 = 1$), $r = 0$.

In terms of individual reproduction, R_0 is an estimate of the number of genes that a female contributes to the next generation. As such, it provides us with a measure that we can use to estimate the evolutionary value of a given strategy, particularly if we calculate an age-specific version, R_x, defined as:

$$R_x = \sum_{y=x} l_y m_y \tag{4.2}$$

Essentially, R_x is the number of same-sexed offspring that a female aged x can expect to produce during the remainder of her life.

An alternative but much less easily calculated measure of future reproductive potential is Fisher's *reproductive value*, v_x. This is defined as:

$$v_x = l_x^{-1} e^{rx} \sum_{y=x} e^{-ry} l_y m_y$$

Reproductive value is an index of a female's future reproductive potential expressed as a ratio of the reproductive value of the average female's expectation of lifetime reproductive output at birth (which is always taken to be $v_0 = 1$). If the population happens to be stationary ($r = 0$), this formidable equation simplifies to:

$$v_x = l_x^{-1} \sum_{y=x} l_y m_y$$

Unfortunately, this happy state probably occurs rather rarely in natural primate populations (for the reasons discussed in Chapter 5, below), so we are generally obliged to use the full equation. Since this is cumbersome and requires us to determine additional variables, it is often more convenient to determine R_x using equation 4.2.

The rhesus data given in Table 4.1 exhibit a number of features that are characteristic of mammals in general and primates in particular. Mortality, for example, is high during the immediate post-natal period, and levels out to a minimum during the juvenile period before beginning to rise at an accelerating pace with age thereafter. Virtually identical patterns have been found in other species of macaques (Dittus 1975, Sugiyama 1976, Fa 1986), gelada (Dunbar 1980b), vervets (Fairbanks and McGuire 1986, Horrocks 1986) and chimpanzees (Teleki *et al.* 1976, Goodall 1983). In each case, the actual values are different, but the general shape of the curve is similar.

Fecundity shows the reverse trend, being zero up to the age of first reproduction, then climbing steeply to reach a maximum during the female's prime and finally falling off steadily as old age sets in (macaques: Dittus 1975, Koyama *et al.* 1975, Masui *et al.* 1975, Takahata 1980, Silk *et al.* 1981a, Fa 1986; baboons: Strum and Western 1982; gelada: Dunbar 1980b; chimpanzees: Teleki *et al.* 1976, Goodall 1983). As a result, reproductive value shows a corresponding inverted-J-shaped trajectory. The initial low value early in life is due partly to the fact that fecundity is zero before puberty and partly to the fact that some females die during infancy yet still contribute proportionately to the v_x for the average female.

A female's future lifetime reproductive output, R_x, depends on four key factors: litter size, age at first reproduction, interbirth interval and lifespan. The majority of primates normally give birth to a single infant, with twins occurring only very rarely. The exceptions amount to a few species of New World monkeys (e.g. marmosets and tamarins) that twin habitually and a few prosimians that produce up to three young at a time (e.g. mouse lemur: Martin 1973).

Age at first reproduction, on the other hand, is more variable, both within and between species. For the rhesus population in Table 4.1, females give birth for the first time at about 4 years of age, some 6–12 months after undergoing puberty. Data from captive populations of a number of Old World monkeys suggest that ages of first reproduction of around 3–5 years are typical (Gautier-Hion and Gautier 1976, Hadidian and Bernstein 1979, Rowell and Richards 1979), with the large-bodied apes being significantly older (*c.* 10 years of age for chimpanzees: Smith *et al.* 1975) and smaller primates rather younger. Age of first reproduction is often several years later in the wild than it is in captivity, mainly due to slower rates of development in the wild where nutritional conditions are generally poor by comparison with laboratory diets. Female baboons, for example, experience their first birth at a mean age of 6.5 years in the wild, about two years later than they do in captivity (Altmann *et al.* 1977). Similarly, Goodall (1983) found that, in the wild, chimpanzees do not usually conceive for the first time until they reach an age of 13 years, some three years later than in captivity, though Harcourt *et al.* (1981a) give 10 years as the mean age of first reproduction in wild gorilla. The values derived from captivity thus represent the minimum age at which females are *capable* of reproducing for the first time under optimum nutritional conditions. The importance of nutritional conditions in triggering the onset of puberty (and thereby determining the age of first reproduction) is demonstrated by the findings that, in provisioned free-living populations, age of first reproduction approximates those recorded in captivity (e.g. macaques: Fa 1986) and that it is delayed if provisioning is withdrawn (macaques: A. Mori 1979a, Sugiyama and Ohsawa 1982a).

The third factor affecting a female's reproductive output is the length of the interbirth interval (in other words, her birth rate). The typical reproductive cycle of a female primate consists of three discrete phases. When a female returns to breeding condition following her previous birth, she undergoes a number of oestrous cycles, each of about one month's duration. Three cycles to conception is probably about the average for most species (see for example van Wagenen 1945), though individual females may be very variable. Once the female has conceived, she enters the second phase, gestation, during which the fetus grows in her womb. Data from both captive and wild populations indicate that pregnancy lasts around 180 days in most species of monkeys (baboons: Altmann *et al.* 1977; macaques: Winkler *et al.* 1984; guenons, colobins and mangabeys: Gautier-Hion and Gautier 1976), rather less in prosimians (Charles-Dominique and Martin 1972; Izard and Simons 1986) and rather more in the great apes (Faiman *et al.* 1981). (For a general discussion of the reproductive characteristics of primates, see Spies and Chappel 1984.) Following the birth of her offspring, the female enters the third and longest stage, the period of lactational amenorrhoea during which lactation inhibits a return to oestrous cycling (Delvoye *et al.* 1978, McNeilly 1979). The one exception to this seems to be the marmosets who are able to conceive again shortly after birth while lactating (Lunn and McNeilly 1982). The mouse lemur may also be able to rear two litters in each breeding season, suggesting that it too may conceive again while lactating (Charles-Dominique and Martin 1972). These may be special cases, however.

By and large, differences in interbirth intervals (and hence in birth rates) between populations largely reflect differences in the length of the period of lactational amenorrhoea. Under the nutritionally benign conditions of captivity, females of many species are capable of reproducing each year. Hadidian and Bernstein (1979), Rowell and Richards (1979) and Fairbanks and McGuire (1984) report interbirth intervals in the order of 12–18 months for a variety of Old World monkey species in captivity (equivalent to birth rates of 1.01–0.68 per female per year). Provisioned populations of free-ranging monkeys generally have comparably high reproductive rates (e.g. macaques: Drickamer 1974a, A. Mori 1979a, Fa 1986). Few natural populations can match such high rates, however: in most cases, interbirth intervals of 18–24 months are more typical (baboons: Altmann *et al.* 1977, Strum and Western 1982; gelada: Dunbar 1980b, Ohsawa and Dunbar 1984; macaques: Takahata 1980; langurs: Winkler *et al.* 1984). Interbirth intervals are considerably longer in the large-bodied apes (5.2 years for one population of wild chimpanzees: Goodall 1983; *c.* 5 years in orang utans: Galdikas 1981) and much shorter in the smaller prosimians (e.g. galago: Doyle *et al.* 1971, Izard and Simons 1986; mouse lemur: Charles-Dominique and Martin 1972) and New World callitrichids (84% of pygmy marmosets twin twice each year even in the wild: Soini 1982; see also Stevenson 1986).

An important exception to this general rule occurs if the infant is stillborn or dies before weaning. In these cases, the mother invariably comes back into oestrus sooner than she would otherwise have done, in many non-seasonal breeders immediately (macaques: Hadidian and Bernstein 1979; gelada: Dunbar 1980b; langurs: Winkler *et al.* 1984; chimpanzees: Goodall 1983). This can have the anomalous effect of elevating birth rates in populations suffering from high neonatal mortality rates in impoverished habitats compared with populations living under more benign conditions where infant survival is greater. (Note that this implies that we must beware of assuming that high birth rates necessarily signify a good quality habitat without corroborative evidence from mortality rates that this is in fact so.)

The length of the lifespan imposes the ultimate limit on the length of the female's reproductive career. This can vary considerably. The smaller prosimians probably live for only about 10 years at most, whereas intermediate-sized Old World monkeys may live to ages of 20 or more and the larger apes may achieve ages as advanced as 45–50 years (Bowden and Williams 1984). It should be noted, however, that while such extremes of longevity do occasionally occur, they are seldom of much significance to evolutionary analyses. In the first place, most female primates enter a non-reproductive post-menopausal phase if they live long enough (Bowden and Williams 1984, Spies and Chappel 1984). Secondly, evolution is in fact a function of what the *average* animal does, so that life expectancy, e_x, or survivorship, l_x, will usually be more relevant. Different strategies may, of course, result in animals incurring radically different survivorship functions as a result of contrasting risks of death. In this case, a comparison of the lifehistory parameters for animals that pursue each of the strategies may be crucial.

Throughout this section, I have concentrated on female lifehistory parameters. Similar analyses can be undertaken for males, with similar constraints operating to influence key variables. The only significant difference is likely to

Table 4.2. Variance in annual birth rates in free-living populations

Species	Population	Birth rate/female		Sample (group-years)	Source
		Minimum	Maximum		
Olive baboon	Gilgil, Kenya	0.427	0.711	9	Strum and Western 1982
Hamadryas baboon	Erer Gota, Ethiopia	0.263	0.950	5	Sigg et al. 1982
Gelada	Simen, Ethiopia	0.143	0.666	40	Dunbar 1980b
Rhesus macaque	Chhatari, India	0.727	1.000	30	Southwick and Siddiqi 1977
Rhesus macaque	Kathmandu, Nepal	0.513	0.725	8	Teas et al. 1981
Rhesus macaque	Cayo Santiago, Caribbean	0.410	0.790	80	Drickamer 1974a
Toque macaque	Polonnaruwa, Sri Lanka	0.590	0.724	4	Dittus 1975
Japanese macaque	Koshima, Japan[a]	0.150	0.750	12	A. Mori 1979a
Japanese macaque	Arashiyama, Japan	0.380	0.870	17	Koyama et al. 1975
Barbary macaque	Gibraltar	0.000	1.000	82	Fa 1986
Hanuman langur	Jodhpur, India	0.091	0.889	12	Winkler et al. 1984
Chimpanzee	Gombe, Tanzania	0.034	0.227	10	Teleki et al. 1976

[a] Unprovisioned population only.

lie in the nature of the reproductive function, m_x. A female's reproductive rate is limited solely by the rate at which she can complete a full reproductive cycle (her reproductive 'turn-round' time). Thus, females tend to be reproductively active throughout most of their lives once puberty has been reached. Among males, on the other hand, active breeding by post-puberty animals may be prevented by social factors like competition among males which effectively allow only the most powerful males to breed. This will have the effect of compressing the male's active reproductive lifespan into a restricted segment of his total lifehistory. Moreover, when he does breed, his reproductive rate will often be very much less variable over time than is the case for females where marked age effects occur.

Variance in Birth Rates

Ecological causes

Not only do birth rates vary over the course of a female's lifetime, but, for females of any given age, they may also vary significantly from one year to the next and from one population to the next within a given species. In a sample of 18 populations of baboons (genus *Papio*), for example, the mean birth rate for individual groups varied from 0.36/female/year to 0.89 (Dunbar and Sharman 1983). Variability of comparable magnitude has been observed within free-living populations of a number of other species (Table 4.2).

This variation in fecundity has a number of possible causes. Environmental factors are particularly important because they affect an animal's physiological condition. In most species of mammals, females have to achieve a minimum nutritional plane before they can ovulate. Frisch (1978, 1982) has suggested that the criterion is a minimum fat:body weight ratio, the implication being that a female can only initiate a new reproductive cycle if she has a significant surplus of energy stored as fat to enable her to undertake gestation without overtaxing herself. Although Frisch's claim has been disputed, it remains a well established fact that, irrespective of the physiological mechanism involved, physical condition is a key factor determining a female's fertility in mammals (Sadleier 1969, Belonje and van Niekerk 1975, Andersen *et al.* 1976). This appears also to be true of primates. A. Mori (1979a), for example, found that the mean birth rate for the Koshima Is. population of Japanese macaques fell from 0.67/female/year during the period when the population was artificially provisioned to 0.32 female/year when provisioning was withdrawn. He attributed the fall in the birth rate to a correlated decline in mean female body weight. Similarly, van Schaik and van Noordwijk (1985a) found a correlation between female condition and birth rate in the longtailed macaque. The importance of nutritional condition is amply attested to by the significantly higher reproductive rates of females living in captivity. There is also evidence of dramatic reductions in the numbers of cycling females following a sudden decline in the food supply for free-living baboons (Hall 1963) and macaques (Loy 1970). In both these cases, females eventually stopped all reproductive activity. Correlations between birth rates and habitat productivity have been reported within a single group over time (baboons: Strum and Western 1982), between groups of the same population inhabiting qualitatively different micro-habitats (vervets: Cheney *et al.* 1986a; colobus: Dunbar in press) and between populations living in different habitats (baboons: Dunbar and Sharman 1983; macaques: Takahata 1980, Ménard *et al.* 1985).

Other environmental factors may also play a significant role in certain habitats. Ohsawa and Dunbar (1984), for example, found that birth rates declined with increasing altitude in the gelada. This was shown to be a consequence of the animals' inability to carry fetuses to term when cold placed heavy energetic demands on them. Declining birth rates with altitude have been documented in a number of other species of mammals, including man. In the latter case, at least, the lower oxygen tension at high altitudes has been implicated as a likely cause (Clegg and Harrison 1971).

Birth rates may also be influenced by a number of demographic factors. Because female fecundity is age-dependent (Table 4.1), the mean birth rate for a group or population will depend on its age structure. The more old females there are, the lower will be the average birth rate, whereas birth rates will be higher if there are more young females (see Table 4.3). Although Strum and Western (1982) found that habitat productivity was the primary factor determining birth rates in their study group of baboons, there was none the less a detectable effect due to changes in the female age structure over time. The correlation between the observed mean birth rate and that predicted by the age structure of the troop was the only one of 10 alternative hypotheses that came anywhere near being significant ($r_s = 0.569$, $p=0.055$). Partial correlation analysis would undoubtedly reveal a significant effect here.

Because interbirth intervals in primates are often longer than 12 months, any tendency towards synchrony in the reproductive cycles of the females will

Table 4.3. Influence of female age structure on mean birth rate in a hypothetical population

| | Number of females | | Mean birth rate |
	Young[a]	Old[b]	
Population A	75	25	$(0.75 \times 0.50) + (0.25 \times 0.25) = 0.438$
Population B	25	75	$(0.25 \times 0.50) + (0.75 \times 0.25) = 0.313$

[a] Birth rate $b_x = 0.50$.
[b] Birth rate $b_x = 0.25$.

result in marked oscillations in the birth rate from year to year. If females reproduce once every two years, a tendency for reproductive cycles to be synchronised will result in years with high and low birth rates alternating. Evidence of within-group synchrony in non-seasonally breeding species has been obtained from baboons (Kummer 1968, Stoltz 1977, Anderson 1982, Patterson 1973, Altmann 1980), gelada (Dunbar and Dunbar 1975, Dunbar 1980b), langurs (Yoshiba 1968, Rudran 1973, Hrdy 1974, 1977), vervets (Rowell and Richards 1979, Kavanagh 1983) and chimpanzees (Nishida 1979, Goodall 1983, Badrian and Badrian 1984).

One major cause of reproductive synchrony is likely to be environmental seasonality. Some species of primates are obligatory seasonal breeders in all habitats (e.g. macaques: Lancaster and Lee 1965), but others appear to be more facultative in this respect. Studies of both langurs (Rudran 1973, Bishop 1979) and howler monkeys (Jones 1985) have shown that populations living in more seasonal habitats have more discrete birth seasons than those living in less seasonal habitats. Reproductive cycles may become entrained to environmental parameters in order to ensure that females have adequate supplies of food during that period of the reproductive cycle when they are under most nutritional stress (Sadleier 1969). In other cases, the onset of reproduction may be triggered by a seasonal flush of vegetation produced by rain (e.g. macaques: Fa 1986; gelada: Dunbar 1980b; patas monkeys: Rowell and Richards 1979). Surprisingly, perhaps, the functional significance and environmental determinants of birth seasonality in primates remain as poorly understood today as they were when they were first reviewed more than two decades ago by Lancaster and Lee (1965).

Social causes

In many cases, however, reproductive synchrony may have little to do with environmental factors. Instead, it may be entirely driven by social factors. In such cases, births may be more or less evenly distributed throughout the months of the year for the population as a whole, but show marked synchrony within individual groups (e.g. hamadryas baboons: Kummer 1968; gelada: Dunbar and Dunbar 1975, Dunbar 1980b). Such synchrony appears to be due to social facilitation among females. In the gelada, for example, females tend to return to oestrus prematurely following the takeover of their units by a new male rather than waiting for the following rainy season when they would normally have returned to reproductive condition (Dunbar 1980b, Mori and Dunbar 1985). This results in birth intervals being shortened by up to 12 months. Rowell and Hartwell (1978) have shown that one female patas monkey returning to oestrus after a period of lactational amenorrhoea triggers the other females in her group to do the same. Challenges to the hegemony of the group's male harem-holder are likely to trigger a return to oestrus in this species, a phenomenon which Rowell (1978) has termed the 'Hoo Haa' effect. Synchrony of menstrual/oestrous cycles is known to be mediated by smell in both humans and rodents (McClintock 1971, 1978). Whether this is also the physiological mechanism that synchronises reproductive cycles over the longer term remains uncertain, however.

Finally, purely social factors may play an important role in determining the birth rates of individual females. In a recent survey of the literature, van Schaik (1983) found a negative correlation between the number of offspring per female and the number of females in a group for 22 out of 27 populations of Old and New World monkeys. Similar results have been reported for gorillas (Harcourt et al. 1981c) and colobus monkeys (Dunbar in press). A number of other studies have reported correlations between an individual female's dominance rank within her group and either her birth rate (gelada: Dunbar 1980a, 1984a; macaques: Drickamer 1974a, Wilson et al. 1978, Dittus 1979, Takahata 1980, Gouzoules et al. 1982; vervets: Whitten 1983, Fairbanks and McGuire 1984) or the number of surviving offspring she has (baboons: Busse 1982; macaques: Silk et al. 1981a). Not all studies that have looked for such a correlation have found one (e.g. vervets: Cheney et al. 1986a; macaques: Wolfe 1984; langurs: Dolhinow et al. 1979), but it is none the less clear that the phenomenon occurs widely.

In some cases, it may be possible to attribute loss of fecundity with declining dominance rank to access to food sources. There is evidence that high-ranking female Japanese macaques are heavier than low-ranking individuals and that they therefore produce infants that are heavier at birth, grow faster and survive better (A. Mori 1979a, Sugiyama and Ohsawa 1982a). Small (1981) found that high-ranking female rhesus macaques had higher fat indices than lower ranking individuals, while Riopelle et al. (1976) found a correlation between infant birth weight and maternal weight at conception in this species.

In other cases, however, loss of fecundity has been found to occur independently of competition for food. Dunbar and Sharman (1983), for example, found no correlation between group size and birth rate in a sample of baboon populations, but did find a correlation between birth rate and adult sex ratio: the more females per male in the group, the lower the mean birth rate. J. Silk (personal communication) also obtained a significant negative correlation between sex ratio (number of females per male) and birth rate in a captive group of bonnet macaques over a 14-year period. Dunbar and Sharman (1983) suggested that the baboon data could be explained by competition among the females for access to male coalition partners as a means of reducing harassment from other group members. Increased harassment was presumed to disrupt a female's reproductive physiology in such a way as to cause increased infertility among females who were unable to acquire suitable allies. An alternative explanation might be that males provide their female partners with access to preferred food sources, so making it possible for these females to breed faster. Though a possible explanation for the baboon data, this is unlikely to be the explanation for Silk's macaques since these had unlimited food, and surplus food was always present at the end of each day.

A decline in fertility with declining dominance rank within groups has been reported for the gelada (Dunbar 1980a). If resource competition alone were responsible, then it would be expected that the variance in fecundity would be greatest *between* units since there will be much greater variation in the food resources in the area covered by the very large herds formed by the gelada (200–500 animals) than in the small areas covered by any one of the constituent units (each of some 10–15 animals). In fact, far from reproductive rates increasing with group size as would be predicted by the larger groups' abilities to monopolise the best resource patches, reproductive rates decline (see Dunbar 1980a). These data can be explained most parsimoniously as an example of partial reproductive suppression resulting from increased rates of harassment on lower ranking females as group size increases.

Reproductive suppression

During the past decade, considerable interest has developed in the suggestion that stress can precipitate reproductive suppression. The physiological processes underlying it have been the subject of a number of clinical and experimental studies in humans (Zacur *et al.* 1976, Carenza and Zichella 1979, Grossman *et al.* 1981, Siebel and Taylor 1982, Howlett *et al.* 1984), talapoin monkeys (Bowman *et al.* 1978, Abbott *et al.* 1986) and marmosets (Abbott 1984). The results can be summarised as follows (see, for example, Yen and Lein 1984). The normal reaction to any form of physical injury, pain or psychological stress is for the brain to pump endogenous opiates (enkephalins and endorphins) into the bloodstream. These act to dampen the pain, probably in order to allow the body to continue functioning long enough for the organism to escape the source of the pain (a predator or aggressive conspecific, for example). One of the costs associated with this benefit is that endorphins are also implicated in the physiological organisation of sexual functions. Release of the opiate blocks the release of luteinising-hormone-releasing-hormone (LHRH) from the hypothalamus. As a result, the pituitary gland is not stimulated to produce LH and, without LH, the gonads remain inactive. The effect can be total suppression of the reproductive system (or, in juveniles, delay in the onset of puberty) or it can be partial (resulting in a high frequency of anovulatory oestrous cycles that otherwise appear to be behaviourally quite normal). In the latter case, the failure to trigger an LH surge once the cycle has been initiated means that the gonads are not 'instructed' to switch off the oestrous phase of the cycle by precipitating ovulation. The result will be oestrous cycles of unusually long duration (as noted in low-ranking female gelada: see Dunbar 1980a).

Total suppression of reproductive function has been noted in a number of monogamous species, including wolves (Zimen 1976), jackals and foxes (Macdonald and Moehlman 1982), klipspringer antelope (personal observation), marmosets (Abbott and Hearn 1978, Abbott 1984) and tamarins

(French *et al.* 1984). In all these cases, daughters do not achieve menarche while they remain with the parental group; but as soon as they leave (or are experimentally removed), they immediately undergo puberty and will conceive within a few weeks of being paired with a male.

Partial suppression leading to high frequencies of anovulatory oestrous cycles has been reported from a wide range of mammalian species (for a review, see Wasser and Barash 1983). Among primates, several studies provide circumstantial evidence of a sudden loss of fertility following physical injury, severe aggression or psychological trauma (humans: Matsumato *et al.* 1968; baboons: Rowell 1970, Samuels *et al.* 1987; macaques: Dittus 1986, Ehardt and Bernstein 1986; gelada: Alvarez 1973). Sackett *et al.* (1975) found high rates of abortion and stillbirth in low-ranking female macaques and attributed this to harassment by other group members, while Wolfe (1979) reported that delayed maturation correlated with high stress levels in a group of Japanese macaques. Fairbanks and McGuire (1986) found low conception and high abortion rates among captive vervet females who were subjected to high levels of aggression because they lacked coalition partners. Fa (1986) found significant negative correlations between annual fecundity and the total number of human visitors for each of two troops of Barbary macaques on Gibraltar: he argued that feeding by visitors induced high levels of aggression among the monkeys and the resulting stress caused reproductive failure. Finally, data from four colobus populations show that increased levels of tension created by the presence of several adult males competing for access to the females produced high rates of reproductive suppression among the females (Dunbar in press).

Taken together, these results suggest that high stress levels can arise from a number of different causes. One of them may even be competition for access to limited food resources.

Though these effects are most conspicuous in females, similar effects are likely to occur in males. Schilling *et al.* (1984), for example, have demonstrated experimentally that reproductive suppression can be induced in low-ranking male mouse lemurs purely by the smell of more dominant males. More recently, Martensz *et al.* (1986) have shown that endorphin levels correlate negatively with dominance rank in male talapoins: although they did not determine whether high endorphin levels caused infertility in low-ranking males, it would be surprising if male physiology was so different from female physiology in this respect.

An interesting example of the way in which behaviour could mediate reproductive suppression in males is given by Henzi (1981). He found that subadult male vervet monkeys are particularly likely to adduct their testes up into the inguinal cavity when involved in interactions with adult males. This is probably intended to reduce the risk of damage to the testes during fights which the animal is intrinsically likely to lose, but it may have the unintended consequence of reducing the male's fertility because sperm production is particularly sensitive to heat (Vandermark and Free 1970). From a low-ranking male's point of view, it may be preferable to suffer some temporary loss of fertility now (when its chances of mating are, in any case, poor) in order to be able to survive and live to breed another day.

Other key parameters of female fecundity are also influenced by both environmental and social variables. In many monogamous species, as we have just noted, age at first reproduction is delayed so long as the offspring remains on its parent's territory. In addition, reduced food availability has been found to delay age at first reproduction in macaques (A. Mori 1979a, Sugiyama and Ohsawa 1982a), baboons (Strum and Western 1982) and vervets (Cheney *et al.* 1986a). Age at first reproduction is also significantly later in the daughters of low-ranking females than in those of high-ranking females (macaques: Drickamer 1974a, Gouzoules *et al.* 1982; baboons: Altmann *et al.* 1986; vervets: Fairbanks and McGuire 1984). Altmann *et al.* (1986), for example, found that daughters of high-ranking females underwent puberty up to 300 days earlier than those of low-ranking females, and as a result they conceived their first offspring up to 150 days earlier. It is not entirely clear what proximate mechanism is responsible in these cases: daughters of high-ranking females might breed earlier because their mothers provided them with access to better food sources *or* because their mothers are able to protect them from harassment by other members of the group. The available evidence cannot distinguish between these two hypotheses, although conventionally there has been a tendency to assume that access to food is the more important.

Aside from the classical ecological emphasis on food as the most important resource limiting primate growth rates, one reason why the causal role of social stress has been doubted is that rates of aggression in wild groups often seem to be too low to have any significant physiological effect (see, for example, Deag 1977). In fact, only very low rates of aggression are required to induce reproductive suppression: the subordinate's *perceived* self-status within the group is sufficient to do so providing it is reinforced by attacks at least occasionally (see Rowell 1966b). In the case of the gelada, the average female received only 3.5 'attacks' per day from other members of her unit, of which 95% were passive displacements or simple threats. Encounters that involved physical contact or a prolonged exchange of threats occurred only about once every five days (Dunbar 1980a). Yet this is enough to reduce a tenth-ranking female's reproductive output by 50% compared with the dominant female of her unit.

In summary, at least two key factors affect the birth rates of individual females as a consequence of their dominance rank, namely access to food and stress. Which of these two emerges as the more important, if either, will depend on the particular ecological and demographic contexts in which the animals happen to live. In some groups, neither may be sufficiently prominent to have a detectable effect; in others, one may be active and the other not, so that some reproductive suppression results; and in yet others both factors will be at work, and their combined effects will presumably give rise to very poor levels of reproductive performance. One of the challenges of the future clearly lies in sorting out exactly what is happening here at the proximate level.

Mortality Rates

Sources of variance

Most species of mammals, including primates, show a characteristic pattern of mortality over the course of a lifetime. Mortality is commonly high during the post-natal period when the unweaned infant is vulnerable to disease and a number of other risks. Once the animal is an independent juvenile, mortality is reduced to a relatively low level, and then it begins to climb with increasing steepness as the animal passes its physical prime (Table 4.1). A number of factors can, however, increase the levels of mortality experienced at any given age.

Mortality during the post-natal period is influenced partly by the mother's ability to provide adequate nutrition for her offspring through lactation and partly by her ability to protect it. A number of studies have found a correlation between infant survival and the mother's dominance rank (macaques: Drickamer 1974a, Wilson et al. 1978, Dittus 1979, Sugiyama and Ohsawa 1982a, Silk 1983; vervets: Whitten 1983, Fairbanks and McGuire 1984; baboons: Busse 1982). Although the precise reasons for this difference remain speculative in most cases, there is clear evidence to implicate access to food in at least some instances. In one population of Japanese macaques, the offspring of high-ranking females were able to achieve higher growth rates than those of low-ranking individuals because their mothers were more effective at providing them with access to the best food sources (A. Mori 1979a). Small and Smith (1986) found that an infant's chances of surviving correlated with its birth weight for male (but not female) rhesus macaques in captivity. Similar results suggesting a correlation between birth weight and post-natal survival have been reported for other mammalian species (e.g. deer: Thorne et al. 1976, Guinness et al. 1978). Other studies have demonstrated increased post-natal and juvenile mortality under conditions of nutritional stress for both primates (macaques: Dittus 1975; gelada: Ohsawa and Dunbar 1984) and other mammals (deer: Smith and Lecount 1979).

Juvenile mortality may, in some cases, have a marked sex bias. Silk (1983), for example, found that the female offspring of low-ranking female bonnet macaques suffered a disproportionately low survival rate compared with both the male infants of low-ranking mothers and infants of either sex of high-ranking mothers. Similarly, Altmann et al. (1986) found that the daughters of high-ranking female baboons were more likely to survive to maturity than either the sons of high-ranking mothers or the daughters of low-ranking mothers, whereas the sons of low-ranking mothers had better survival chances than the daughters of low-ranking mothers or the sons of high-ranking mothers.

Among mature animals, the most important factor influencing mortality rates is sex. In most species, males suffer higher mortality than females from puberty onwards. Consequently, males invariably have shorter life expectancies than females. Among the gelada, for example, males can expect to live for about 12 years compared with about 14 years for females (Dunbar 1980b). Other field studies suggest comparable differences in the life expectancies of the two sexes (macaques: Sugiyama 1976, Dittus 1979; vervets: Cheney *et al.* 1986a).

One of the main causes of higher male mortality is that, in most species, it is males rather than females that migrate between groups (Harcourt 1978). While moving between groups, animals are especially susceptible to predation. It is no doubt significant that the only primate species in which males appear to have a significantly longer life expectancy than females (the chimpanzee: Teleki *et al.* 1976) is also one of the very few species in which the females rather than the males migrate. In addition to the costs of migration, males are likely to suffer higher mortality because of their more frequent involvement in fights, either when trying to enter a new group or when contesting access to oestrous females during the mating season (see, for example, A. Jolly 1966a, Wilson and Boelkins 1970, Dittus 1977). Injuries received during these fights can be fatal, but in any case impose an additional energetic load on males through the costs of tissue repair.

Within sexes, mortality rates may also vary in relation to status. Several studies have recently suggested that low-ranking animals are especially prone to mortality during periods of environmental stress (macaques: Dittus 1977, Silk 1983, van Noordwijk and van Schaik, in press; vervets: Wrangham 1981). Cheney *et al.* (1981) found that in the Amboseli population of vervets low-ranking females were particularly susceptible to disease, but high-ranking ones were more likely to die from predation. There is no obvious explanation as to why this should be the case. One possibility is that high-ranking females have unrestricted access to the best food sources and are therefore unlikely to die from starvation and disease, leaving only predation as a likely alternative.

Survivorship curves may also vary widely both between species and, within species, between populations (Table 4.4). Life expectancy at birth, e_0, was estimated to be 7.5 years for female rhesus macaques on Cayo Santiago (Sade *et al.* 1976), 4.5 years for female Japanese macaques at Takasakiyama (Masui *et al.* 1975), but only 0.6 years for female toque macaques at Polonnaruwa (Dittus 1975). Female gelada living at Sankaber in Ethiopia's Simen Mountains achieved $e_0 = 13.8$ years, whereas those in the nearby Gich population were only able to manage 10.3 years (Ohsawa and Dunbar 1984).

Causes of mortality
This wide variance in survivorship reflects the influence of a number of

Table 4.4. Life expectancy at birth and survival to maturity for female primates

Species	Population	Life expectancy e_0 (years)	Survival to 4 years (l_4)	Source
Yellow baboons	Amboseli, Kenya	4.0	0.50	Altmann 1980
Hamadryas baboons	Erer Gota, Ethiopia	—	0.74	Sigg et al. 1982
Gelada	Simen, Ethiopia	13.8	0.88	Dunbar 1980b
Rhesus macaque	Cayo Santiago, Caribbean	7.5	0.73	Sade et al. 1976
Japanese macaque	Takasakiyama, Japan	4.5	0.54	Masui et al. 1975
Toque macaque	Polonnaruwa, Sri Lanka	0.6	0.20	Dittus 1975
Vervet	Amboseli, Kenya	1.6	0.27	Cheney et al. 1986a
Vervet	CPRC, Davis, USA	9.0	0.71	Smith 1982
Chimpanzee	Gombe, Tanzania	10.9	0.46[a]	Teleki et al. 1976

[a] Survival to 9 years (considered to be approximately equivalent in developmental terms).

environmental factors. Four main causes of mortality have been identified: starvation, disease, temperature stress and predation.

Food shortage seems to be the main cause of mortality in populations of macaques (Dittus 1977), howler monkeys (Milton 1982) and baboons (Hamilton 1986), although death by starvation is particularly common only during exceptional years. Bramblett (1967) suggested that tooth wear due to the ingestion of large quantities of grit in the diet resulted in death by starvation at relatively early ages in one population of baboons inhabiting a thorn-scrub habitat in Kenya. A comparison of two populations of colobus monkeys living in adjoining areas showed that the population living in the poorer quality habitat suffered very much higher mortality rates (Dunbar in press). It is clear from Table 4.4 that most of the populations characterised by high survival rates for immatures and long life expectancy are either provisioned populations (e.g. Cayo Santiago rhesus, Japanese macaques) or ones that live in relatively rich habitats (e.g. the Simen gelada: the gelada habitats have the highest rainfall of any of the study areas listed in Table 4.4).

Disease has been implicated as an important cause of death in chimpanzees (Goodall 1983), gelada (Ohsawa and Dunbar 1984) and howler monkeys (Smith 1977, Milton 1982). One of the difficulties with identifying disease as a cause of mortality is that it is rarely possible to carry out autopsies in the field. Moreover, the visible effects of disease are often obscured by starvation once an animal becomes too ill to feed, so that death may in the end be attributed to starvation rather than disease. Sick animals are also more likely to be caught by predators.

Temperature stress, though an important consideration in certain habitats, does not normally cause mortality directly. Rather, death will result when animals are unable to maintain critical metabolic processes under conditions that would not normally kill them. Ohsawa and Dunbar (1984) noted that persistent low temperatures combined with high rainfall result in high frequencies of respiratory infections in gelada living at high altitudes (above 3000 m asl). At the other extreme, high ambient temperatures may result in death when animals are unable to gain access to water (e.g. vervets: Cheney et al. 1981, Wrangham 1981). Although temperature stress is seldom the direct cause of death in any environment where primates live, its role in precipitating death by other causes should not be overlooked on that account.

Finally, predation, as we noted in Chapter 3, can constitute an important cause of mortality in some populations (e.g. vervets: Cheney et al. 1986a), but has not generally featured so prominently as a cause of death in most studies.

The difficulty with identifying cause of death in the field, combined with the inevitably small sample sizes, makes it difficult to deduce any firm conclusions on the relative importance of these sources of mortality. Only two studies provide sufficient data to give a useful picture (Table 4.5). These underline the important point that causes of mortality are likely to vary widely from species to species (and even population to population), depending

Table 4.5: Causes of mortality in two populations of African primates

	Number of cases recorded	
Cause of death	Vervets[a]	Chimpanzees[b]
Predation	54[c]	0
Injury, cannibalism	0	17
Illness	19	23
Orphaned	5	6
Old age	0	3
Unknown	24	8

[a] Based on data for the Amboseli, Kenya, population (Cheney *et al.* 1986a).
[b] Based on data for the Gombe, Tanzania, population (Goodall 1983).
[c] Includes both known ($n = 14$) and suspected ($n = 40$) cases.

on both the ways in which the biological state variables interact and the relative prominence of key factors like predation and food availability. The vervet data demonstrate the significant impact that predation can have on a small primate. In contrast, animals as large as chimpanzees are all but immune to predation, and other factors (in this case, disease) come to the fore. Note that, in both studies, being orphaned is a significant cause of death during infancy. In general, infants whose mothers die before weaning seldom survive (see review by Thierry and Anderson 1986).

One of the most important findings to emerge from demographic studies over the last decade is the fact that when populations suffer heavy mortality (as during famines or droughts), mortality does not necessarily fall evenly on all age–sex classes. Under extreme conditions, the young bear a disproportionate share of the mortality (vervet: Struhsaker 1973; howler monkeys: Milton 1982; macaques: Dittus 1977). Dittus (1977) found that juvenile females were particularly likely to die when his toque macaque population underwent a period of severe food shortage. He argued that this was because juvenile females are least able to defend themselves or to attract coalitionary support against other group members when competition for access to limited resources becomes severe. Silk (1983) was able to confirm this in an experimental study of bonnet macaques.

★

An understanding of both the causes of mortality and its distribution over an animal's lifetime is important at two distinct levels. First, survivorship patterns are a key factor determining the effectiveness of reproductive strategies in the individual. Strategy-dependent mortality that reduces longevity necessarily reduces lifetime reproductive success and hence fitness. Thus, to determine the evolutionary value of a given behavioural strategy, we need to be able to determine the average lifespan that an animal can expect under normal conditions and the extent to which that lifespan is shortened (or lengthened) as a result of pursuing a particular alternative strategy. Secondly, mortality is a key factor determining population dynamics. An understanding of how environmental and behavioural factors are likely to affect the future structure of the population is crucial to any understanding of the value of particular strategies because the population provides the pool of interactees from which the individual has to choose its rivals or allies.

Chapter 5

Demographic Processes: (2) Population Parameters

This chapter is concerned with some of the population-level consequences of lifehistory processes, in particular with population growth rates and migration. In the final section, I discuss sex ratios, a variable that frequently plays a crucial role in determining the costs and benefits of different behavioural strategies, but whose significance has often been overlooked.

Population Dynamics

The manner and rate in which populations or groups increase or decrease over time is determined by the birth and death rates to which their members are subject. Many populations of primates are, of course, undergoing a dramatic decline: this is largely a consequence of human interference in natural ecosystems, notably through deforestation. None the less, some of the more adaptable species and those that are less directly affected by human interference may be increasing steadily. No population can grow indefinitely, however. Sooner or later, the number of animals in the group or population will reach the maximum that its ranging area can sustain. (This limit is conventionally known as the *carrying capacity* of the area.) At this point, populations either stabilise, begin to crash or lose animals by emigration.

In theory, a population can reach a state of equilibrium in which births exactly balance deaths. It is then said to be demographically *stationary*. For this to be the case, however, fecundity and mortality schedules must remain constant over a prolonged period of time. With large vertebrates that reproduce slowly and have long lifespans, this is likely to occur only in exceptional cases, mainly because natural disasters tend to occur at intervals that are shorter than the average lifespan of the animals. Since these catastrophes tend to disrupt fecundity and mortality schedules, the populations are pushed away from their natural growth trajectories. The long generation time and the presence of several generations at the same time make it difficult for the populations to return to the natural trajectory before the next catastrophe overtakes them.

Baboons, for example, are relatively long-lived: most individuals that survive to puberty can probably expect to reach the age of 15 years and a significant number will live to be 20 or so. However, severe droughts occur on a cyclical basis in eastern Africa with a periodicity of 10–11 years (Wood and Lovett 1974). Droughts of such depth place severe stress on natural populations, often decimating them (see Dittus 1977, Hamilton 1986). In other cases, long-term changes in vegetation associated with macro-climatic events can have similar consequences (e.g. baboons and vervets in the Amboseli region of Kenya: Altmann and Altmann 1970, Struhsaker 1973, Cheney *et al*. 1986a). In addition to the regular 10-year cycle of droughts, eastern Africa appears to be subject to other longer-term weather cycles, including an 80-year cycle (see Caughley 1976, Nicholson 1981), which in turn create long-term changes in vegetational profiles. Winkless and Browning (1975) and Pearson (1978) provide succinct reviews of the variety of short- and long-term weather cycles that have been identified and their biological effects.

Many populations of primates exhibit alternating periods of growth and decline, in some cases involving a catastrophic crash (e.g. macaques: Southwick and Siddiqi 1977; langurs: Winkler *et al*. 1984; howler monkeys: Milton 1982). In most cases, population crashes are brought about by starvation. Milton (1982) has argued that the population crashes that occur in the Barro Colorado population of howler monkeys occur because of periodic shortfalls in the availability of essential nutrients in the ecosystem. Hamilton (1986) found that when most of the water sources in their habitat dried up during a severe drought, baboons were unable to range widely enough in search of food: many animals died of starvation as a result.

Traditionally, food availability and predation pressure have been considered the most important factors limiting the growth of populations. The first can act directly through both birth and death rates to determine population growth rates, whereas the second can only act on mortality. Other environmental variables may also exert considerable impact in certain areas. Ohsawa and Dunbar (1984), for example, found that cold stress affected both birth and death rates in wild gelada, such that population growth rates tended towards zero as altitude increased, despite a steady increase in the amount of food available. Such cases are probably exceptional outside of temperate latitudes and high-altitude habitats in the tropics.

Food availability is usually considered to be a more important factor than predation, partly because its effects are more obvious. There is, for example, ample evidence to suggest that a population will respond rapidly to an increase or decrease in its food supply. Provisioning has allowed populations of both rhesus and Japanese macaques to expand dramatically (Sade *et al*. 1976, A. Mori 1979a, Sugiyama and Ohsawa 1982a). The subsequent withdrawal of provisioning in the Japanese macaque populations led to rapid crashes in population size. Similarly, Dittus (1977) reported a severe crash in his toque macaque population in Sri Lanka following the worst drought in 44 years, while at the same time one troop that had access to a garbage dump continued to grow at the extemely high rate of 12.5% per year. Hamilton (1986) documented a similar collapse in a population of baboons in Botswana following a drought.

Evidence of the converse (an increase in population size following an increase in the amount of food available) has been documented by Harding (1976). He found that the baboon population at Gilgil, Kenya, increased ten-fold when the introduction of cattle drinking troughs allowed the baboons to expand their range into areas where lack of surface water had previously prevented them from making use of the available food resources. Similarly, D.R. Rasmussen (1981a) reported that, in the decade following the creation of the Mikumi National Park, Tanzania, the baboon population expanded rapidly as a direct result of the cessation of agriculture in the area. In a comparative study of three qualitatively different habitats, Ménard et al. (1985) found that groups of Barbary macaques living in richer forested habitats were larger and more stable and had both higher growth rates and a higher proportion of young animals than groups living in poorer quality moorland habitats.

Some habitats, however, appear to be sufficiently rich that food resources do not seem to limit population growth. Iwamoto (1979) concluded that the gelada (a species unique among the primates in being a grazer) had more food available throughout the year than was required to sustain the population. From a comparison of the quantities of food eaten by individual animals and the productivity of the grasslands, he concluded that, on average, the animals only consumed about 5.5% of the net primary production of the habitat during the year despite their large population size. In this case, it appears that population growth rates may be limited by the rate at which animals can throughput food. At very high altitudes (in this case, 4000 m asl), the energy costs of thermoregulation are so great that the animals cannot consume grass fast enough to meet the demands imposed on them by the climate. As a result, this population can only just maintain itself at a stable population size (Ohsawa and Dunbar 1984).

Coelho et al. (1976) also suggested that the energy requirement of a population of howler and spider monkeys in Guatemala was well below what was available in the habitat. This particular claim, however, has been disputed. Cant (1980) has pointed out that not all the food present in a habitat is available for monkeys to use. Large quantities are taken by ecological competitors (particularly fruit-eating birds) and a proportion escapes consumption by rotting or falling to the ground before the animals can eat it. The importance of other species in limiting the amount of primary production available for primates cannot be underestimated, though it is rarely given serious consideration. Strum and Western (1982), however, have demonstrated that a number of demographic parameters for a population of baboons correlate best with a weighted index of the total biomass of all herbivores in the habitat (their 'Food Competition Index').

Dittus (1979, 1980) has argued that there is an in-built mechanism that prevents primate populations from rising to excessive levels. He argued that competition for scarce resources inevitably leads to high mortality among the competitively least effective members of the population. Since these are often juvenile females, growth rates are naturally retarded as recruitment into the cohort of breeding females declines. In addition, the surviving females will be growing older (and so have lower fecundity: see Table 4.1). Other studies have also noted that juveniles tend to bear the brunt of the mortality during population crashes (vervets: Struhsaker 1973; howlers: Milton 1982).

There is, however, compelling evidence to suggest that not all populations behave in this way. Some populations respond simultaneously along a number of dimensions when crashes occur. In studies of Japanese macaque populations by A. Mori (1979a) and Sugiyama and Ohsawa (1982a), population crashes following the cessation of artificial feeding were associated with a reduction in female body weight, lower birth rates, lower infant survival rates and delayed age at first reproduction (Table 5.1). In neither case was mortality among juveniles particularly significant. Other studies have also emphasised the importance of a decline in the birth rate and/or an increase in perinatal mortality as group size increases towards its asymptotic size (baboons: Rowell 1969; many other species: van Schaik 1983). Although increased mortality of juvenile females undoubtedly has some effect, Dittus's suggestion that it is a universal mechanism controlling population growth remains doubtful.

A comparison of lifehistory parameters for a large group of captive rhesus macaques with those of the free-living (but provisioned) Cayo Santiago population revealed only two significant differences. In the presumably optimal environment provided by captivity, the fecundity of young females (< 5 years of age) was higher and the mortality of adults (> 3 years of age) was lower (Smith 1982). These two differences were sufficient to allow the captive

Table 5.1. Consequences of withdrawal of provisioning on two populations of Japanese macaques

	Koshima[a]		Ryozenyama[b]	
	Before	After	Before	After
Age at first birth (years)	6.2	6.8	5.2	6.7
Birth rate/female/year	0.67	0.32	0.59	0.34
Survivorship to 2 years (l_2)	0.851	0.312	0.854	0.723
Population growth rate (%)	15.4	−4.0	13.4	4.9

[a] Based on Mori (1979a).
[b] Based on Sugiyama and Ohsawa (1982a).

population to grow at double the rate of the free-living population. Because groups in poorer quality habitats grow more slowly, we can expect to find smaller group sizes in these habitats. This has been confirmed by studies of Japanese and Barbary macaque populations (Takasaki 1981, Ménard *et al.* 1985).

Rudran (1979) has suggested that in some species of howler monkeys infanticide may regulate population size when densities reach critical limits. Although providing some support for those who view infanticide as a pathological aberration (see p. 258), it seems unlikely that infanticide could do more than scratch the surface, the more so given that it takes several years for the reduction in the female cohort to work its way through into the breeding population so as to reduce birth rates. In any case, it is clear that even among howler monkeys the various populations respond in different ways to food shortage, some by emigration (Jones 1980), others by increased juvenile mortality (Milton 1982), so that infanticide is unlikely to be a widely applicable mechanism.

Migration and Fission

In a number of well documented cases, population growth has continued to maintain rates as high as 10–15% per year without any evidence of adverse demographic effects. In some cases, this has been possible because the food supply has increased proportionately as the population has grown (e.g. provisioned populations of macaques: Drickamer 1974a, Masui *et al.* 1975), but in other cases populations have been completely natural (e.g. gelada: Ohsawa and Dunbar 1984). (It is worth pointing out here that, despite what seems to be unrestricted growth, none of these populations is demographically stable, let alone stationary. Indeed, individual groups commonly differ radically in their gross birth, death and growth rates: see Sade *et al.* 1976, Ohsawa and Dunbar 1984.) In all these cases, emigration is the most important process regulating local population size. This may involve the emigration of individuals in what amounts to a continuous trickle or it may involve entire groups (or parts of groups) leaving to find new ranging areas elsewhere.

Emigration of individuals occurs as a natural process in all species, though most of this consists simply of individuals moving out of one group into another one nearby, a process for which the term *transfer* is perhaps more appropriate. In a number of species, the majority of males transfer between groups at least once during their lives, usually between puberty and full physical maturity (macaques: Boelkins and Wilson 1972, Sugiyama 1976, van Noordwijk and van Schaik 1985, Caldecott 1986a; baboons: Packer 1979a; gelada: Dunbar 1980b, 1984a; vervets: Henzi and Lucas 1980, Cheney and Seyfarth 1983, Horrocks 1986). In a few exceptional cases, males tend to remain in their natal groups and females are the ones that transfer (e.g. chimpanzees: Pusey 1980) or both sexes change groups with more or less equal frequency (gorilla: Harcourt 1978; red colobus: Marsh 1979b). Although it has been claimed that migration is primarily a male phenomenon in the majority of species, this has been disputed.

Moore (1984), in an extensive review of female transfer, insists that females are far more active than has previously been supposed, even in those species that have been considered as archetypal male-transfer species. He argues that female transfer is particularly common among folivores, largely because there is much less pressure on folivores to form integrated groups in the defence of scarce resource patches (cf. Wrangham 1980). Moore's (1984) data on the occurrence of female transfer can be interpreted in other ways, however, for diet is confounded by habitus: most frugivores are terrestrial and most folivores are arboreal. Since arboreal species experience less predation risk than terrestrial species (when body weight is taken into account), we could argue with equal plausibility that female transfer tends to occur in those species whose way of life minimises predation risk and does not occur in those species subject to high levels of predation risk. Of 25 species in Moore's (1984) sample that are definitely known to be transfer or non-transfer species,[1] 8/12 frugivores are non-transfer as against only 1/13 folivores ($\chi^2 =$ 7.107, df=1, $p < 0.01$). On the other hand, 9/13 terrestrial species are non-transfer compared with 0/12 arboreal species ($\chi^2 = 10.067$, df=1, $p < 0.005$). Of the four exceptions among the terrestrial species, two are the chimpanzee and the gorilla whose susceptibility to predation despite a terrestrial lifestyle is likely to be significantly less than that experienced by primates of more conventional body size. If we include them with the arboreal species in a 'low-risk' group, then 0/14 are non-transfer species, as against 9/11 ' high-risk' species ($\chi^2 = 14.255$, df=1, $p \ll 0.001$). Because arboreality and folivory are so highly confounded in this sample of species, it is difficult to test conclusively between the two hypotheses. None the less, the weight of evidence would seem to be in favour of predation risk as the more important factor militating against female transfer between groups. A more conclusive test, however, will depend on the availability of data from species whose habits diverge on the two dimensions (i.e. arboreal frugivores and terrestrial folivores).

Hitherto, the consensus has been that emigration of this type is largely concerned with the avoidance of inbreeding (see, for example, Wilson 1975). This view has not gone unchallenged: both Harcourt (1978) and Moore and Ali (1984) have pointed out that emigration from natal groups often seems to occur in order to provide access to better reproductive conditions (e.g. easier access to mates or better territories) and that migration tends to be limited to one sex because only one sex normally faces limitations in that particular respect. Packer (1979a) and D.R. Rasmussen (1979), for example, have shown that male baboons tend to move from groups with few oestrous females into groups with higher ratios of oestrous females to adult males.

Similarly, Harcourt *et al.* (1976) noted that gorilla females seem especially likely to change groups when they have failed to breed successfully. Marsh (1979a) argued that female red colobus may change groups in order to avoid the risk of infanticide following a male takeover of their group.

Further evidence against the inbreeding-avoidance hypothesis derives from observations that male transfer between groups is often far from random. In some cases, males of the same natal group migrate together (macaques: Meikle and Vessey 1981, van Noordwijk and van Schaik 1985), whereas in others there are marked tendencies for most transfers to take place between a very restricted set of groups (macaques: Colvin 1983a; vervets: Henzi and Lucas 1980, Cheney and Seyfarth 1983).

In practice, the frequently quoted risks of inbreeding have probably been overestimated. For one thing, the likelihood of inbreeding is actually quite small in most primate populations even if animals mate at random because the majority of individuals will be dead or post-reproductive by the time their opposite-sexed offspring reach sexual maturity (Altmann and Altmann 1979). Only in expanding populations with short interbirth intervals and long life expectancies will the risk of parent–offspring matings become significant. But here, of course, the pool of potential mates is very much greater so that the likelihood of encountering a close relative is greatly reduced. Sibling incest, by contrast, is more likely, but because the male's active breeding career is short in most cases, the probability of a brother and sister fathered by the same male surviving to breed together is again quite small. In any case, it has been argued that a degree of inbreeding may in fact be beneficial (Bateson 1983; see also Rowley *et al.* 1986). While it remains clear that emigration helps to minimise any costs that inbreeding might incur, it is probably fair to say that inbreeding is not a significant cause of emigration in primates (see also Waser *et al.* 1986).

This kind of trickle migration, important though it is in social and genetic terms, has rather little impact on group size or local population density. Aside from the fact that the numbers of animals involved are often quite small, much of it is bidirectional, so that, in the long run, a group may gain as many animals by immigration as it loses by emigration. For population density to be significantly reduced, mass migration is usually necessary.

Although entire groups do occasionally move out of an area in search of better ranging conditions, the more common process involves the fission of a group and the emigration of one of the subgroups. Group fission is a well documented phenomenon, having been observed not only in provisioned troops of macaques (rhesus: Sade *et al.* 1976, Chepko-Sade and Sade 1979; Japanese macaques: Koyama *et al.* 1975, Masui *et al.* 1975), but also in wild populations of baboons (Nash 1976), gelada (Ohsawa and Dunbar 1984), colobus monkeys (Dunbar in press), langurs (Hrdy 1977, Mohnot *et al.* 1981), guenons (Bourlière *et al.* 1969, Cords and Rowell 1986) and chimpanzees (Goodall 1983).

There is evidence to suggest that, when groups do undergo fission, they often split along lines of 'least genetic resistance': related animals tend to stay together so that the group splits where animals are least closely related to each other (macaques: Koyama 1970, Chepko-Sade and Sade 1979; baboons: Nash 1976). As a result, the two halves generally have higher mean co-efficients of relatedness (at least through the female line) after fission than the original group did beforehand (Chepko-Sade and Olivier 1979). In other cases, individuals who groomed with each other frequently prior to the fission tended to stay together afterwards (macaques: Missakian 1973; gelada: Dunbar 1984a; guenons: Cords and Rowell 1986).

Melnick and Kidd (1983), however, found no evidence for a higher degree of maternal relatedness in the two halves of a troop of wild rhesus macaques that underwent fission (at least in terms of the distribution of serum poly-morphisms). They argued that, in wild populations, the small family sizes due to high post-natal mortality rates, combined with the fact that only one or two males are normally responsible for the conceptions that occur during any given mating season, could result in the animals being more closely related through the paternal line (see also Altmann 1979). Melnick and Kidd suggest that when group fission occurs in these cases, it may tend to occur along lines of 'least paternity', whereas 'least maternity' divisions may be more likely in expanding populations which can generate large matrilines. An alternative explanation is that, in the absence of extended kinship groups, animals may prefer to form coalitions with unrelated individuals (see Chapter 10) and will prefer to remain with these individuals rather than go with their relatives when fission occurs.

In some cases, the growth rate of a population may be very high, yet the population density within a given area may remain more or less constant as a result of group fission and emigration. *Net* stability of population size in populations with high intrinsic growth rates has been documented in gelada (Ohsawa and Dunbar 1984) and colobus monkeys (Dunbar in press). In both these cases, high growth rates within a confined population resulted in the regular emigration of groups of animals into adjacent areas where population growth rates were very much lower (or even negative). These appeared to be areas of high natural mortality where the population was only able to maintain itself through immigration. Such habitats act as 'demographic sinks' that absorb excess population from adjacent areas of high growth, thus acting so as to maintain global population densities below the critical limit.

Group size itself is determined by a combination of four demographic processes, namely births, deaths, immigrations and emigrations. Cohen (1971) has developed a family of models (the BIDE models) designed to predict the distribution of group sizes within a population. While generally valuable, his models do assume that demographic processes remain stationary over time, something that seems unlikely in most primate populations. None the less, the models have proved quite successful in predicting the sizes of various kinds of temporary subgroups whose determinants are essentially social and hence independent of environmentally driven birth and death rates. These have included the size distributions of interacting groups of preschool children (Cohen 1971) and the party sizes of semi-solitary orang utans (Cohen 1975).

The importance of these models lies in the fact that they allow us to predict future distributions of group size and to determine their likely stability. When an animal's choice of social and reproductive strategies depends crucially on the availability of suitable partners or groups of a certain size, these models may allow us to determine the way in which a group's (or population's) demographic structure is likely to change over time. An example of this is provided by the gelada. In this species, fission plays a particularly important role in regulating group size. Units undergo fission when their size becomes so large that conflicts of interest occur among the members, thus giving rise to increasing social fragmentation as the group's size grows (Dunbar 1984a). Demographic modelling has suggested that fission is likely to occur, on average, twice during a female's reproductive life in the particular population under study. The point at which a unit actually undergoes fission, however, is independent of its size and depends primarily on the rates at which units acquire secondary males, this in turn being dependent on the level of competition among males for access to breeding females (see Dunbar 1984a).

Note that the fission of a particular group does not occur instantaneously. It depends on the accumulation of costs being borne by the members and these change gradually with time rather than in large jumps. In addition, fission may depend on appropriate demographic or social conditions being realised even though the ecological or reproductive costs may make fission a desirable course of action. Thus, in the gelada, fission can occur only after a unit has acquired a second male. If no male joins, the females of the group may continue to incur increasingly severe costs as the group's size grows with time without being able to do anything about it. Because of these 'frictional' factors that slow down the rate at which demographic processes respond to their precipitating factors, group sizes will often exceed the ideal size at which they ought to undergo fission. It is for this reason that we generally find a distribution of group sizes in a population with relatively high variance around the mean rather than a tightly clustered set of groups at the mean size.

It is also worth emphasising here that fission can occur as a result of either ecological costs or social costs reaching unacceptable levels. In the gelada, again, bands undergo fission because the density of animals exceeds the carrying capacity of their ranging area (Ohsawa and Dunbar 1984). But ecological factors of this kind are obviously irrelevant to whether or not the constituent reproductive units of the band themselves undergo fission. In their case, fission occurs because the size of the unit results in intolerable levels of harassment (and hence reproductive suppression through stress) for low-ranking females.

Sex Ratio

Aside from the sheer number of animals in the population or group, sex ratio is the most important demographic variable that influences the choice of strategies open to an individual animal. In this respect, several different aspects of sex ratio need to be distinguished. The *primary* sex ratio is the ratio of males to females at conception, while the *secondary* (or neonatal) sex ratio is that at birth and the *tertiary* (or adult) sex ratio is that for the adult component of the population. These invariably differ significantly, partly because of differential mortality rates between the two sexes and partly because of differences in the rates of maturation. In most primates, for example, males are not considered to be adult until they have completed physical growth (usually between the age of 6 and 8 years in most monkeys), whereas females are usually considered to be adult shortly after puberty (usually at around the time they produce their first offspring, some time between 4 and 6 years old). These definitions are intended to reflect the respective sexes' abilities to contribute to the species' gene pool, and thus to be a rough measure of the level of sexual competition. Since not all males are members of reproductive units (in some species, a proportion of the adult males are found in all-male groups), it is sometimes more convenient to refer to the *socionomic* sex ratio (the ratio of adult females to adult males *within* reproductive groups). However, in primates, the long interbirth intervals mean that, at any one time, the majority of females are not sexually active, and it may sometimes be useful to determine the *operational* sex ratio (defined as the number of sexually active females per adult male at any given moment: Emlen and Oring 1977). Whereas the more conventional sex ratios are fixed, the operational sex ratio can obviously vary enormously from day to day as females come into and go out of oestrus.

These ratios are related to each other as a hierarchical sequence in which each ratio is the outcome of an interaction between the preceding ratio in the series and certain extrinsic variables (Figure 5.1).

For most primate species, the sex ratio at birth approximates 50:50. Only two studies have found neonatal sex ratios that are significantly biased in favour of one sex: one population of greater galago had a sex ratio of 57:43 in favour of males (Clark 1978) and one band of hamadryas baboons produced a ratio of 32:68 in favour of females in 82 births over a seven-year period (see Sigg *et al.* 1982).

Although in general 50:50 is the norm at birth, this applies only when sample sizes are large. Most primates in fact live in rather small groups and, within these, sex ratios in any given year may often be heavily biased by small-sample statistical effects. These are a consequence of the fact that, when small samples are taken from a population with a given mean and variance, extreme values are more likely, and the smaller the sample the more extreme they will be. In large samples, these extreme values tend to cancel

Table 5.2. Variance in the neonatal sex ratio within individual groups of primates over time

Species	Population	Percentage of males			Sample (group-years)	Source
		Min.	Mean	Max.		
Yellow baboon	Amboseli, Kenya	37.5	50.0	66.7	8	Altmann 1980
Hamadryas baboon	Erer Gota, Ethiopia	7.1	32.0	60.0	7	Sigg et al. 1982
Gelada	Simen, Ethiopia	25.0	51.0	61.0	8	Dunbar 1980b
Japanese macaque	Koshima, Japan	25.0	50.0	75.0	7	A. Mori 1975
Japanese macaque	Ryozenyama, Japan	37.5	49.1	70.0	6	Kurland 1977
Toque macaque	Polonnaruwa, Sri Lanka	36.4	51.4	60.0	6	Dittus 1975
Barbary macaque	Gibraltar	0.0	54.2	100.0	75	Fa 1986
Chimpanzee	Gombe, Tanzania	0.0	47.1	100.0	16	Goodall 1983

Figure 5.1. Relationships between the four different sex-ratio variables

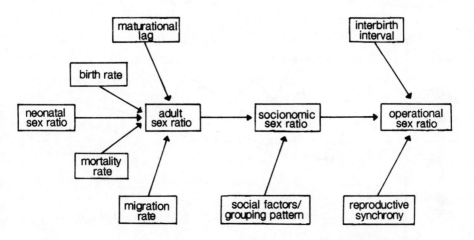

each other out to yield values that are closer to the true population mean. Table 5.2 illustrates just how variable neonatal sex ratios can be within individual birth cohorts in a number of different species. Yet, in all cases except one, the mean value for each of the sampled populations is very close to 50:50.

This variance in the neonatal sex ratio will obviously filter through to influence adult sex ratios in later years. In addition, there are a number of other variables that affect the adult sex ratio, including (1) variations in the birth rate (see Ohsawa and Dunbar 1984), (2) environmentally determined differences in the mortality rates for the two sexes (e.g. Dittus 1977, Hamilton 1986; see also Dunbar and Sharman 1983), and (3) variations in the rates with which males and females migrate between groups or populations. The interaction of these factors can cause dramatic oscillations in the adult sex ratio in small populations. Figure 5.2 gives two examples of the extent to which the adult sex ratio can vary over time. Similar changes in the adult sex ratio over extended periods of time have been noted in baboons (Rowell 1969), gelada (Ohsawa and Dunbar 1984) and howler monkeys (Milton 1982).

Birth rate can affect the adult sex ratio by a curious quirk in the mathematics of the demographic processes in those species that have a significant difference in the ages at which the two sexes mature (see Ohsawa and Dunbar 1984). Because females enter the adult cohort several years earlier than males do, the relative number of males and females entering in a given year will reflect the relative birth rates in those two years when the two cohorts were born. If the population is growing, the number of breeding females will be

Figure 5.2. Variation in the adult sex ratio over time within two small populations of macaques

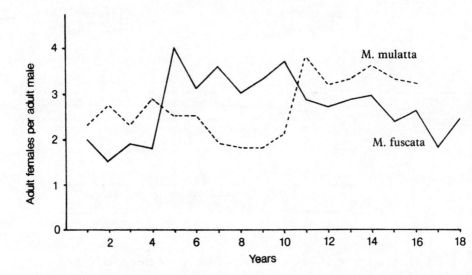

Sources: Japanese macaque (M. fuscata) from Koyama et al. (1975); rhesus macaque (M. mulatta) from Southwick and Siddiqi (1977). (Redrawn with permission of the publishers from Dunbar 1987 in B. Smuts et al. eds, Primate Societies. © *1987 by The University of Chicago Press. All rights reserved.)*

increasing year by year. As a result, the females entering the adult cohort in any given year will derive from a later (and therefore larger) birth pool, whereas the males will have come from an earlier (and therefore smaller) one. Adult sex ratios will become increasingly biased in favour of females as the birth rate rises. Conversely, in declining populations, the cohort of breeding females will be declining with time, so that males, having been born earlier, will be drawn from a larger birth pool than the females; the adult sex ratio will then be biased more in favour of males.

Finally, because of the maturation lag, environmental events that reduce the birth rate in any given year will cause fluctuations in the future adult sex ratio as those cohorts work their way through the adult segment of the population. Modelling shows that a sudden drop in the birth rate (as might be caused by a severe drought in one year) causes the adult sex ratio to move down in favour of males some four years later (when the females 'born' that year mature) and then to rise in favour of females two to three years later (when the males of that cohort 'mature'), after which it oscillates in either direction until the effects are finally damped (Figure 5.3). Significantly, the

Figure 5.3. Effect on future adult sex ratio of a year in which no infants were born (year 0) in a population with a stable age distribution and reproductive characteristics similar to those of many Old World monkeys (i.e. females mature at 4 years of age, males at 6; both sexes die at 14 years of age and females give birth once every two years, with neonatal sex ratios at exactly 50:50).

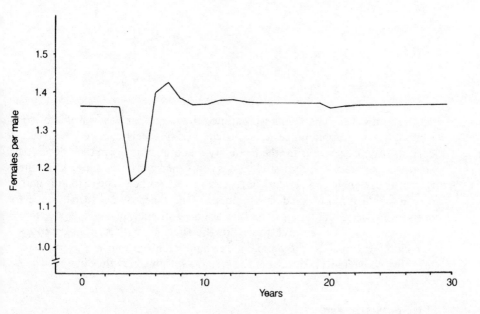

effects of a single bad year can still be detected some 15–20 years later, well after many of the animals involved would have died. It is this effect in particular that makes it unlikely that primate populations will often achieve demographic stationarity (or even stable age distributions).

Note

1. In addition to the 23 species in Moore's (1984) table, I have included black-and-white colobus among the transfer species on the evidence of Dunbar (in press), and the gelada among the non-transfer species on the evidence of Ohsawa and Dunbar (1984). I have also counted the hamadryas baboon as a non-transfer species: following Wrangham (1980), Moore seems to have misinterpreted the evidence that female hamadryas transfer between one-male units as implying transfer between ecological groups. In fact, it is clear from Sigg *et al.* (1982) that female hamadryas transfer only between one-male units of the same band.

Chapter 6
Time Budgets and Other Constraints

The previous three chapters have outlined the main features of primate ecology and demographic processes. This chapter considers some of the consequences of those features for the flexibility that animals have in their decisions about how to behave. The first section is concerned with time budgets and the way in which their components interact with each other. The second is concerned with ways in which demography can limit the choice of social partners open to an individual on particular occasions. Although in the long run subject to evolutionary selective forces, in the short term these considerations provide constraints on the animal's freedom of choice that are imposed on them by the biological features of the systems within which they live.

Time Budgets

A time budget is a classification of the way in which an animal distributes its available time among those categories of activity that are important for its survival and reproduction. In principle, time budgets can be drawn up with an infinite number of categories of behaviour, but conventionally only four main ones are used. These are *resting, moving, feeding* and *social activity.* A fifth category *other activity* (covering such things as drinking, mating, territorial behaviour, etc.), is sometimes included so that the total adds up to 100% of available time. The category 'other activity', however, seldom accounts for more than 2% of the time budget and can usually be ignored.

There are a number of problems associated with the definitions of the main activity categories, particularly feeding and social activity. In different studies, feeding has been defined as (1) ingesting food, (2) harvesting food, (3) harvesting or ingesting food, (4) any of these, plus moving between nearby feeding sites, and (5) all of these as well as long-distance progressions from one feeding area to another. Social activity has sometimes been taken to mean only social grooming, in other cases all friendly interactions, and in yet others any kind of interaction (including agonistic ones). Some studies have used the category 'grooming' rather than social activity to make this more specific, but confusion has then been introduced by the fact that some investigators have included autogrooming (self-grooming) under this rubric. On the other hand, in some socially inactive species (e.g. the gorilla), sitting in physical contact may serve the same function as grooming as an expression of social preference: some studies have therefore included 'sitting in contact' either as a separate category or as part of the heading 'social activity'. In general, the problems of definition are less acute in the larger, more social species where grooming accounts for more than 90% of all social activity.

Although there are obvious methodological problems about using data collected by investigators using definitions, the inherent biases are unlikely to be in any consistent direction. Hence, they will tend to reduce the significance of any relationships that might exist between activity categories and environmental variables rather than to inflate the significance of any statistical tests. If we get a significant result from such a test, it is likely to be a very significant one. There is, in fact, a greater risk of overlooking relationships which are borderline.

We know from the discussion in Chapter 3 that the amount of time spent feeding depends on body size, the habitat-specific costs of thermoregulation, the animal's active metabolic requirement, the reproductive metabolic requirement and the nutritional quality of the available food. Time spent moving, on the other hand, is determined primarily by the dispersion of food in the habitat and by the size of the foraging group. Surprisingly, perhaps, time spent moving does seem to be quite unrelated to the amount of time spent feeding (baboons: Altmann 1980, Dunbar and Sharman 1984; gelada: Iwamoto and Dunbar 1983, Dunbar and Dunbar in press). Social time seems to be independent of both of these. Analyses of several species' time budgets, for example, suggest that animals make every effort to conserve social time whenever additional demands are made on their time budgets by environmental conditions (baboons: Dunbar and Sharman 1984; gelada: Dunbar and Dunbar in press; macaques: Fa 1986). Social time is probably determined largely by the extent to which the animal has to invest time in interacting with other individuals in order to secure their cooperation as allies (see Chapter 11). This will be dependent primarily on (a) the minimum amount of time required to ensure an ally's loyalty, and (b) the number of allies that an individual needs to maintain.

Analyses of data from 11 baboon populations yielded the following relationships between the four main activity categories (see Dunbar and Sharman 1984):

$$R = 46.05 - 0.57F$$
$$M = 33.48 - 0.26F$$
$$S = 10.8$$

subject to the constraint that:

$$F + R + M + S = 100$$

where F is the percentage of time spent feeding, R the percentage spent resting, M the percentage spent moving, and S the percentage spent in social activity. The relationship between moving and feeding time was, however, weak, suggesting that moving time might either be a constraint like social time or that moving time is largely dependent on other extrinsic factors (e.g. group size, length of day journey). This set of system equations will not necessarily apply to other species with different dietary strategies, but given that we can establish such a set of relationships in any given case we can then use it to explore the consequences of changes in an animal's feeding time requirements that result from specified changes in its energetic requirements. We shall make use of this to explore maternal time budgets in particular in Chapters 9 and 12.

The most important point to note from these equations in the present context is that resting seems to act as a catch-all category that absorbs whatever time is not required for other activities. It thus provides a source of uncommitted time that can be drawn on whenever additional time is required for some other activity. Data from a number of species besides baboons confirm that this is in fact the case (gelada: Dunbar and Dunbar in press; macaques: Fa 1986, Seth and Seth 1986; vervets: Lee 1984). A similar conclusion is suggested by a comparison of the time budgets of two South American tamarin species: both time spent moving and time in social activity remain constant, but as feeding time increases with increasing body size the additional feeding-time requirement is taken out of resting (see Terborgh 1983).

Resting time, however, is not an unlimited source of additional time. Not only is there an absolute limit imposed by the fact that there are only 12 hours of daylight in a tropical day, but some resting time may be essential for other purposes (e.g. vigilance, allowing tired muscles to recover, digestion or simply as time gaps betwen successive bouts of activity). The use of resting time for digestion may be especially important for folivores: in order to extract nutrients from plant structural elements, leaves may have to be fermented for a prolonged period and this kind of digestive activity seems to be incompatible with other activities. Bovids and cervids that 'chew the cud' have to devote 10–15% of their time to rumination, an activity that can only be undertaken while resting quietly (see, for example, Jarman and Jarman 1973). Pseudo-ruminants like colobines and howler monkeys that actively ferment what they eat may also need to devote time to inactivity in order to allow this process to occur. There is evidence to suggest that folivorous primates do suffer a cost in this respect. The more folivorous black-and-white colobus spends more time feeding and resting and less in social activity than the more frugivorous red colobus (compare Dunbar and Dunbar 1974c with Clutton-Brock 1972), and the same is true for the more folivorous gibbons (the siamang) compared with more frugivorous species such as the white-handed gibbon (Raemakers 1979). Among primates in general, there appears to be an inverse relationship between time spent resting and time spent in social activity (Figure 6.1). Where complexity of social organisation depends on the time available to service social relationships, species that are committed by their digestive strategy to spending a great deal of time resting may find the extent to which they can evolve complex societies limited by the amount of time they can afford to devote to social interaction.

In addition to time required for digestion, environmental conditions may impose limitations on the amount of time that the animals can be active. This

Figure 6.1. Relationship between time spent resting and time spent in social activity in species of primates

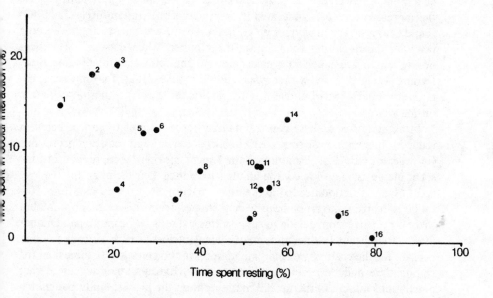

Time spent resting (%)

Sources: 1, rhesus macaque (Teas et al. 1980); 2, gelada (Dunbar 1977); 3, Guinea baboon (Sharman 1981); 4, yellow baboon (mean of three populations, see Dunbar and Sharman 1984); 5, chacma baboon (mean of two populations, see Dunbar and Sharman 1984); 6, olive baboon (mean of four populations, see Dunbar and Sharman 1984); 7, capped langur (Islam and Husein 1982); 8, ringtailed lemur (mean of two populations: Sussman 1977); 9, gorilla (Fossey and Harcourt 1977); 10, brown lemur (mean of two populations: Sussman 1977); 11, red colobus (Clutton-Brock 1972); 12, spider monkey (Richard 1970); 13, Tana River red colobus (Marsh 1981b); 14, black colobus (McKey and Waterman 1982); 15, black-and-white colobus monkey (Dunbar and Dunbar 1974c); 16, mantled howler monkey (Richard 1970). Spearman correlation: $r_s = -0.594$, p < 0.05.

is particularly relevant in habitats charaterised by very high ambient temperatures where animals may be obliged to rest in the shade during the hottest part of the day. So far, little attention has been given to this problem, other than to note that some populations of primates often rest during the middle of the day when temperatures are at their highest (e.g. patas: Hall 1965a; guenons: Aldrich-Blake 1970; macaques: Southwick *et al.* 1965). In antelope, temperature is known to have a marked effect on both behaviour and resting patterns (Dunbar 1979b). Time that the animals are forced to spend resting to avoid high heat loads is time that is effectively removed from their overall time budget. This means that they have to pack all their essential time-budget requirements into a shorter active day, with the inevitable result that conflicts of priority will become more acute. Comparable problems face animals living in temperate latitudes where day length varies significantly over the course of the year. With a shorter period of daylight in winter, time budgets must be compressed if the animals are to do all the feeding they need to do — and that at a time when energy requirements may be elevated significantly by the thermoregulatory costs generated by lower ambient temperatures. Fa (1986) found that Barbary macaques on Gibraltar spent twice as much time feeding per hour during winter than during the summer; when seasonal differences in day length are taken into consideration, the total amount of time spent feeding each day in winter was still 1.5 times that for summer, the difference being attributable to the additional costs of thermoregulation during winter.

In extreme cases, these constraints may force animals to choose between reducing the amount of time spent in social activity and reducing group size (in order to reduce the amount of time spent moving). A reduction in time spent moving will also allow the animals to reduce their feeding time requirement by that fraction needed to fuel that travel.

Time budgets have not been widely studied from this point of view, but there is evidence from gelada to support this suggestion. Iwamoto and Dunbar (1983) found that feeding time requirement increased with altitude to compensate for the costs of thermoregulation up to the point where almost all the resting time had been committed to feeding. Extrapolation suggested that populations living as little as 200 m higher than the highest study population would exhaust all their resting time. The only way in which further increases in altitude could be accommodated would be to reduce group size in order to reduce the distance the animals had to travel each day, so as to effect savings on moving time and feeding time. There was evidence that this was already beginning to happen in the highest altitude study area: group sizes here were significantly smaller than those at lower altitudes (Ohsawa and Dunbar 1984).

Lee (1983a, 1984) found that vervet groups living in better quality territories obtained more food and spent less time feeding and more time in social activity than groups living in poorer territories. Similar results have been reported for rhesus macaques (Seth and Seth 1986). The quality of relationships was also affected by energetic constraints: when feeding conditions deteriorated in the dry season, immature vervets spent less time playing and more time engaged in energetically less expensive social activities like grooming (Lee 1984). Müller-Schwarze *et al.* (1982) also found that a reduction in energy intake of as little as 1% was sufficient to cause the frequency of play behaviour to drop by 35% in white-tailed deer fawns. Such effects are bound to have a significant impact on the structure of social groups. If animals have less time to spend interacting, then they must either spend less time interacting with each of their social partners or reduce the number of individuals they interact with. Either way, groups are likely to become socially fragmented and break up into smaller units.

Another conspicuous influence on time budgets is reproduction. Female mammals in general require significant additional quantities of nutrients to make gestation and lactation possible (see Chapter 3). These impose extra demands on a female's feeding-time allocation. Dunbar (1983e) found that lactating female gelada spent about 30% more time feeding each day than non-lactating females; much of this additional feeding time was achieved by starting to feed significantly earlier in the day than other individuals. The additional time cost to the female is largely a function of the infant's energy demands on her. Data from gelada and yellow baboons reveal that the amount of time spent feeding by mothers increases steadily up to the time that the infant begins to fend for itself (see Altmann 1980, Dunbar and Dunbar in press). At its peak, the amount of time spent feeding by female gelada was 75% greater than it had been prior to the birth of the infant. This massive increase in the time spent feeding necessarily detracts significantly from the time the female has available for social activity. She has little choice but to withdraw from some of her less important relationships (see Chapter 9). Altmann's (1980) data from yellow baboons can even be interpreted as suggesting that mothers try to shift some of the burden of servicing relationships on to their friends, repaying this 'debt' later once lactation is over and they have more time available.

The main reproductive problem faced by most male primates is how to maximise the number of fertilisations achieved, the problem of rearing the infants so produced being left largely to the females concerned. Generally, this means ensuring that fertilisation occurs on every opportunity a male has to mate with a female. Since the timing of ovulation cannot be ascertained with total certainty, some degree of mate-guarding may be necessary to ensure that other males do not also mate with the female and fertilise her instead. (For an introduction to mate-guarding, see Parker 1976.) Among primates, mate-guarding is a serious problem only in species like baboons, macaques and chimpanzees that live in large multimale social groups. In these species, males may have to devote a considerable amount of time and energy to forming consortships with oestrous females in order to prevent competitors from mating with them (see, for example, Kaufmann 1965, Hausfater 1975, Tutin 1979).

K.R.L. Rasmussen (1985) has analysed the time budgets of male and female yellow baboons in relation to whether or not they were in consort at the time. Although there were complicating factors, her results generally show that both males and females spent less time feeding while in consort than they did when not consorting. This was particularly true of higher ranking individuals of both sexes, who showed marked declines in feeding time while in consort. In the case of males, this reduction in feeding time was associated with an increase in time spent interacting (mainly with the oestrous female) and time spent moving (a consequence of having to follow closely on the female's heels in order not to lose her). On average, males in consort spent about 36% less time feeding than they probably ought to have done. Berkovitch (1983) also found that the proportion of time spent feeding by female olive baboons fell from 52% under normal conditions to 41% during sexual consortships (a drop of 21%). Packer (1979b) found that olive baboon males spent around 27.4% of their time feeding under normal conditions, but only 17.5% while in consort with an oestrous female (a drop of 36% in their feeding time allocation).

Demographic Constraints on Behaviour

Demographic structure affects the behaviour of animals in two ways. First, it imposes constraints on the choice of social partners open to them. Secondly, it may influence the level of competition for access to specific resources, thereby increasing the costs of obtaining that resource or limiting the animal's access to them.

Family size and partner availability
Partner availability is not something that has attracted a great deal of attention hitherto. There has, rather, been a tendency in ethology and its derivatives to assume that an animal's options in partner choice are unrestricted. This may be the case in species that live in large unstructured swarms, but in those species that live in small semi-closed social groups, demographic structure itself can severely limit the options available, thereby colouring the nature of the social relationships in which the animals are involved.

Altmann and Altmann (1979) pointed out that, in their Amboseli baboon population, the probability of an infant surviving to the age of 2 years is only 0.45. With a birth rate of 0.5/female/year, a neonatal sex ratio of 50:50 and only 17 females in the group, the number of juveniles of any given 6-month age block of the same sex is only:

$$17 \times 0.25 \times 0.5 \times 0.45 = 0.96$$

In other words, an infant that survives to become a juvenile is unlikely to have another juvenile of the same age and sex to play with in the group. A male infant, for example, would either have to play with a female if it wanted to play with an individual of the same age or with an older or younger animal if it wanted to play with a male. Since both the nature and intensity of play and social relationships differ considerably between the sexes at any given age and, within sexes, over time (macaques: Simonds 1965; baboons: Kummer 1968; gelada: Dunbar and Dunbar 1975; squirrel monkeys: Baldwin 1969), this is likely to have a significant effect on the style of behaviour that an individual adopts as an adult. It may, for example, significantly affect an individual's aggressiveness and its skill at manipulating conspecifics.

Unfortunately, the long period of immaturity that characterises primates means that we have little idea as yet how early social experience influences style of behaviour as an adult. None the less, experimental studies have suggested that early social experience can have significant and lasting effects. Spencer-Booth and Hinde (1971a), for example, found that a temporary separation from the mother as short as seven days can still be detected in the behavioural responses of the infant up to two years later. Though excessively crude in their conception, the studies of maternal deprivation carried out in the 1950s and 1960s by Harlow and his colleagues have shown just how devastating such experiences can be. Animals reared in complete social isolation fail both to interact normally with other adults and to function adequately as mothers (for a recent review, see Swartz and Rosenblum 1981). More recently, Simpson and Howe (1986) have been able to show that small groups of rhesus macaques develop characteristic styles of behaviour which are reflected in the behaviour of their respective infants at 12 months of age. Inter-group differences in this respect were much greater than intra-group differences, and could be attributed to differences in the personalities of the adults and the styles of behaviour that they imposed on their groups.

Demographic processes can also have a dramatic impact on an adult's opportunities for interacting with close kin. This is illustrated in Figure 6.2 which gives pedigrees for three notional females (the matriarchs) living under different lifehistory regimes, assuming that all the infants born into the pedigree are female. Figure 6.2(a) represents a population with a low reproductive rate such as that found in the Amboseli baboon population: the age at first birth is 6 years, the birth rate is 0.5 per year with a survivorship to maturity of $l_{0-4} = 0.5$ (hence, an adjusted interbirth interval of about 3 years when allowance is made for the fact that females come back into oestrus soon after the death of an infant). Figure 6.2(b) represents a medium reproductive rate population (e.g. the Simen Mountains gelada) where the age at first reproduction is 4 years and survivorship is $l_{0-4} = 0.75$, but the birth rate is still only 0.5 per year (giving an adjusted interbirth interval of 2.5 years). Finally, Figure 6.2(c) shows a high reproductive rate population such as the provisioned rhesus macaque population on Cayo Santiago in the Caribbean:

Figure 6.2. Descendants of a female (shown on left in each graph) at the time of her death when all the infants born into her matriline are female. Three populations with different reproductive characteristics are illustrated: (a) a low reproductive rate with the age at first birth at 6 years and an interbirth interval for surviving offspring of 3 years; (b) an intermediate reproductive rate with the age at first birth at 4 years and an interbirth interval for surviving offspring of 2.5 years; (c) a high reproductive rate population with an age at first birth of 4 years and an interbirth interval of 1 year. A thick line corresponds to the period of sexual maturity

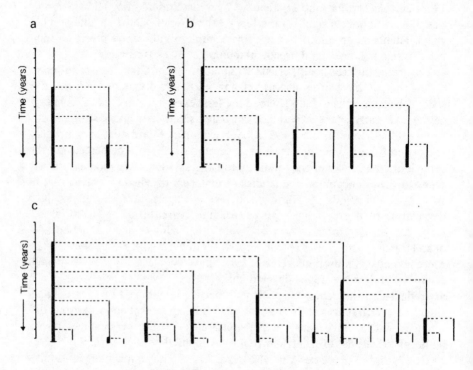

here, age at first reproduction is still 4 years and survivorship is $l_{0\text{-}4} = 0.75$, but the birth rate approaches 1.0 per year (hence an adjusted interbirth interval of 1.5 years). In all cases, females are assumed to die at age 14 years (probably a fairly typical average life expectancy for animals that achieve adulthood).

While the diagrams in Figure 6.2 represent the extreme case in which all the infants born are female, the fact that the sex ratio at birth is 50:50 means that a wide variety of different pedigrees are actually possible. By working out all the possible combinations of male and female offspring, we can determine the probability that the matriarch will have various numbers of mature (i.e. post-puberty) female descendants at the end of her own life. The resulting probabilities for different sized kin groups (including the matriarch) are given in Table 6.1, together with the mean kin group size in each case. What is

Table 6.1. Frequency distribution for kin group size for populations with different reproductive rates (see Figure 6.2)

Number of post-puberty females	Proportion of kin groups		
	Low	Intermediate	High
		Reproductive rate	
1	0.500	0.028	0.006
2	0.500	0.167	0.028
3	—	0.333	0.072
4	—	0.333	0.150
5	—	0.139	0.244
6	—	—	0.261
7	—	—	0.183
8	—	—	0.050
9+	—	—	0.006
Mean size	1.5	3.4	5.4

immediately striking about these distributions is that age at first reproduction, birth rate and survivorship all have a significant effect on the range of kin group sizes. The higher the net reproductive rate (after mortality has been taken into account), the larger will be the mean group size, the greater the variety of group sizes and the smaller will be the proportion of matriarchs that have no mature female kin. It is in this last respect that the differences are particularly striking. As many as half of the 'Amboseli' females will have no female relatives, whereas only six in every 1000 will be in this position among 'Cayo Santiago' females.

Table 6.2 gives the actual family sizes for two baboon groups: main troop at Amboseli (based on pedigrees given by Altmann 1980) and EC troop at Gilgil (based on data given by Johnson 1984). These correspond roughly to the low and medium reproductive rate populations, respectively. Though the mean kin group size is smaller than predicted by Table 6.1 in both cases, the ranges of the two distributions are exactly as would be expected. The more skewed distributions in the observed cases (hence the lower mean family size) can be attributed to the fact that in constructing the model kin groups of Figure 6.2, I assumed for simplicity that all females died at the age of 14 years, whereas in reality many females will die much earlier and much later than this. In addition, the models plot family sizes at the end of the matriarch's life, whereas many of those in the two baboon populations will include matriarchs that have not yet reached the ends of their lives: they will have smaller families than they would otherwise have. The general trend, however, is clearly in line with the theoretical demographic predictions. We shall see in Chapter 10 that this can have profound consequences for females' decisions about whom they should form alliances with.

Table 6.2. Frequency distribution of female kin group sizes in two baboon groups inhabiting different quality habitats

Number of post-puberty females	Number of kin groups[a]	
	Amboseli[b]	Gilgil[c]
1	10	8
2	3	6
3	—	5
4	—	2
5	—	2
Total groups	13	23
Mean size	1.2	2.3

[a] Defined as the female descendants of a living matriarch who are older than 4 years of age; the matriarch herself is included.
[b] Based on pedigrees given by Altmann (1980); mean annual rainfall is 225 mm.
[c] Based on pedigrees given by Johnson (1984); mean annual rainfall is 690 mm.

A more specific example of the way in which demographic variables can limit the options open to an animal is provided by the gelada. In this species, females form coalitions in order to buffer themselves against harassment by other group members. Analysis of the costs and benefits of various possible allies suggests that the most profitable coalition is one formed between a mother and her daughter, with the daughter's value in this respect declining progressively with her birth rank (Dunbar 1984a). (By birth rank, I mean the position in the order in which the mother's offspring were born.) Because of the vagaries of the lifehistory processes, not all females will find themselves in a position to form such a coalition: half the females will have a son as the first-born offspring and a significant proportion (12.5%) will have only sons for their first three offspring (after which it makes little difference what sex the offspring are since the mother will normally be dead before they have time to mature into useful allies). Such females simply do not have the option of forming coalitions with daughters and must pursue strategies that, in general terms, are less than ideal. One such strategy is to join a new male when one attaches himself to the unit. But, again, this is an opportunistic strategy: males join reproductive units on average only once every five years (or twice during the average female's reproductive lifespan). The ideal, in other words, is not always attainable because of the way demographic events happen to turn out.

Level of competition

Aside from creating tactical opportunities for animals, the demographic structure of a group may also affect the level of competition for access to limited resources, be these resources physical (e.g. feeding or sleeping sites) or social (e.g. allies). Both sex ratio and absolute group size will be important considerations in this respect.

De Waal (1977) provides an example from longtailed macaques that illustrates rather nicely the way in which sex ratio can affect intra-sex fighting (Table 6.3). When the sex ratio was heavily biased in favour of females, a high proportion of the agonistic events were exchanged between females, but when the sex ratio approached unity, most of the aggression was between males. These data suggest that competition was occurring within the commoner sex for access to (or control over) the less common sex. Note that competition here need not be for sexual partners, but for access to social partners and/or allies.

Table 6.3. Frequencies of intra-sex aggressive acts in two groups of longtailed macaques

Group	Adult sex ratio (females/male)	Mean frequency of acts/dyad/hour	
		Between males	Between females
A	1.3	0.389	0.173
B	3.0	0.175	0.448

Source: de Waal (1977).

In a sexual context, the operational sex ratio may be more relevant. In baboons, Hausfater (1975) found that the frequency of fighting and wounding among males increased when one of the group's females was in oestrus, but then decreased steadily as more and more females were in oestrus together. Since a male can only consort with one female at a time, the more females there are in oestrus in the group the more males will be able to acquire consort partners and hence the fewer males there will be to compete for access to them. Packer (1979a), D.R. Rasmussen (1979) and Manzolillo (1986) all report that male baboons tend to emigrate from groups with lower operational sex ratios and move into groups with higher ratios.

Similar effects will be encountered where males compete for control over whole groups of females. In this case, the amount of competition for groups will depend on the ratio of the number of groups available in the population (i.e. the number of resource patches) to the number of males waiting to gain control over one. Dunbar (1984a) found a positive correlation between the adult sex ratio in gelada bands and the proportion of adult males that held harems. In addition, there was a negative correlation between the proportion of all males that held harems and the proportion of groups that contained two or more adult males. Gelada males acquire harems either by taking over a group from the encumbent harem-holder or by joining a group as a subordinate follower (Dunbar and Dunbar 1975): in either case, the unit becomes a multimale group for a time. Consequently, the frequency of multimale groups in the population mirrors the rate at which males acquire harems, and hence reflects the level of competition for access to harems by males that do not yet have their own. The two correlations can be interpreted as a causal chain connecting low adult sex ratios (i.e. ratios that approach equality) with high frequencies of multimale units via high rates of harem entry by males (these being generated by high levels of competition among males for control over the limited number of female groups).

The composition of a group can also have a dramatic impact on the quality of social relationships. Seyfarth (1977) has used simple models to show how competition for access to preferred social partners can result in the tendency for animals to form just the kind of discrete dyads that characterise the social relationships of so many groups of monkeys. Experimental manipulations of group composition have also demonstrated that the quality of interactions between two individuals can change quite markedly when a third individual is present (hamadryas baboons: Stammbach and Kummer 1982; squirrel monkeys: Vaitl 1978). In the presence of a competitor, individuals become significantly more assiduous in pursuing and maintaining preferred social relationships, apparently in an attempt to reduce the risk that the partner will be lost to the competitor.

Group size itself may be an important factor for two reasons. First, the more animals there are performing some behaviour, the more occurrences of that behaviour there will be. One example is provided by the way threats and harassment tend to accumulate on the lowest ranking members of a group simply because there are more individuals to harass them (Figure 6.3). The second aspect arises from the fact that an individual animal has only a limited amount of time in which to interact with other members of its group. As a result, its social experiences of other group members will necessarily decline as group size increases. Figure 6.4 shows that the mean diversity of an individual's interactions with its fellow group members declines as the number of potential interactees increases. Evidently, the animals do not attempt to interact with each new member of the group as the size increases. One consequence of this is that gelada groups become increasingly fragmented as their size grows. There appears to be a threshold at a unit size of six adults (five reproductive females and a breeding male) at which the unit suddenly becomes socially unstable (Dunbar 1984a). At this point, units are particularly easy for other males to take over.

Another example is the way in which dominance hierarchies become increasingly unstable (as measured by the number of rank reversals) as they get larger (macaques: A. Mori 1977a; gelada: Dunbar 1984a). Part of the problem here seems to be that as group size increases individuals do not have the time available to interact with all members of their group (or even to observe them interacting with individuals that they themselves interact regularly with). Because their relationships with these individuals are less certain,

Figure 6.3. Median frequency of threats received per hour by female gelada in relation to dominance rank (vertical bars are interquartile ranges). Sample sizes are 7, 7, 7, 6 and 3 individuals, respectively

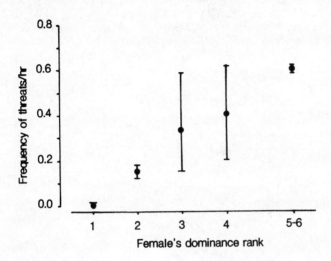

Source: Dunbar (1984a).

they need to test the situation when they do meet. As a result, more of their interactions will be influenced by day-to-day fluctuations in relative status than is normally the case. (Note that this has nothing to do with the statistical effect that linear hierarchies are more likely to be found by chance in small groups: see Appleby 1983a. We are here concerned not with rank orders within hierarchies but with the stability of individual dyadic relationships.) Badrian and Badrian (1984a) have also commented on the importance of opportunity to interact for the stability of relationships among pygmy chimpanzees (or bonobos), though the context in this case is rather different. Common chimpanzees typically occur in groups of only about four individuals, so that their opportunity to interact with other members of their community is relatively limited. Bonobos, in contrast, are found in much larger groups (typically 8–16 individuals). Interactions are more common among bonobos than among common chimpanzees, and their relationships seem to be more stable in consequence.

Altmann and Altmann (1979) provide one last example of the way in which demographic variables can influence the stability of dominance hierarchies. Mortality rates in the Amboseli baboon population are very high: half the infants born into the group fail to survive through to adulthood. As a result, the cohorts of infants that do mature into the adult complement of the group each year are quite small in number and are therefore more likely to be biased

Figure 6.4. Mean diversity of interactions among the adult members of gelada reproductive units, plotted against the number of adults in the unit, for a sample of 25 reproductive units. Each point is the mean of the Buzas–Gibson indices of the diversity of interactions for individual members of each unit: the index measures the relative evenness with which an individual distributes its social time among the various members of its unit, adjusted for unit size (with 1.0 indicating a completely even distribution). For further details, see Dunbar (1984a)

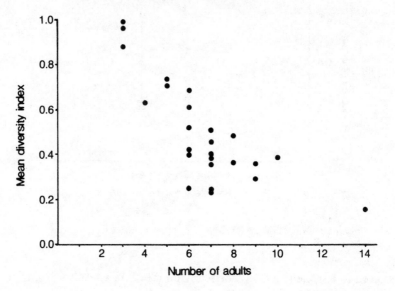

Source: redrawn from Dunbar (1984a, Figure 37) with the permission of the publisher.

towards one sex or the other by small-sample statistical effects. As it happened, no females survived from the 1971 and 1972 birth cohorts. The result was that the female dominance hierarchy remained very stable for several years with few challenges or rank reversals. However, six of the seven surviving infants from the 1973 birth cohort were female. These began to challenge for high rank as a group in 1976 when they were 3 years old. The result was a period of unprecedented chaos with a dramatic increase in the rate of rank reversals among the females and a very high frequency of fights. As the Altmanns note, it would have been impossible to understand what was happening in this group (or why it was happening) without a knowledge of the group's past demographic history. Samuels and Henrickson (1983) report a very similar case in a captive colony of rhesus macaques: here, the simultaneous maturation of a large cohort of juvenile females resulted in a sudden outbreak of aggression and the consequent instability of rank relationships in the female dominance hierarchy.

One lesson to be drawn from this is that, with most field studies lasting only 1–2 years, an investigator's impression of life in a primate group will inevitably be coloured by the particular situation pertaining at the time of his/her study. Thus, observers studying the Amboseli baboons during 1976 would have come away with a view of baboon society radically different from that of observers who had studied the group a couple of years before. I suspect that many of the disputes that have surfaced from time to time over just what counts as the 'normal' behaviour of a species can probably be attributed to differences in the demographic structure of different study populations.

Chapter 7
Evolution of Grouping Patterns

In Chapter 2, I argued that primate societies should be viewed as multi-layered sets of coalitions based on relationships that differ in intensity, character and function. The most obvious and perhaps important respect in which these clusters of relationships differ from each other is, of course, their spatial localisation and temporal stability. Traditionally, observers have always recognised that animals of many species spend much of their time in the physical company of conspecifics rather than wandering alone. In this chapter, I shall concentrate mainly on the evolution of group-living.

In exploring this issue, we need to draw a clear distinction between the reasons why primates form groups at all and the reasons why, given that they live in groups, they prefer to live in ones that have a particular composition. I suspect that a tendency to confuse these two questions lies at the heart of many of the disagreements about the adaptive significance of primate societies. I shall therefore begin by examining the various possible reasons why primates might live in groups rather than alone. The benefits of grouping cannot, of course, be viewed in isolation from the costs of doing so, so that only when both the advantages and disadvantages of grouping have been considered can we proceed to assess the relative importance of the possible functions and the role they play in determining group size. Once we have answered these questions, I shall go in the final section to ask why particular kinds of groups might have evolved.

Why Form Groups?

Four main reasons have been offered to explain why primates might live in groups. These are: (1) protection against predators, (2) defence of resources, (3) foraging efficiency, and (4) improved care-giving opportunities. In the following subsections, I summarise the evidence that has been adduced in support of each of these hypotheses in turn. Note that in many cases this evidence is open to interpretation and could be used to support more than one hypothesis. At this stage, my intention is really to present the best case I can for each suggestion rather than to evaluate their relative importance. I shall postpone any such attempts at testing between them until after we have had a chance to discuss the costs of group-living.

Protection against predators

Several sources of evidence have been presented in support of the hypothesis that primates form groups as a means of defence against predators. There have, for instance, been a number of reported incidents in which large-bodied primates have driven off dogs, leopards, cheetah or lions (baboons: DeVore and Washburn 1963, Smuts 1985; guenons: Gautier-Hion and Gautier 1985; chimpanzees: Tutin and McGinnis 1981, Boesch and Boesch 1984). Busse (1977) presents evidence to show that attacks on red colobus by chimpanzees are significantly less likely to be successful if a colobus male makes an active attempt to drive them off. Ransom (1981) has described a case in which a male baboon acted protectively towards an orphaned infant in the presence of chimpanzees (known to be active predators of the baboons). There are in addition a number of other features of primate biology that can be interpreted as reflecting predation as a selection pressure. Males of the more terrestrial species, for example, have significantly larger canines relative to body weight than those of species that are more arboreal in habit (Harvey et al. 1978): terrestrial animals are presumed to be more susceptible to predation than arboreal animals (but for a different interpretation, see Wrangham 1980).

The deterrence value of groups need not, however, depend on the presence of large powerful males. The size of a group alone will often be sufficient to deter a predator from attacking, either because it cannot single out a victim so easily or because it is not prepared to risk a concerted attack by many animals even if those animals are individually small. Female open-country antelope, for example, commonly form all-female groups for reasons of predation even though they can do little to defend themselves in an active sense (Jarman 1974).

There is considerable evidence to suggest that animals respond to high predation risk by forming larger or more compact groups. Baboon groups, for example, become more compact when moving through high-risk areas such as dense vegetation where predators can easily surprise them, or across open areas where there are few trees to climb for safety (Altmann and Altmann 1970, Byrne 1981, D.R. Rasmussen 1983). Sigg (1980) recorded the response of a band of hamadryas baboons to the sudden appearance of potential danger (humans, dogs or natural predators) while they were foraging. Out of a total of 34 sampled occasions, the baboons responded by a significant closing of ranks (i.e. reduced inter-individual distances) in 29 and by an increase in speed of travel in 26. Neither response can be convincingly explained away.

Group size may also be influenced by predation risk. Van Schaik and van Noordwijk (1985b), for example, found that macaques living on a predator-free island formed significantly smaller groups than those of a population living on the adjacent mainland where predators were abundant. Anderson (1981) noted a similar tendency for baboon groups to fragment and disperse in predator-free habitats but to forage in more cohesive groups in habitats where predators were common. Kummer *et al.* (1985) provide a further instructive example. Hamadryas baboons living in Ethiopia where predators occur at low densities normally forage as bands that consist of several reproductive one-male units, but in Saudi Arabia, where predators have long since been eradicated, the one-male units commonly sleep and forage on their own. In addition, there is evidence from a number of non-primate taxa to suggest that solitary individuals suffer higher rates of mortality (mainly because of an increased risk of being caught by predators) than individuals who live in groups (hartebeest: Gosling 1974; buffalo: Sinclair 1977; African wild dog: Frame and Frame 1976).

It has also been suggested that primates form polyspecific associations in response to predation pressure (West African cercopithecines: Gartlan and Struhsaker 1972; New World platyrhines: Terborgh 1983). Polyspecific associations are presumed to have the advantage of increasing group size without increasing the level of competition for particular resources because different species are more likely to occupy different niches. This interpretation has been questioned by Waser (1982) who showed that many reported polyspecific associations actually occurred no more often than would be expected if animals foraged independently on similar food sources. In addition, he argued that where polyspecific associations do occur more often than expected, the benefits are invariably one-sided and that, in at least some cases, these benefits are related to foraging rather than predation risk. If this is the case, then we still need to ask why the species that is being exploited is prepared to tolerate the presence of the exploiting species. It is possible, for example, that while one species benefits in terms of foraging advantages, the other gains through reducing predation risk. Barnard and Thompson (1985) provide a detailed analysis of just such a trading of different benefits in mixed-species flocks of birds. Plovers are prepared to tolerate the presence of gulls despite the latters' habit of stealing food from them because the gulls' greater visual acuity allows the plovers to concentrate on foraging without having to bear the high costs of vigilance.

This raises another important aspect of the anti-predation functions of grouping, namely that detection rather than deterrence may be the key factor in some cases. A very large theoretical and experimental literature has developed over the past decade and a half in this area (see, for example, Hamilton 1971, Vine 1971, Siegfried and Underhill 1975, Treisman 1975a, b, Hoogland and Sherman 1976, Lazarus 1979, Bertram 1980). There are three quite separate issues involved here, namely (1) the ease with which predators can locate prey when searching randomly in a habitat, (2) the ease with which prey can detect approaching predators in sufficient time to take evasive action, and (3) an individual prey animal's risk of being caught by a predator once it has been detected.

Several theoretical analyses have shown that predators searching at random are less likely to find prey if they are clumped into groups than if they are distributed more evenly around the habitat. Early detection of an approaching predator may not only allow prey to take evasive action but may also allow them to notify the predator (either visually or vocally) that it has been detected. A predator whose hunting strategy depends on getting close enough to a prey for a final rush (or pounce) will gain little in continuing to stalk a prey that has already seen it and so is more likely to give up and try elsewhere. (For a recent review of this aspect of predator–prey relationships, see Caro 1986). The third possibility (originally suggested by Hamilton 1971) is that it will pay a prey animal to become part of a group because the presence of other individuals reduces the risk that it will be caught by a predator, given that once the predator has located some prey it will usually capture only one animal.

On balance, however, grouping as a means of reducing the risk of either detection or capture by predators seems rather unconvincing in the case of primates. They tend to assume that prey cannot take evasive action (or defend themselves) and have little warning of a predator's attack. This would seem to be inconsistent with both primates' visual acuity and their widespread use of alarm calls and visual displays. Moreover, those primates that are likely to have least warning of approaching predators (i.e. nocturnal species) are almost all solitary. On the other hand, there is recent evidence to support the vigilance hypothesis. Van Schaik et al. (1983a), for example, found positive correlations between the distance at which a predator was detected and party size in four species of south-east Asian primates, though in only two cases were the correlations significant. De Ruiter (1986) has also shown that capuchin monkeys living in a small group spent significantly more time in visual scanning of the environment than those living in a large group. Wirtz and Wawra (1986) present data from humans to show that scanning rates decline with increasing group size.

Two main criticisms can be levelled against the predation hypothesis. One is that predation is seldom observed and thus cannot constitute a serious threat to the animals (e.g. Wrangham 1983, Cheney and Wrangham 1987). The other is that the presence of many males in multimale groups does not correlate well with the levels of predation faced by natural populations as originally claimed by, for example, Crook and Gartlan (1966) and Busse (1977) (e.g. Andelman 1986). The force of both these observations is weakened, however, by the fact that they confuse radically different questions. The first confuses predation as a cause of *mortality* with predation risk as a determinant of *evasive behaviour* (see p. 53). By analogy, it is rather like insisting that foraging behaviour could not possibly have evolved to allow animals to acquire energy because most animals die of disease and injury rather than starvation (a claim that is clearly substantiated by the data in Table 4.5). The second fails to distinguish between the decision to form a group (i.e. to aggregate) and the decision to form groups of a particular demographic type. While the lack of correlation is certainly evidence against the hypothesis that *multimale* groups evolve as a defence against predators, it has little bearing on the question of whether *large* groups have evolved for this reason even if large groups often happen to be multimale.

Resource defence

The defence of a year-round supply of food to ensure the survival of oneself and one's offspring has traditionally been considered one of the most important factors militating in favour of group-living, particularly in birds, where it is often associated with territorial defence. The most obvious supporting evidence from primates comes from the fact that, in many species, groups do defend territories (e.g. lemurs: A. Jolly 1966a, Pollock 1975; howler monkeys: Chivers 1969; colobus: Oates 1977b; guenons: Struhsaker 1967c, Aldrich-Blake 1970; gibbons: Ellefson 1968, Chivers 1974). In addition, among non-territorial species, group size is often an important factor determining a group's ability to displace other groups from key resources (e.g. baboons: Hamilton *et al.* 1975, 1976; macaques: Dittus 1986; orang utans: Mackinnon 1974).

The primary justification for resource defence also seems to be well substantiated. There is incontrovertible evidence that, in the final analysis, primate populations are limited by food availability (e.g. Dittus 1977, A. Mori 1979a). Given this, it seems reasonable to conclude that groups that can monopolise an area for their own exclusive use must gain a significant advantage. Moreover, within groups, animals do compete for priority of access to resources and success in that competition is known to determine an animal's survival and reproductive success (Dittus 1977, Wrangham 1981).

Because of these considerations, some authors (e.g. Wrangham 1980, 1983, Andelman 1986, Dittus 1986) have argued that resource defence is the single most important factor militating in favour of group-living in primates. The hypothesis is not without a certain amount of counter-evidence, however. Thus, Janson (1985) found that intra-group competition for food was about ten times greater than inter-group competition in capuchin monkeys. Conversely, Lee (1983b) found that juvenile vervet monkeys did not suffer any loss of feeding time when they lost their primary source of protection as a result of being orphaned, suggesting that coalitions are not always formed in order to gain access to food. Neither set of observations can be considered conclusive evidence against the hypothesis, however, because they only consider a possible short-term cost to grouping. We would need to know whether other long-term benefits to grouping or coalition formation in terms of access to resources outweigh these short-term costs before we could be sure that resouce defence plays no role.

Foraging efficiency

In certain species of large carnivores, foraging efficiency improves significantly with increasing group size: larger groups can capture and subdue larger prey (see Schaller 1972, Kruuk 1972, Macdonald 1983). Although some primates have become proficient hunters, no population derives a significant proportion of its diet from hunting large prey so that this explanation as it stands is unlikely to be important in the case of non-human primates (though it might possibly be relevant to hominids). A modified version, however, might apply to the more insectivorous species: prey acquisition rates might improve with group size if there are more individuals to help locate and/or disturb concealed prey (see Schoener 1971). Laboratory studies of foraging efficiency in tamarins have suggested that the animals forage very effectively in groups because they keep close watch on where other group members have foraged, thereby saving time by not repeatedly searching places that have already been visited by other individuals (Menzel and Juno 1985).

An alternative possibility is that animals form groups in order to facilitate more efficient harvesting of slowly renewing resources or because they are obliged to assemble at resource patches that are large and highly clumped. Where key food species have a relatively slow rate of renewal (e.g. fruiting trees), it may pay animals to exhaust one patch completely and then exploit successive patches in turn until sufficient time has elapsed for the first patch to have regenerated. Such a strategy has been noted in certain species of birds (finches: Cody 1971; geese: Prins *et al.* 1980) and at least some grazing mammals (see McNaughton 1979). Terborgh (1983) has suggested that squirrel monkeys form large groups in order to exploit the few large fruiting fig trees that are available during the dry season. Since individual trees can support large numbers of animals, grouping by the animals from a large area gives them access to a sufficient number of trees to see them through the dry season. Were a smaller group to attempt to defend a single tree, they would presumably find that much of the fruit would have rotted long before the dry season ended.

An alternative suggestion is that animals form groups in order to be able to monitor food production over a wider area. Such groups must, of course, be unstable so as to permit individuals to travel widely, but must also assemble regularly in order to make the transfer of information concerning ephemeral food sources possible. This kind of communal exchange of foraging information may occur in birds (Ward 1965, Ward and Zahavi 1973). A possible primate example is offered by the hamadryas baboon. Hamadryas bands regularly disperse into smaller sub-units (one-male groups) during the course of the day's foraging, but they regularly assemble at a waterhole at midday and at the sleeping cliff each evening. Moreover, the process by which the day's route is decided each morning at the sleeping cliff (see below p. 242) provides precisely the opportunity for animals to exchange information on the relative value of different routes (see Kummer 1968, Stolba 1979). Sigg and

Stolba (1981) have also pointed out that the resident band makes much more efficient use of its ranging area than do visiting bands, precisely because the animals are more familiar with the location of ephemeral resources. Many other species have fission–fusion social systems that could function in similar ways (e.g. Guinea baboons, pigtail macaques, spider monkeys and chimpanzees: see p. 10). Indeed, there is evidence for some of these species to suggest that scattered parties do converge on particularly good food sources as a result of special calls given by individuals that locate them (chimpanzees: Reynolds and Reynolds 1965; toque macaques: Dittus 1984).

Data from a number of other species do confirm that animals may congregate in unusually large numbers at particularly good food sources (macaques: Dittus 1984; orang utans: Mackinnon 1974; chimpanzees: Sugiyama 1973, Wrangham 1977; bonobos: Kano and Mulavwa 1984, Badrian and Badrian 1984a). Indeed, Andelman (1986) has specifically proposed that food patch size is the main factor determining group size (and thus grouping tendencies) in Old World cercopithecine monkeys, and there is ample evidence (reviewed in Chapter 5) showing that, within species, group size does increase as habitat quality improves.

Unfortunately, although animals might well be forced to separate out when patch size declines, this explanation rather begs the question as to why they should form large groups when food is superabundant or patch size very large since there is no reason why large groups should be preferable to small groups under these conditions. However, if the resource is *very* patchily distributed so that there are only a few super-rich patches, then animals might be forced to convene in very large groups. An example of this might be the squirrel monkeys discussed above. Kummer (1968) noted that a chronic shortage of safe sleeping sites obliges hamadryas baboons to form very large sleeping troops on those few cliffs that are available. Struhsaker and Gartlan (1970) also noted that groups of patas monkeys sometime form temporary associations at waterholes during the dry season when only a few still contain water.

Care-giving

Finally, primates might form groups in order to ensure that they will have help in rearing their offspring. This could include (1) direct help in carrying and caring for their infants, (2) indirect help in the form of territorial defence or protection from harassment by other individuals, or (3) the adoption of infants should the mother die. Help with rearing has been a particularly important factor promoting sociality in birds, but among mammals help with rearing is common only among canids (Kleiman 1977, Macdonald and Moehlman 1982). Among primates, cooperative rearing in any form is found only among the South American callitrichids and the siamang (Kleiman 1977).

The value of allies in protecting both individuals and their offspring from harassment has been widely documented in primates (e.g. Vaitl 1978, Silk 1982, Smuts 1985). In addition, a number of other studies have shown that females are more likely to leave a group if they have no access to powerful allies (macaques: Grewal 1980, Sugiyama and Ohsawa 1982b, Yamagiwa 1985). Finally, adoption of infants, though not common, has certainly been documented in a number of species (macaques: Hasegawa and Hiraiwa 1980, Berman 1983; baboons: Hamilton *et al.* 1982, Smuts 1985). All of these might provide reasons why individual females might choose to live in groups. The main problem which they all face, however, is that none of them requires groups sizes of more than two or three adults. Thus, while they might explain the evolution of certain *types* of group (e.g. monogamous pairs), they cannot account for the evolution of the large groups containing more than 10 adults that figure so prominently in primates as a whole.

Costs of Group-living

The benefits that an animal presumably derives from living in a group are off-set by the disadvantages that are inevitably incurred by living in close prox-imity to conspecifics. These costs come in two forms. Direct costs arise as a result of competition between members of the same group for access to limited resources. Indirect costs arise as a result of group-members being obliged to coordinate their activities in order to be able to remain together. Of these, the first is probably the more serious.

Since we have discussed indirect costs in some detail in the preceding chapter, I shall confine my remarks here to two observations. The first is that animals in a social group have to coordinate their activities so as to be able to remain together in order to benefit from the advantages of group life. Because lactating females, for example, need to feed more than other individuals, they may force other members of the group to move more often than they would otherwise do (see, for example, Dunbar 1983e). Alternatively, they may them-selves be prevented from moving to as many new feeding sites as they would like to visit by the inertia imposed by the rest of the group's refusal to move. The second point is that larger groups have to cover greater distances each day to meet their nutritional requirements (p. 48). In longtailed macaques, for example, increases in group size are directly reflected in increases in the length of the day journey and, consequently, in the amount of time spent moving (van Schaik *et al.* 1983a). The more time an animal has to spend moving, the more energy it will consume and hence the more feeding it will have to do to balance its energy budget. It will therefore have proportionately less time in which to rest or take part in social activity. In addition, longer day journeys inevitably mean that the animals will spend more time exposed to the risk of predation.

The direct costs of grouping, on the other hand, arise largely as a result of harassment by other individuals. This can come in the form of direct competition for access to resources or in the form of more generalised harassment intended to reinforce dominance relationships. A number of studies of provisioned groups, for example, have demonstrated that rates of aggression increase dramatically when food resources become highly clumped (macaques: Southwick 1967, Loy 1970, A. Mori 1977b, Fa 1986; baboons: Balzamo *et al.* 1973; talapoin monkeys: Gautier-Hion 1970; chimpanzees: Wrangham 1974, Kuroda 1984). Wrangham (1974) also noted that in wild chimpanzees the frequency of attacks per animal increased roughly in line with the number of individuals attending feeding sites. What may be important here is not so much the scarcity of food (the patches often contain a superabundance) as in the fact that the animals are crowded together and consequently get in each other's way. Some evidence to support this is provided by Koyama *et al.* (1981) who found that as groups of Japanese macaques became more compacted on artificial feeding sites, so the frequency of vocalisations (which was interpreted as an index of social tension) increased. Harcourt (1987) has reviewed the evidence on dominance behaviour in primates and concluded that animals are generally most often able to exert their dominant status only when resources are highly clumped in space. He also noted that large group sizes may have the same effect by forcing animals to crowd together.

In addition, increases in group size are likely to result in more time being spent searching for food by all individuals, partly because of an increase in the frequency of displacements. Stacey (1986) compared estimated energy intake and time spent feeding by members of three baboon groups in the same habitat. He found that although energy intake was constant across group size, animals in the smallest group spent only half as much time feeding as those in the largest group. Van Schaik *et al.* (1983a) also found that an increase in group size was associated with an increase in the time spent searching for food by all individuals, partly because of an increased frequency of displacements during feeding. Individuals that find a good feeding place and lose it to another animal waste all the time it took to locate that food source. Correlations between group size and levels of competition and/or aggression within the group have been reported from studies of capuchin monkeys (de Ruiter 1986), vervets (Fairbanks and Bird 1978), chimpanzees (Wrangham 1977, Wrangham and Smuts 1980) and gorilla (Watts 1985).

These costs need not fall evenly on all members of a group. Access to food sources and/or time spent feeding have been found to correlate with dominance ranks within wild groups of baboons (Sigg 1980, K.R.L. Rasmussen 1985) and capuchin monkeys (Janson 1985). Although Post *et al.* (1980) found no significant differences in the amount of time spent feeding by individual baboons, they did find that low-ranking individuals had significantly shorter feeding bouts and that their feeding bouts were interrupted more frequently than those of higher ranking animals. They also noted that the different age–sex classes had diets that differed in significant ways, suggesting perhaps that the animals attempted to minimise the adverse effects of interference while feeding by separating out ecologically. By feeding on less preferred food items, for example, low-ranking animals would reduce the likelihood that a higher ranking individual would displace them and take over the site.

In fact, direct harassment may not always be necessary to disrupt the feeding bouts of low-ranking individuals. Keverne *et al.* (1978) have shown that, in captive groups of talapoins, the frequency of visual monitoring is inversely related to dominance rank. The amount of visual monitoring that an animal does is primarily a function of its nervousness, and reflects the animal's need to keep track of the movements of more dominant individuals in order to avoid being attacked unawares. Such monitoring inevitably interferes with the flow of an animal's activity, thus reducing its feeding efficiency.

Taken together, these results suggest that although being a member of a group does not necessarily affect an individual's energy intake, it does affect the costs it incurs in acquiring what it needs. The fact that it has to spend longer feeding means that it has less time available for rest and social interaction. In baboons, gelada and vervets, for example, an increase in time spent feeding results in a reduction in resting time (Dunbar and Sharman 1984, Dunbar and Dunbar in press, Lee 1983a). Although social time is usually conserved whenever possible, conditions can deteriorate so badly that animals are forced to sacrifice some of their social time too (baboons: Hall 1963; gelada: Dunbar and Dunbar in press; macaques: Loy 1970).

Loss of social time will be of critical significance because social interactions are essential for the coordination and cohesion of groups through time. Enforced reductions in social time may result in reduced meshing of social relationships within the group, and a consequential tendency for groups to become socially fragmented due to the emergence of clearly defined cliques. Such problems undoubtedly contribute to the likelihood of group fission among gelada (Kummer 1975, Dunbar 1984a). In rhesus macaques, fission is most likely to occur when subgroups cease to interact with each other: this is usually precipitated by the death of an elderly female that had retained relationships with some members of both subgroups (Chepko-Sade and Sade 1979). In Japanese macaques, females are more likely to emigrate out of their natal groups if they have become socially peripheralised and no longer have close grooming and kinship ties with the core of the group (Grewal 1980, Sugiyama and Ohsawa 1982b, Yamagiwa 1985).

Perhaps the most direct reproductive cost of group-living, however, is the fact that some individuals may be subjected to intense harassment by other members of the group. Harassment received by a female has been shown to increase progressively as her rank declines in baboons (Altmann 1980), gelada (Dunbar 1984a) and talapoins (Bowman *et al.* 1978). In addition, Altmann (1980) found that not only were female baboons displaced much more fre-

quently following the birth of an infant than in the month preceding parturition, but the rates with which they were displaced correlated far more
strongly with the females' dominance ranks after parturition than they did
before. Among gelada, harassment of females increased significantly when
they came into oestrus (Dunbar 1980a). Similar results have been reported
for captive rhesus macaques (Wilson 1981).

Since such harassment can lead to reproductive suppression (p. 67), living
in a group can have a very serious impact on a low-ranking female's reproductive performance. Among gelada, for example, the effect is sufficient to
halve birth rates over a 10-rank span (Dunbar 1980a, 1984a). In monogamous species such as marmosets and tamarins, suppression may be total
(Abbott 1984, French et al. 1984). The benefits of living in a group must therefore be considerable for a female to be willing to bear such heavy costs in terms of
lost reproductive opportunities. In the next section, I attempt to integrate these
costs and benefits into a single framework in such a way as to be able to assess
their impact on the evolution of sociality.

Evolution of Groups

In the first section, I outlined four main reasons why primates might form
groups, namely (1) protection against predators, (2) defence of exclusive
access to resources, (3) increased foraging efficiency, and (4) more effective
rearing practices. In this section, I shall try to evaluate their relative
importance. Since, in discussing these causes of grouping, we noted that the
second pair are unlikely to apply widely, I shall begin by briefly examining
these two hypotheses before going on in the following subsection to a more
detailed test of the first pair of hypotheses. In the final subsection, I examine
the more general question of how these variables interact to limit group size.

Foraging and rearing hypotheses
The arguments for the increased efficiency of resource utilisation fall into two
kinds. Those for which there seems to be most evidence generally refer to
tendencies for animals to congregate on scarce resources (patchily distributed
food, sleeping sites or waterholes). This is largely a non-functional reason for
group formation, and seems to result from the fact that, when competition for
a severely limited resource becomes too fierce, individuals or groups are
simply unable to prevent competitors joining them on the resource (providing
the resource patch is large enough to accommodate all the animals): the best
they can do is to defend their own personal space. Such groupings are likely to
be unstable and to disintegrate as soon as the more dominant members can re-
assert themselves. As such, it fails to explain why many species live in more
stable groups and so must be less interesting as a general explanation for the

evolution of stable social groups than the alternative suggestion that grouping actively benefits foraging efficiency. As noted on p. 111, however, foraging efficiency is only likely to apply in the case of insectivores (most of which are in fact solitary: e.g. tarsiers, galagos, mouse lemur, aye-aye) or those species that live under marginal conditions where plant foods are highly unpredictable (e.g. hamadryas baboons). The South American callitrichids are the only insectivorous primates that are habitually social in this sense, and these live in small (mainly monogamous) groups for reasons that almost certainly concern care-giving rather than foraging efficiency (see p. 281). As a general explanation for the evolution of sociality in primates, foraging efficiency is thus either inconsistent with the biology of many of the species to which it might apply or begs the question as to why animals should need to form groups at all.

Similarly, the relevance of care-giving must also be doubted. On the one hand, group-living is itself the cause of much of the harassment and stress that animals seek allies to buffer themselves against, so that harassment *per se* can hardly rank as the primary factor promoting sociality. Despite cases of adoption in wild primates, the evidence is quite clear that, in the majority of cases, those immatures that most need care (i.e. dependent infants) die if they are orphaned (Rhine *et al.* 1980, Goodall 1983, Lee 1983b; for a general review, see Thierry and Anderson 1986). Clearly, the value of conspecifics in this respect is at best unpredictable and patchy in its distribution. The third possibility (direct help in rearing dependent offspring) occurs predominantly in the callitrichids, a taxonomic group that is in many different respects rather specialised (Ford 1980). In any case, the extensive literature on parental care makes it clear that only one or two helpers are needed to maximise the likelihood of rearing offspring to maturity. Thus although it may account for the evolution of monogamy, it is unlikely that parental care can explain the evolution of large groups. I shall discuss the relationship between monogamy and parental care in more detail in Chapter 12.

Neither of these two alternatives, then, commands the wide taxonomic distribution that would warrant its being considered a major factor in the evolution of grouping, though they may well be relevant in particular cases. This leaves us with only two serious contenders: defence against predators and resource defence.

Predator-defence and resource-defence hypotheses

Hitherto, evaluation of the relative importance of these two hypotheses has tended to rest primarily on intuitive assessment of their strengths and weaknesses and the particular penchant of the individual commentator. The arguments that have been advanced for and against the hypotheses have thus done little to resolve an already ambiguous situation. What we urgently need are formal tests that force the two hypotheses into direct competition with each other such that what the animals do must favour one hypothesis at the expense of the other. This effectively makes the animals choose between the competing hypotheses and so provides a more powerful test.

Van Schaik's tests.   The first attempt to do this was made by van Schaik (1983). He pointed out that these two hypotheses affect key behavioural and demographic variables in qualitatively different ways. He argued that the resource-defence hypothesis, if it is to work effectively, must overcome the fact that group-living inevitably generates severe costs for the animals. In other words, inter-group competition for resources must be greater than intra-group competition so that individuals do better when in groups than when foraging alone, at least up to the point where group size is so large that the disruption begins to outweigh the advantages. This should result, he suggests, in an inverted-U-shaped relationship between rate of food acquisition and group size which should translate directly into a similar relationship between group size and female reproductive rates since these are assumed to be limited primarily by the rate at which females can acquire nutrients (Goss-Custard *et al.* 1972, Wrangham 1980). In contrast, there is nothing to counteract the effect of within-group competition for access to resources under the predator-defence hypothesis since the hypothesis itself is concerned mainly with *adult* survival. Thus, behavioural variables such as feeding rates can be expected to decline with group size due to the conventional costs of grouping. As a result, female reproductive rates should decline linearly with increasing group size. Only when reproduction is viewed in the very longest term (i.e. as *lifetime* reproductive output) will a positive effect due to group size be found, reflecting the increased life expectancy of adults as increasing group size reduces the risk of predation. (Note that lifetime reproductive rate will also be an inverted-U-shaped function of group size under this hypothesis for exactly the same reasons as the resource-defence hypothesis predicts such a relationship: measuring lifetime reproductive output would not, therefore, allow us to distinguish between the two hypotheses.) The relationships between group size and female reproductive rates predicted by the two hypotheses are graphed in Figure 7.1.

In order to test this prediction, van Schaik (1983) examined data on the numbers of offspring per female in relation to group size in 27 populations of Old and New World monkeys. Twenty-two of the regression equations were negative, and though in most cases the slopes were close to zero the probability of so many negative coefficients being due to chance alone is very small (binomial test: $p=0.001$).

Moore (1984) has criticised van Schaik's test on the grounds that his sample is heavily biased in favour of species in which females commonly transfer between groups. He argues that Wrangham's (1980) hypothesis on resource defence applies only to species where females do not migrate because only in these species will groups consist of closely related females for the reasons that Wrangham's hypothesis presupposes. If the sample is split on this

Figure 7.1. Predicted effect on reproductive rate of two competing theories of group formation: (a) groups allow females to reproduce more successfully by defending areas of exclusive resource use (solid line), and (b) groups allow females to avoid predation and so survive longer (dotted line)

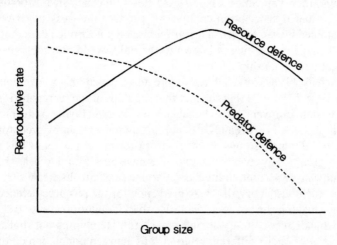

Source: redrawn from van Schaik (1983) with the permission of the publisher.

basis, Moore claims that the data for species in which females are not known to migrate yield a mean *positive* correlation between group size and number of offspring per female that is more consistent with the resource-defence hypothesis. Unfortunately, Moore's argument here rests on only three populations (out of the 27 in van Schaik's sample), two of these being for the same species (vervets). In fact, when taken together, the data for the two vervet populations tend to support the predation-defence hypothesis. If the resource-defence hypothesis were true, then we would expect larger groups to be able to compete more effectively when conditions deteriorated so that the effect on reproductive rates should be exaggerated: the regression coefficient ought, therefore, to become more positive as habitat conditions deteriorate, but in fact it becomes *less* positive. In any case, as a defence of Wrangham's hypothesis, Moore's argument is a weak one, for it effectively implies that the hypothesis is much less widely relevant than Wrangham (1980) originally supposed. Essentially, Wrangham had argued that if females form groups only for reasons of resource defence, then it will always pay them to form groups with closely related individuals (which would tend to reduce the rates with which females migrate). Presumably, then, some other hypothesis is necessary to account for the fact that all other species form groups even though the females migrate regularly and are unrelated to each other.

A more serious criticism of van Schaik's test is that only a positive correlation would conclusively distinguish between the two hypotheses. As can be seen from Figure 7.1, both hypotheses predict a negative relationship between group size and mean reproductive output on the right-hand side of the graph. Since these graphs are only notional, we have no way of knowing where the hump in the resource-defence hypothesis occurs. If it occurs at relatively small group sizes, then van Shaik's results are clearly inconclusive. Moreover, an inverted-U-shaped distribution could easily produce the weak negative correlations obtained by van Schaik: a linear regression put through data that are an inverted-J shape with a long tail down to the right would inevitably produce a weak negative relationship.

Van Schaik (1983) offered two more tests which might overcome this problem. First, he argued that the fact that the slope of the regression is steeper for infants than for juveniles can be taken as evidence in favour of the predation hypothesis because (1) juveniles are more likely to be taken by predators than infants, and (2) small groups are likely to suffer more heavily in this respect than large groups. However, since juveniles have been found to bear the brunt of mortality in some food-limited populations (e.g. Struhsaker 1973, Dittus 1977), we might also make the same prediction for the resource-defence hypothesis, thus making the test less conclusive than it might otherwise be.

Van Schaik's second test was more powerful. He pointed out that the two hypotheses also predict different relationships between population density and mean group size. The resource-defence hypothesis is specifically concerned with an individual's attempts to overcome the problem of competition for access to scarce resources. Since the hypothesis argues that large groups compete more successfully in this respect (Wrangham 1980), group size ought to increase steadily with population density as groups are forced into more intense competition with each other. The predator-defence hypothesis, conversely, assumes that within-group competition will increase as group size increases so that a trade-off will occur at the point where the disadvantages of large group size begin to outweigh the anti-predator advantages: at this point, groups will undergo fission rather than grow larger. Van Schaik assumes that density largely reflects group size, mainly on the grounds that ranging areas remain constant. This allows him to argue that, as density increases, the costs of group living will result in the very largest groups splitting into two smaller ones. Since the proportion of very large goups in the population will be greater at high densities, the proportion of groups undergoing fission should increase linearly with increasing density. Consequently, while mean group size will initially increase with density as groups grow in size, once past the critical threshold group size should decline again as more of them undergo fission. Mean group size should thus be an inverted-U-shaped function of population density. These predictions are graphed in Figure 7.2.

Van Schaik's argument in the case of the predator-defence hypothesis is based on the questionable assumption that range size remains constant as group size increases. Evidence from many different species shows that range size actually correlates with group size (see p. 47), suggesting that groups adjust the size of their ranging areas as they change in size. Although his

Figure 7.2. Predicted effect on group size of the two competing theories of group formation as population density rises: resource-defence hypothesis (solid line) and predator-defence hypothesis (dotted line) (see text for details)

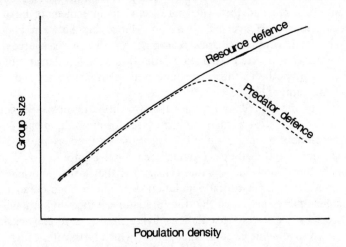

Population density

Source: redrawn from van Schaik (1983), with the permission of the publisher

explanation is debatable, however, van Schaik's prediction can be rescued on the grounds that other evidence reviewed in Chapter 3 shows that range sizes become compressed as population density rises to high levels. Since groups will then be unable to adjust range size in response to increases in the number of animals, they will in fact be forced to undergo fission. There is unequivocal evidence of this process in colobus, for example (see Dunbar in press). There will thus tend to be a 'settling down' effect at high densities, just as van Schaik suggested. Note that this explanation itself need have nothing to do with predation risk, but it does provide a clear alternative based on the assumption that the resource-defence hypothesis exerts no influence at all.

Van Schaik (1983) was able to test these predictions using data from only two species, his own study of longtailed macaques over a four-year period and data derived from seven studies of the Barro Colorado howler monkey population. In both cases, the data showed a closer fit to the inverted-U shape predicted by the predator-defence hypothesis.

Some further tests. We can use van Schaik's two sets of predictions to devise a still more powerful test that yields diametrically opposite predictions for the two hypotheses. The logic of the test goes like this. If we map the graphs from Figure 7.1 on to those for Figure 7.2, we obtain a causal relationship between population density as the independent variable and birth rate as the dependent variable that is mediated by group size. Thus, for the predator-defence hypothesis, Figure 7.2 tells us that when population density is low, group size will be small, which (from Figure 7.1) should mean that birth rates will be high; at intermediate densities, group sizes will be relatively large, which will mean low reproductive rates; finally, at high densities, group sizes will be intermediate and this will be reflected in intermediate birth rates. Thus, we arrive at a (reversed) J-shaped relationship between population density and birth rate. Analogous reasoning for the two resource-defence graphs predicts an inverted-J-shaped relationship between density and birth rate. The two hypotheses thus yield predicted relationships that are mirror images of each other (Figure 7.3).

As a test of these predictions, data for two different taxa are plotted in Figure 7.4. In Figure 7.4(a), data on birth rates for nine *Papio* baboon populations are plotted against population density (measured as the number of animals per km² of the troop's ranging area), and Figure 7.4(b) plots estimated birth rates from nine separate censuses of the Barro Colorado howler monkey population against total population size. The baboon data clearly fit the J-shaped curve predicted by the predator-defence hypothesis. Even though we lack data from populations living at intermediate densities, it would be hard to imagine circumstances under which birth rates could be high enough to transform this distribution into the inverted-J shape required by the resource-defence hypothesis. Birth rates well in excess of one infant per female per year would be necessary, and even captive baboons are unable to sustain such high rates for any length of time. The howler data in Figure 7.4(b) also tend to support the predator-defence hypothesis. If the lower left-

Figure 7.3. Predictions of the two competing hypotheses of group formation for the effect of population density on reproductive rate: resource-defence hypothesis (solid line) and predator-defence hypothesis (dotted line). The graphs are obtained by mapping the graphs from Figure 7.1 on to the graphs for Figure 7.2

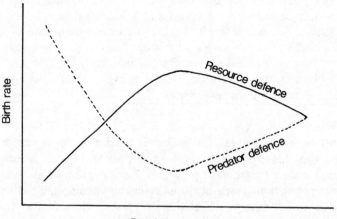

Figure 7.4. Tests of the predictions shown in Figure 7.3. (a) Birth rate plotted against population density for Papio *baboon populations: 1, Amboseli, Kenya (estimated from DeVore and Hall 1965); 2, Suikerbosrand, South Africa (Anderson 1982); 3, Shai Hills, Ghana (Depew 1983); 4, Kuiseb, Namibia (estimated from Hamilton et al. 1975, 1976); 5, Honnet Reserve, South Africa (estimated from Stoltz and Saayman 1970); 6, Mulu, Ethiopia (Dunbar and Dunbar 1974b, unpublished data); 7, Bole, Ethiopia (Dunbar and Dunbar 1974b, unpublished data); 8, Gombe, Tanzania (J. Oliver, personal communication); 9, Budongo, Uganda (Patterson 1976). (b) Number of infants per female plotted against total population size for the mantled howler monkey population on Barro Colorado, Panama, during different censuses between 1932 and 1980, as given by Milton (1982). Regression lines fitted by eye in both cases.*

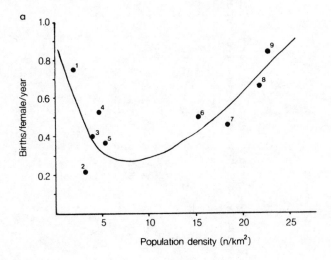

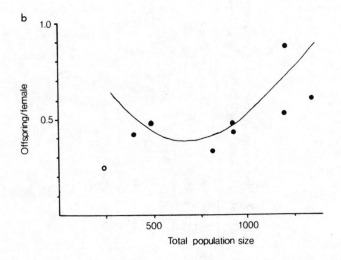

Table 7.1. Re-assigned birth rate values from the baboon data of Figure 7.4(a) to give idealised U-shaped and inverted-U-shaped relationships to population density, together with the observed order (see text for details)

Study number	1	2	3	4	5	6	7	8	9
Density (n/km^2)	1.9	3.2	4.0	4.7	5.4	15.3	18.4	21.9	22.8
Birth rate/year:									
observed	0.750	0.215	0.396	0.522	0.358	0.500	0.462	0.667	0.845
U-shaped	0.845	0.667	0.500	0.396	0.215	0.358	0.462	0.522	0.750
inverted-U	0.215	0.396	0.500	0.667	0.845	0.750	0.522	0.462	0.358

Sources: as for Figure 7.4.

hand point is ignored, the data clearly fall on a J-shaped curve. This point is exceptional for several reasons which suggest that ignoring it may not be entirely unjustified. First, it was a census taken shortly after a yellow fever epidemic swept through the area and apparently decimated the population (Milton 1982). Secondly, and perhaps as a result, mean group sizes were significantly lower than in any other census ever taken on the island (a mean of 8.0 compared with the minimum mean of 13.8 for the other censuses taken before or afterwards). Thirdly, the distribution of infant age classes differs from all other years: whereas in all other years the number of infants in the three age classes distinguished by Milton (1982) has an inverted-U shape, in this one year it is completely even. In any case, even if we do not discount this census, the rising birth rate at higher population densities seems to conform better with the predictions of the predator-defence hypothesis than those of the resource-defence hypothesis. (The latter would imply that birth rates would only begin to decline again at densities far in excess of the island's carrying capacity and this would be logically inconsistent.)

We can test the hypotheses more precisely by fitting curvilinear regressions to each of the distributions in Figure 7.4 and asking whether the data fit U-shaped or inverted-U-shaped curves better. The least squares best fit equations for log–log transformations of the data are:

$$\text{for baboons: } \ln y = 0.372 - 1.602\,(\ln x) + 0.441\,(\ln x)^2 \qquad (7.1)$$
$$\text{for howlers: } \ln y = 0.173 - 0.635\,(\ln x) + 0.075\,(\ln x)^2$$

(For the reasons outlined above, I have excluded the point marked by an open circle from the howler data-set for this analysis.) There is no *a priori* way of deciding what the form of the theoretical curves should be, but one solution is to redistribute the actual data in such a way as to approximate a perfect curve most closely (for a simpler example of this procedure, see Dunbar and Sharman 1984). To do this, we re-cast the data by assigning the values for birth rate in arithmetic order to the outermost points on the observed distribution of population density, working sequentially inwards by alternating left and right arms of the distribution. Working from highest to lowest birth rates gives a U-shaped distribution, while lowest to highest gives an inverted-U. The resulting distributions are given in Table 7.1 for the baboon data. We can then set curvilinear regressions through these distributions and compare the parameters of the equations with those of equations 7.1. (Methods for calculating curvilinear regressions and for testing the significance of their parameters are given by Pollard 1977.) The results for the second (linear) parameters of the two observed regressions are given in Table 7.2. In both cases, the observed parameters are clearly more similar to those of the respective U-shaped distributions demanded by the predation-defence hypothesis than the inverted-U of the resource-defence hypothesis. Although the difference between observed and expected values of the resource-defence

Table 7.2. Comparison of the observed regression parameter for linear slope for the distributions in Figure 7.4 with the slopes for the idealised U-shaped and inverted-U-shaped distributions predicted by the predator-defence and resource-defence hypotheses (see text for details)

| Species | Observed slope | Predicted slope[a] | | Goodness-of-fit | | | | 95% Confidence limits on observed slope | |
| | | | | Predation hypothesis | | Resource hypothesis | | | |
		Predation	Resource	t	p(1-tailed)	t	p(1-tailed)	lower	upper
Baboons	−1.602	−2.593	2.966	1.150	0.292	−5.302	0.0009	−3.419	0.215
Howlers	−0.635	−1.442	1.116	0.379	0.360	−0.823	0.227	−1.473	0.203

[a] Determined from Table 7.1.

hypothesis are only significant in the case of the baboons, the expected value lies within the 95% confidence limits of the observed value on the predation hypothesis and outside the 95% confidence limits on the resource-defence hypothesis in both species. Moreover, we can effect a more general test by using Fisher's formula (see Sokal and Rolf 1969, p.621) to combine the significance levels of the two tests to search for underlying trends. The results indicate a significantly better fit to the predation-defence hypothesis ($\chi_4^2 =$ 17.018, $p < 0.01$) than to the resource-defence hypothesis ($\chi_4^2 = 4.505$, $p < 0.30$). An analysis for the third (quadratic) parameters of the equations yields substantially similar results. These tests thus come firmly down in favour of the predation-risk hypothesis.

Three other attempts have been made to test between the two hypotheses. Dittus (1986), for example, noted that when two groups of toque macaques fused, all the females of the subordinate group fell to the bottom of the joint dominance hierarchy. As a result, they suffered reproductive suppression and produced few or no offspring. In contrast, females of the dominant group maintained the same fecundity and survival rates as they had prior to the fusion, while at the same time they presumably gained by acquiring the ranging area of the rival group. Dittus interprets these observations as being incompatible with the predator-defence hypothesis (on the grounds that the females of the subordinate group did not benefit from being in a larger group) and as being consistent with the resource-defence hypothesis (on the grounds that the females of the dominant group benefited at the expense of their rivals). It is not immediately clear to me that these events can be used to support either hypothesis, let alone test conclusively between them. It is not clear, for example, why groups that form essentially for reasons of predation should not also exploit their advantages of size to gain access to more profitable resources. Moreover, the context of the observations makes it impossible to distinguish unequivocally between the two hypotheses because they are necessarily confounded in the fused group. Thus, the predator-defence hypothesis expects females at the bottom of a large hierarchy to do rather badly. The fact that the groups have fused is not itself a test of the predator-defence hypothesis since, given the dynamics of most primate groups, there is no reason to suppose that small groups will attempt to form larger groups. Rather, the normal expectation under the hypothesis would be that small groups do rather badly and that group size will therefore tend to drift upwards because animals are reluctant to leave them (in other words, fission does not occur).

A more direct test was made by de Ruiter (1986) who compared aspects of the behavioural ecology of capuchin monkeys living in a small (8-animal) group and a large (25-animal) group in the same habitat. He found (1) that animals in the smaller group spent significantly more time scanning the environment and foraged at greater heights, and (2) that those in the larger group travelled significantly further each day, fed on riskier and/or less nutritious foods and were involved in more agonistic encounters. He argued that these

Table 7.3. A test of two hypotheses to account for herd formation in gelada: the percentage of lone reproductive units in seven different habitats in three localities would correlate positively with mean slope of the ground from the horizontal if predation risk was the determining factor or negatively with grass density if defence of scarce patchily distributed resources was critical

	Bole gorge	Gich escarpment	Sankaber escarpment	Sankaber gorge	Bole plateau	Gich plateau	Sankaber plateau	Correlation to herd size
Herds of one unit (%)	92.4	73.3	72.3	31.5	16.7	13.1	12.4	τ = 0.715[a]
Mean slope of ground (°)	50	75	49	33	5	15	10	τ = −0.333[b]
Ground cover (%)	39.4	—	91.4	79.3	82.0	81.4	88.9	

[a] p = 0.015 (1-tailed); with Gich escarpment omitted, τ = 0.733, n=6, p=0.028 (1-tailed).
[b] p = 0.235 (1-tailed).

results are only compatible with the predator-defence hypothesis, for the results in (1) imply that the animals were concerned about greater predation risk in small groups, while those in (2) imply that animals in large groups suffer significant foraging costs.

Finally, Dunbar (1986) noted that three different hypotheses might explain why gelada reproductive units form herds. These were that herds would form (1) when food was superabundant (as originally proposed by Crook 1966), (2) when visibility conditions were good enough to allow units to maintain spatial coordination, or (3) when the risk of predation was high. A comparison of the frequencies of lone units in three adjacent sectors of habitat that differed in these respects suggested that predation risk was the most likely explanation. We can use this same approach, with additional data from two other study areas, to make a more direct test between the predator-defence and resource-defence hypotheses. As before, the proportion of herds that consist of a single reproductive unit is used as a measure of the units' willingness to travel alone in a given habitat. The resource-defence hypothesis would then make the following predictions. Where grass is abundant, it will usually be too common and too evenly distributed a resource to be worth defending and there will be no reason why units should not travel alone. However, as grass density declines, the grass is likely to become more patchy in its distribution; in addition, its scarcity will render other food sources (e.g. fruiting bushes or areas where bulbs are particularly abundant) of greater dietary significance. We would then expect an ability to defend resource patches to be more important: units should then prefer to travel in herds in order to be able to monopolise desirable patches more effectively. (All other resources, such as waterholes and sleeping sites, are superabundant throughout the gelada's present range and will not therefore be a confounding source of competition between units.) The resource-defence hypothesis thus predicts a *negative* correlation between habitat quality and the frequency of lone units. Conversely, if herds form in order to reduce predation risk, we would expect a negative relationship between predation risk and the frequency of lone units. Since predation risk is a function of the availability of steep cliffs down which the animals can escape, we can again use the slope of the habitat as an index of relative predation risk: the steeper the slope, the less easily predators can get at them. Hence, the frequency of single units should correlate *positively* with the mean slope.

Table 7.3 gives the frequency of single units, mean slope and mean proportion of ground-level vegetation cover in seven habitats in three areas. Although the correlation between the environmental variable and the percentage of single units is in the predicted direction in both cases, only that for mean slope is significant. (A one-tailed test is appropriate here since we are testing a specific prediction: a correlation of opposite sign would be just as much evidence against the hypothesis as no correlation at all.) One possible problem with this test is that grass density is likely to be influenced by the slope of the ground: hence, a negative correlation between these two variables may confound the results. If we remove the effect of this relationship by calculating partial correlations that take the third variable into account, we find that ground cover has a negligible effect on the correlation between mean slope and the percentage of single units ($\tau_{ab \cdot c} = 0.700$, $n=6$), but that taking slope into account greatly reduces the apparent correlation between ground cover and the percentage of single units ($\tau_{ac \cdot b} = -0.139$, $n=6$). Although it is not possible to assign significance levels to Kendall partial correlation coefficients, none the less a comparison of the magnitude of the change in each case is sufficient to warrant the conclusion that resource defence plays a negligible role in herd formation among the gelada.

Some criticisms.   Two general criticisms might be levelled against these tests. One is that all the tests focus on the value of groups of different *size*, whereas our real concern should be with why animals do not forage solitarily. In other words, we should be comparing the behaviour of animals on their own with that of animals living in small groups. The other is that the only two unequivocal tests of the hypotheses (i.e. Figure 7.4a and Table 7.3) relate to open-country species where predation is most likely to be relevant: resource defence may yet turn out to be the reason why groups occur in species whose niche makes them less susceptible to predation (i.e. arboreal forest-dwelling species and large-bodied apes).

The suggestion that we might be testing the wrong hypothesis cannot, in fact, be taken seriously. For one thing, we are interested primarily in the factors that promote grouping in living primates. Most of these species are already predisposed to social life so that our interest must focus on the question of what *maintains* grouping in these species. This can only be explored by an analysis of the costs and benefits of living in groups of different sizes. It would, of course, be a legitimate question to ask what factors first prompted the ancestral solitary primates to live in groups, but this is a different question and not one that we can answer directly with tests on living species. Even if this were not the case, we would still have considerable difficulty trying to test between the two competing hypotheses by comparing the behaviour of solitary and group-living individuals of the same species because both hypotheses would yield exactly the same prediction: group-living individuals will be more successful than solitary ones (albeit for different reasons). A solitary animal might forage less successfully than one in a group, for example, *either* because it lacked the support to defend access to a resource patch *or* because it spent too much time scanning for predators. Moreover, evidence that groups can displace solitary animals from resource patches would not of itself tell us anything because, even if the predator-defence hypothesis is correct, there is no reason why groups should not use the advantage conferred by their size for other purposes. In short, there are just too many confounding variables for a simple two-way comparison to tell us very much.

The second criticism is both more serious and more interesting. Grouping as a means of resource defence clearly remains a candidate explanation in at least some cases, just as foraging efficiency and care-giving remain appropriate in certain specific species. But it is also interesting because it suggests that grouping might be multifactorially determined and that different factors might emerge into ascendancy in different contexts. Given that we are primarily concerned here with species whose ancestors have been social for many millions of years, there is no real reason why grouping might not evolve for one reason and subsequently be taken over for some other purpose. This then raises a question: did the earliest primates first become social for reasons associated with predation or for reasons associated with resource defence, with the alternative reason for grouping coming to the fore at a later date once a change of niche had altered the impact of other selection forces?

Much, of course, depends on whether evidence can be adduced to show that resource defence is a relevant factor in the grouping strategies of particular species. At present, the balance of the evidence from a wide range of species clearly favours the predator-defence hypothesis. Given that this is so, I shall rely most heavily on the significance of predation risk in my subsequent account of primate social evolution. None the less, it is worth pointing out that the advocacy of the resource-defence hypothesis has played an important role in focusing attention on the factors promoting grouping in primates. Hitherto, most of the claims that have been made on behalf of the predator-defence hypothesis (e.g. Alexander 1974) have largely been based on assumption or analogy with other taxa (particularly the birds). The arrival of a competing theory has forced us to generate tests of sufficient power to discriminate between them, thereby greatly enhancing our understanding of the factors involved.

Determinants of group size

The preceding analyses suggest that there are two important factors influencing group size in primates: predation risk promotes the formation of large groups while the costs of group-living give rise to tensions that lead to the disintegration of large groups. The balance between these two countervailing forces yields the optimum group size and this will obviously be habitat-specific (see also Wittenberger 1980, Rutberg 1984, van Schaik and van Hooff 1983). This optimum group size will, however, be subject to the constraint imposed by a third factor: the availability and dispersion of food in a given habitat will place an upper limit on the size to which groups can grow if every member is to be able to acquire the food it needs. Thus, we arrive at a four-dimensional relationship in which reproductive success (the criterion that the animals are trying to optimise) is plotted against group size (the behavioural response that the animals adjust in order to optimise the criterion variable) and two different state variables (habitat quality and predation risk). This raises problems, because we cannot easily represent a four-dimensional space graphically. To circumvent this problem in order to be able to illustrate how and why these variables might interact to influence group size, I shall begin by considering each of the two state variables in isolation to show how each one affects lifetime reproductive success as a function of group size.

The value of grouping as an anti-predator strategy will depend closely on both the local density of predators and the availability of safe refuges: in other words, it is predation *risk* that is the key factor. Lifetime reproductive success, R_0, will thus depend not only on the size of the group but also on the level of predation risk in any given habitat. The interaction between these three variables can be represented as a family of curves, each showing how R_0 increases with group size in a habitat with a specific level of predation risk (Figure 7.5). These curves rise to an asymptote whose location depends on the habitat-specific level of predation risk: R_0 is maximised at lower group sizes in habitats with lower predation risk. Conversely, we can see that for any given group size, R_0 will decline as predation risk increases.[1] We have thus represented a three-dimensional space in two-dimensions by representing the third dimension as a family of curves: these graphs are actually a three-dimensional surface.

The costs of group-living can be represented similarly. These are of two kinds: those that are due to competition for access to limited food or other essential resources and those that are due to the physiological stresses of living in close proximity to other individuals. The latter costs will be a simple function of group size, so that we can represent their effect on R_0 as a straight line sloping down to the right: as group size increases, so the average lifetime reproductive success of the members declines. (Note that this is only the *average* cost: in some cases, this may be borne equally by all group members, but in other cases low-ranking individuals may bear a disproportionate share of the costs.) We can view this graph either as the proportion by which R_0 is reduced by living in a group of a certain size (relative to that for groups of size $n=1$) or as an additive component of R_0 that is summed with the component derived from predation-risk. I prefer the first of these, but the same results are obtained in either case. Competition for access to food, on the other hand, will have the effect of steepening the slope of the size-dependent cost graph as habitat quality deteriorates (or the habitat's patchiness increases). The influence of the costs of group-living can thus be represented as a family of curves that slope downwards to the right from approximately the same origin in the upper left corner of Figure 7.5.

The optimum group size will be that which maximises R_0 given both the predator-risk and the resource quality of the habitat.[2] This should mean that as a habitat's value on one state variable alters, the optimum group size will move up or down the gradient on the other state variable. Thus, if predation risk increases due to the loss of forest cover (a shift from predation curve A to curve B, for example), the optimum group size will move upwards (from n_A to n_B). Figure 7.6 suggests that, on average, baboons do live in significantly larger groups in more risky (i.e. more open) habitats. Conversely, if habitat

Figure 7.5. Optimal group size in primates. The costs of group-living in terms of reproductive suppression and interference in feeding efficiency cause lifetime reproductive output, R_0, to decline as group size increases, with the rate of decline being steeper in poorer quality habitats (narrow lines). Conversely, group size buffers females against the risk of predation so that reproductive output will increase with group size up to a maximum determined by the female's reproductive capacity, with that maximum being reached at lower group sizes in habitats with low predation risk (heavy lines rising to the right). The optimum group size in any given habitat is given as the intersection of the habitat-specific predation-risk and cost graphs. If predator-risk increases (as from curve A to curve B), the optimal group size will move down the appropriate cost gradient to a new equilibrium point (i.e. group size will increase from n_A to n_B). If habitat quality then deteriorates, the optimum group size will move down the relevant predation-risk curve to a new equilibrium, n_C. If habitat quality also determines predation risk (e.g. rich habitats have many trees which provide refuges in which the animals can escape from predators), then each habitat will have its own pair of habitat-specific cost and benefit graphs whose intersection uniquely determines the optimum group size for that type of habitat. In this event, the equilibrium group size may well be non-monotonically related to habitat quality (dotted line). The shape of this relationship will, however, depend on the precise shapes of the cost and benefit graphs and on the relationship between them

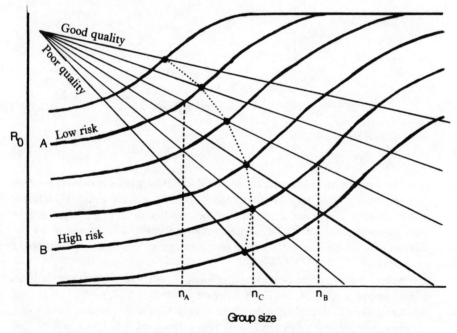

quality deteriorates, so the optimal group size will move along the relevant predation-risk gradient to settle at a new size, n_C. Precisely such a shift in group size has been documented over the course of a decade and a half in the Amboseli (Kenya) baboon population (Figure 7.7). As the habitat has deteriorated with time following changes at the macro-climatic level (see Western and van Praet 1973), so group sizes have declined. Note that this decline in

Figure 7.6. Mean group size for populations of Papio baboons living in habitats of high and low predation risk. Predation risk was assessed purely in relation to the availability of trees as refuges from predators: high risk habitats are open grassland and savannah habitats while low risk habitats are woodland or forest habitats. The horizontal bars indicate the median group size in each case. The two distributions are significantly different (Mann–Whitney test: z = 2.058, p = 0.039 2-tailed)

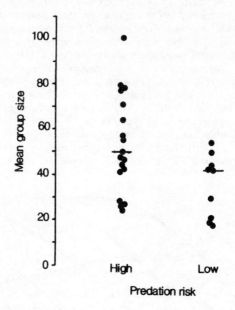

Sources: see Appendix, p. 326

group size has occurred despite the demand for large groups presumably being imposed by predation risk (assuming that this has remained roughly constant, if not actually increased as tree cover has declined). This provides a nice example of the way in which animals may be forced to compromise on the ideal solution to one problem by the constraints imposed by other features of their ecology.

Figure 7.5 reveals one other result of some significance. Given that habitat quality and predator risk are likely to be correlated (at least to the extent that rich habitats will also have many trees that the animals can use as refuges, whereas poor habitats will tend to be more open), rich habitats will be less risky than poor habitats for any given density of predators. If this is so, then each cost graph in Figure 7.5 will correspond to one (and only one) predator-risk graph. On plotting the intersection of the cost and predator-risk graphs for individual habitats (indicated by the dotted line linking certain inter-sections on Figure 7.5), we find that optimal group size can be a curvilinear function of habitat quality: in other words, it can increase and then decrease

Figure 7.7. Distribution of troop sizes for the Amboseli (Kenya) population of yellow baboons at three periods during the continuing decline in habitat quality caused by changes in the macro-climate of the region. The horizontal bars indicate median group sizes in each case. Differences between successive censuses are significant (Mann–Whitney tests, p < 0.05 in each case)

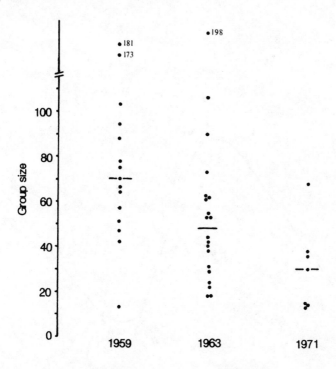

Sources: 1959 from DeVore and Washburn (1963); 1963 from Altmann and Altmann (1970); 1971 from Hausfater (1975).

again as habitat quality declines. The particular set of curves drawn on Figure 7.5 suggest that group size will be a J-shaped function of habitat quality, but the actual shape will in fact depend critically on the precise relationship between the habitat-specific cost and benefit curves. None the less, it remains clear that, whatever this relationship might be, the distribution of optimal group sizes may well be a non-monotonic (i.e. U-shaped) function of habitat quality.

That this may in fact be the case is indicated by Figure 7.8 which plots mean group size for 31 *Papio* baboon populations against mean annual rainfall (as a readily available measure of habitat quality). I offer no comment on the apparent N-shaped distribution[3] except to say that it emphasises my point by implying that the situation may be even more complex than I have suggested. Nor, it should be added, is this curvilinear relationship unique to

Figure 7.8. Mean group size for 31 populations of Papio *baboons plotted against mean annual rainfall for each habitat (rainfall being an index of habitat productivity). Open circles indicate mean band size for populations of hamadryas baboons: note that they are not significantly smaller than the groups of other baboon species living in comparable habitats. The regression line is set by eye*

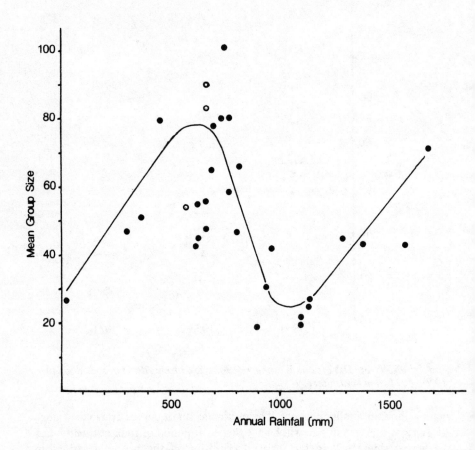

Sources: see Appendix, p. 326

baboons. Figure 13.2 (p. 316) suggests that mean party size for chimpanzees may also be a U-shaped function of annual rainfall. I do not wish to place too much weight on the particular shape of the relationship implied by Figure 7.5, since it depends on the exact relationship between the habitat-specific cost and benefit curves and these cannot be determined theoretically. None the less, Figure 7.8 emphasises the fact that variables like group size will not always be simple linear functions of some unitary environmental variable like habitat quality in the way that we have tended to assume. Clearly, one priority for future research must be to sample more populations at the extremes of the rainfall distribution in order to try to determine the shape of these curves empirically.

In conclusion, one note of caution should be sounded. We need to draw a clear distinction between two very different components of a species' anti-predator strategy, namely detection and avoidance. Group size is less important in terms of detection (the improvement in detection rate falls off rapidly as group size increases: Pulliam 1973) than in terms of avoidance (where the deterrance effect is probably proportional to group size over a very wide range). If visibility conditions are good, animals may be able to minimise group size by relying on early detection followed by evasion (e.g. patas monkeys: Hall 1965a). But where visibility is poor and the risks of being surprised by a predator at close quarters are high, they may have to rely on the deterrent effects of large group size (e.g. forest-living baboons like the mandrill) or large physical size (e.g. the orang utan and the gorilla) if they are to be able to exploit a terrestrial niche. High predation risk need not, therefore, inevitably select for large group size if the animals can minimise risks in some other way.

Evolution of Social Structure

In most species, males and females maximise their respective lifetime reproductive outputs by maximising different criteria in the biological process of reproduction (Goss-Custard et al. 1972, Trivers 1972). The reproductive output of a female is primarily limited by the rate at which she can produce and rear offspring (her reproductive turn-round time), and this is mainly determined by the quantity and quality of food available in the habitat (see Chapter 4). A male, on the other hand, is limited mainly by the rate at which he can fertilise females, so that his reproductive output is partly a function of the rate at which females themselves can reproduce and partly a function of his own ability to monopolise females as they become available.

Because female mammals are obliged to reproduce more slowly and to have more at stake on each reproductive event than males, they will generally be more concerned to ensure that they gain access to the best possible conditions in which to rear their offspring. This means that female reproductive considerations will generally take precedence over those of the males (Goss-Custard *et al.* 1972, Trivers 1972, Wittenberger 1980, Rutberg 1983). Females will therefore tend to distribute themselves around the habitat in the way that suits them best, given their particular ecological strategies and the distribution of essential resources. The males will then compete among themselves in order to map their own distribution on to that of the females. Charles-Dominique (1977) has been able to show that this is exactly what happens when dwarf galago are released experimentally into a new habitat.

This raises two questions. First, given that it pays a female to live in a group rather than alone, whom should she choose to form such a group with? Secondly, how will the distribution of males be influenced by the females' decisions about grouping?

Female grouping strategies

Among primates, the female's preference is likely to be for female group companions for two reasons. First, the difference between reproductive strategies of the two sexes means that males are likely to be less dependable as group companions because it will always be in their interests to desert the female to search for other females with whom to mate. Unless the female can coerce the male into investing heavily in parental care (generally unlikely because there is little help the male can give), she will do better to fall back on the company of other females. Secondly, many studies have noted that males and females often feed on different species of plants or in different places (baboons: Post *et al.* 1980, Kummer 1971; mangabeys: Waser 1977a; colobus: Clutton-Brock 1973; guenons: Cords 1986; lemurs: Pollock 1977; orang utans: Rodman 1977). These differences reflect contrasts both in the two sexes' nutritional requirements and in their abilities to exploit different micro-habitats due to differences in body size. Differences of this kind will inevitably generate conflicts over ranging patterns and the timing of progressions that will lead to the fragmentation of groups. In Chapter 6 we saw just how disruptive of time budgets the need to maintain close coordination with another individual could be (p. 96). Although conventional wisdom assumes that similarity of ecological niches creates increased competition (e.g. Wrangham 1979), the significance of this competition is attenuated (and may even be offset) by the advantages that accrue from being able to minimise conflicts about where and when to forage. In reality, we have no idea just how serious feeding competition actually is in terms of lost reproduction for most of these species. Nor, more importantly, do we know what proportion of the total competition in any given case is due to the various causes (i.e. group size, other females, males, etc.).

Given that females will have a preference for living in groups with other females, two further considerations will prompt them to choose to live with relatives rather than with unrelated individuals.

First, relatives tend to grow up in close social proximity and will therefore be more familiar with each other than non-relatives are. The importance of this lies in the fact that group-living depends on trust: by living in a group and providing reproductive benefits to other individuals, an animal incurs certain costs which it can only offset if it gains reciprocal benefits in kind at a later date. If the beneficiary of its cooperative actions reneges on this implicit agreement, then the very basis of social life is undermined. Willingness to trust another individual depends largely on past experience of that individual's reliability under the appropriate circumstances. An important source of such experience may be observation of the individual(s) concerned interacting with third parties. Close relatives are likely to be able to make more accurate assessments (and therefore afford to take bigger investment risks) by virtue of the greater familiarity that they have with each other through having grown up together. We also need to bear in mind that social groups do not materialise out of nowhere but have to be built up one step at a time. The easiest route to sociality is therefore for offspring to remain with their parent(s) rather than dispersing as they would normally do in a semi-solitary species.

The second consideration is that relatives can contribute to each other's inclusive fitness through kin selection. Helping relatives who share genes by virtue of descent from a recent common ancestor can significantly increase the likelihood that specific genes will be represented in future generations (see p. 19). In evolutionary terms, animals can afford to risk such an investment in a relative because, even if the relative later reneges, some return will accrue through kin selection. Some authors have suggested that kin selection is the main (or even the only) reason why animals form groups with relatives. This need not be the case, since cooperation with relatives may simply be more efficient or more convenient than cooperation with non-relatives (for an example, see Dunbar 1984a). For the reasons outlined in Chapter 2 (p. 21), evidence that animals distribute their altruistic behaviour in proportion to their relatives' relatedness to them cannot of itself be taken as evidence that such behaviour has evolved through kin selection. Nor can group-living itself be justified by appeal to kin selection alone since group-living requires a predisposition for cooperation before it can evolve. Kin selection can reinforce the selective advantages of group-living, but cannot itself promote it.

Male grouping strategies

Given that females prefer to live with female relatives if they are going to live in groups at all, and given that the size and dispersion of these groups is determined by the interaction of food availability and predation risk, how should males map themselves on to the female distribution?

Modelling male behaviour.   We can assume, from a Darwinian standpoint, that a male's primary concern will be to gain exclusive control over the largest possible group of breeding females. His success in doing so, however, will depend primarily on the operational sex ratio (i.e. the number of females that are in oestrus at any given moment relative to the number of breeding males). If the male's ability to maintain control over a group of females depends on his ability to control access to oestrous females (because competing males will not be interested in anoestrous females), then his ability to resist intruding males will depend on the probability that more than one female will be in oestrus on the same day (Emlen and Oring 1977). (Remember that a male cannot pinpoint the time of ovulation with total certainty, so he must stay with a female in order to prevent other males from mating with her at the crucial moment. I discuss this issue in more detail in the next chapter.)

Knowing the reproductive characteristics of the females, we can easily determine the probability that two or more females will be in oestrus together from the binomial expansion:

$$P(x \geqslant 2) = \sum_{x=2}^{n} \binom{n}{x} p^x (1-p)^{n-x}$$

where $P(x \geqslant 2)$ is the probability that two or more females will be in oestrus at the same time, n is the number of females in the group, x is the number of females in oestrus at the same time, and p is the probability that any one female will be in oestrus on any given day. The expression $\binom{n}{x}$ directs us to determine the number of ways in which we can select x individuals from a group of n: this can either be looked up in a table of binomial coefficients or calculated directly as $n!/[x!(n-x)!]$ where $n!$ signifies the factorial of n.

For females that breed randomly throughout the year, the probability that a given female will be in oestrus on any given day, p, is given by the proportion of days that the female spends in oestrus during a full reproductive cycle. Assuming that a female takes an average of three menstrual cycles to conceive and that the oestrous phase of each cycle (i.e. the period during which ovulation is likely to occur) is about 10 days (van Wagenen 1945), the probability that a female will be in oestrus on any given day will be $p = 30/365 = 0.082$ if females give birth every year, and $p = 30/730 = 0.041$ if the interbirth interval is normally two years. If breeding is seasonal, the degree of synchrony between the cycles of different females will be greater: this will have the effect of reducing the length of time for which females might be in oestrus and will be exactly equivalent to reducing the length of the reproductive cycle because males are presumed only to be interested in females during the time when they *might* be in oestrus. Thus, if the interbirth interval is two years, but breeding occurs only during a six-month period each year, then half the females will come into oestrus during the first breeding season and half during the second: this is equivalent to all the females breeding continuously with an interbirth interval of just one year (i.e. $p = 0.082$ as above). Similarly, if females take fewer or more oestrous cycles to conceive, on average, this will have the effect of increasing or decreasing p proportionately.

The probability of two or more females being in oestrus together for groups of different size is graphed in Figure 7.9 for several different reproductive patterns. The first point to notice about Figure 7.9 is that the probability of two or more females being in oestrus together (and hence the male's ability to maintain control over the group as a whole) depends on how synchronised the females' reproductive cycles are as well as on the number of females in the group. The precise point at which a harem-holding male can no longer prevent rival males from gaining access to his females depends on how the number of co-cycling females relates to the frequency of intrusions by other males and on how the harem-holder's ability to block these intrusions relates to the rate of intrusion. Assuming that the frequency of intrusions is constant over time (but it need not always be: see, for example, Dunbar 1984a), the point of transition will appear as a horizontal threshold running across the graph in Figure 7.9. Once the probability of co-cycling females rises above this critical threshold, the harem-holder will be unable to prevent other males entering his group and mating with his females. Data presented below (Figure 7.10) suggest that, in the gelada, the transition from one male to multimale groups occurs at about eight reproductive females. Interpolating this value into the graph on Figure 7.9 that corresponds most closely to the gelada's reproductive characteristics (a four-month breeding season due to reproductive synchrony among the females and a 2.5-year interbirth interval) suggests that the crisis point lies at $p(x \geqslant 2) = 0.187$. I have assumed a value $p = 0.2$ for simplicity in drawing the horizontal line across Figure 7.9, but this is obviously only an approximation and further data from a variety of species would be needed to fix it with any certainty.

Some striking evidence to support the general principle of this analysis comes from field studies of redtail monkeys by Cords (1984) and blue monkeys by Tsingalia and Rowell (1984). Both species normally live in one-male groups, and the single breeding male is usually able to monopolise matings with the females as these come intermittently into oestrus. In both studies, however, there were rare occasions when up to six females were cycling together. During these periods, the harem-holder was unable to prevent other males from joining the group and mating with the females. Once the females ceased cycling, the extra males left the group, which then reverted once more to being a conventional one-male group. Another test of the model is provided by Janson's (1984) data on brown capuchin monkeys since these differ significantly in terms of the female's reproductive characteristics from those assumed for the model. According to Janson's (1984) findings, a female capuchin is in oestrus for six days on each menstrual cycle (his Figure 5) with the probability of conception on each cycle being 0.14. Since the expected number of cycles to conception is the reciprocal of the rate of conception

*Figure 7.9. Influence of interbirth intervals and reproductive synchrony on the
probability of two or more females being in oestrus together in groups of different size.
The probability of one female being in oestrus on any given day, p, is a composite
function of the length of the oestrous period, the mean number of cycles to conception
and the degree of synchrony in the reproductive cycles of the females. The graphed
values of p correspond to: (a) a two-year interbirth interval with no synchrony between
females (i.e. females give birth randomly throughout the year) (p = 0.041); (b) a
one-year interbirth interval with no synchrony or a two-year interval with a six-month
breeding season (p = 0.082); (c) a one-year interbirth interval with a six-month
breeding season (p = 0.164); (d) a two-year interbirth interval with a two-month
breeding season (p = 0.246); (e) a one-year interbirth interval with a two-month
breeding season (p = 0.492). The horizontal line across the graph represents the
threshold above which males are unable to prevent other males entering the group and
mating with the females in oestrus (see text)*

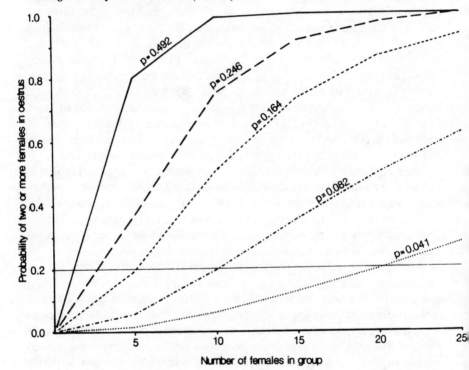

(Meyer 1970), the average number of cycles that a female will have each
breeding season is $(1/0.14) = 7.1$. Hence, during each breeding season she
will be in oestrus for $7 \times 6 = 42$ days. With a seven-month breeding season
(see Janson's Table 1), the probability of a female being in oestrus on any one
day is $p = 42/210 = 0.200$. With an average of four females in the group, the
probability of two or more females being in oestrus simultaneously is $p(x \geqslant 2)$
$= 0.181$. This is close enough to the threshold shown in Figure 7.9 to suggest
that capuchin groups should be multimale, as in fact they are.

Figure 7.10. Proportion of gelada reproductive units that have two or more adult males, plotted against the number of females in the unit. Sample sizes are 1, 16, 17, 29, 16, 14, 11, 9, 4 and 4 units, respectively

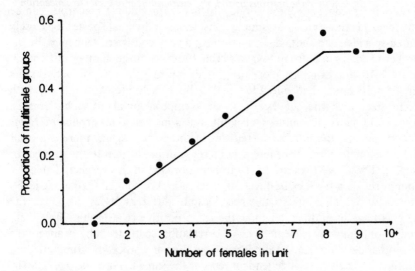

Source: Dunbar and Dunbar (1975), Dunbar (1984a) and unpublished data.

Irrespective of precisely where the threshold falls, Figure 7.9 generates two significant sets of results. First, seasonal breeders are unlikely to have single-male harem-type social systems even if the females live in small groups. Conversely, non-seasonal breeders may be able to hold quite large harems without serious risk of being unable to resist rival males' entry attempts. This, of course, is exactly the contrast we find between macaques (most of which are obligate seasonal breeders and live in multimale groups: see Caldecott 1986b) and a species like the patas monkey (which breeds randomly in time and lives in large one-male groups that contain about the same number of females as most macaque groups do: see Hall 1965a, Rowell and Richards 1979). In addition, Ridley (1986) has shown that, in a sample of 33 primate species, multimale groups tend to occur significantly more often in species that have short breeding seasons (i.e. high reproductive synchrony) whereas one-male groups occur significantly more often in non-seasonal breeders.

The second implication is that, in general, the number of males in a group is likely to be a function of the number of breeding females. This is confirmed by a number of sets of data. Andelman (1986) found a linear relationship between the mean number of males and the mean number of females in the groups of 18 species of cercopithecine monkeys (baboons, gelada, macaques and guenons). Although there are many conspicuous exceptions (notably among the guenons), species with an average of less than six females per group are invariably one-male grouping and those with more than ten females are always multimale.

Figure 7.9 suggests that the exceptions in these cases probably reflect differences in breeding synchrony. The effects of reproductive synchrony can be controlled for more easily by simply considering individual species (or closely related species), and these do in fact reveal a very consistent tendency for the number of males in a group to be a simple function of the number of females. Figure 7.10, for example, plots the proportion of gelada reproductive units that contain more than one adult male: this clearly increases linearly with harem size, with the suggestion of an asymptote at about eight females. Figure 7.11 plots the number of adult males in individual groups of *Papio* baboons against the number of females in the group. Again, there is a clear tendency for the number of males to increase in proportion to the number of females. These two taxa are far from being unique in this respect. Table 7.4 summarises data from eight different species drawn from all taxonomic levels from New World monkeys to apes. Despite different bases on which the analyses have been done, all show the same positive relationship. The likelihood of obtaining as extreme a set of correlation coefficients by chance were there no underlying relationship is very low (Fisher's procedure for combining significance levels from independent tests of the same hypothesis: $p < 0.001$).

Figure 7.11 exhibits two interesting trends that deserve comment. One is that very small baboon groups are likely to contain only a single male. The limit seems to be about 10 females — much the same as the maximum size of gelada one-male units. The second is that there is a suggestion that the number of males in a group reaches an asymptote of 12–15 males at group sizes of around 30 females: this emerges particularly clearly if grouped medians are calculated (given by the solid line in the figure). Such an asymptote might arise because, once this number of males cooperate, they form a sufficiently powerful coalition to be able to prevent other males from joining, no matter how large the number of females may become. This is intuitively plausible given that male baboons normally migrate between groups as individuals rather than in groups.

There are two general points to note about these results. First, the formal social system seems to be irrelevant to the fact that the number of males in a group is a simple function of the number of females. The same relationship occurs not only among species that live in unstable parties in a fission–fusion society, but also in species that typically live in one-male groups and those that live in multimale groups. Thus, it seems that the conventional distinction between species that form one-male groups and those that form multimale groups is in reality only a distinction between the size of group that females typically live in. Although many forest guenons, for example, live in one-male groups, few of them in fact seem to be obligate harem-forming species.

The second point is that these results seem to apply even though females are not always in oestrus: obligate seasonal breeders do not seem to differ

Figure 7.11. Number of adult males in individual groups of Papio *baboons, plotted against the number of adult females in the group*

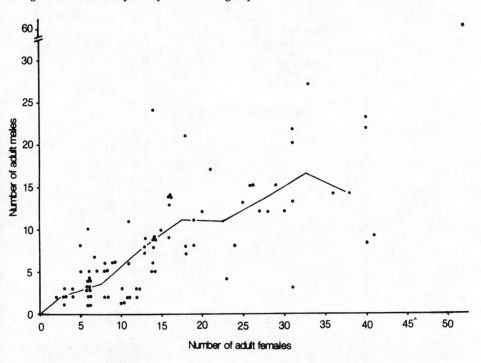

Sources: see Appendix, with additional data from Seyfarth (1976), Post (1978), Popp (1978) and Johnson (1984). Hamadryas are not included in this graph. Solid line gives the median values for grouped data in 5-female blocks.

from species that breed year round. This seems puzzling since males presumably cannot compete for access to oestrous females in the non-breeding season. Why should such groups continue to be multimale groups? There are several possible reasons, none of which can be evaluated in any detail at present. One obvious possibility is that the correlations in Table 7.2 are a fortuitous consequence of some other factor and have nothing to do with the males' access to females. The males may, for example, provide a benefit to the females, perhaps through resource defence (see below). However, while this might explain why females tolerate (or encourage) males joining them, it does not explain why males should be prepared to remain in a group. To be worth their while doing so, they must receive some benefit in return, and access to cycling females as and when these occur seems an obvious candidate. A second reason why males might remain in a multimale group is that it is simply not worth their while leaving because females are so highly clumped in their distribution. Once in a group, the costs of trying to join another group in

Table 7.4. Correlation between number of adult males and females in groups of various primate species

Species	Population	Comparison[a]	Total number of groups sampled	r_s	n	p (1-tailed)	Source
Baboons	Various	A	86	0.821	34	<0.001	Figure 7.11
Gelada	Simen, Ethiopia	B	121	0.888	10	<0.001	Figure 7.10
Hanuman langur	Various	C	105	0.524	8	0.091	Hrdy 1977
Red colobus	Various	D	30	0.800	4	0.100	Marsh 1979a
Black-and-white colobus	Various	B	46	0.462[b]	5	0.217	Dunbar, in press
Howler monkey	Barro Colorado, Panama	E	27	0.527	27	0.002	Smith 1977
Chimpanzee	Mahale, Tanzania	A	113	0.833	9	0.003	Table 13.6
Bonobo	Lamako, Zaire	A	190	0.933	9	<0.001	Table 13.6

[a] Bases of analysis: A: mean number of males in groups containing x females; B: proportion of multimale groups in groups with x females; C: percentage of multimale groups versus mean group size in individual populations; D: mean number of males versus mean number of females in groups for individual populations; E: number of males versus number of females in individual groups.

[b] A median test reveals a significant difference between small and large groups (χ^2=9.690, df=1, $p < 0.01$).

the face of resistance by the resident males may outweigh the disadvantages of remaining. Moreover, if the males of a group are related to each other (having migrated from the same natal group: see p. 82), then there may be consider-able advantages in staying in a group where you can count on the support of a coalition of related individuals. In addition, the risks of predation may make it desirable for males to be in a group rather than becoming solitary even when there are no oestrus females to compete for. A third possible reason is that males cannot be certain when females are going to come into oestrus. Since the simultaneous cycling of several females is a purely statistical event, males cannot predict exactly when they will occur. They can only 'know' that such events are increasingly more likely as groups become larger. In such a situ-ation, the best strategy may be to join a group and wait rather than searching randomly through the habitat since a male who searches in this way may risk missing the critical days when several females do come into oestrus together. The bigger the groups that females form, the more advantageous it is for males to remain with them rather than searching randomly (see below, p. 309).

Female choice.   So far, I have assumed that females are disinterested by-standers in the competition among the males. Up to a point, this is a reason-able assumption. Males need not confer significant advantages *or* disadvantages on a group of females, so the number of males present in a group may be irrelevant to the females so long as their intra-sexual fighting does not become too disruptive for the females themselves. In some species, at least, the females are relatively aggressive and can force the males to remain on the periphery of the group during the non-breeding season (e.g. squirrel monkeys: Baldwin 1971, Baldwin and Baldwin 1972; talapoins: Gautier-Hion 1970, Wolfheim 1977).

None the less, it has been suggested that the females of some, if not all, species exert a significant level of control over the number of males they are prepared to tolerate in their group. Both Wrangham (1980) and Caldecott (1986b), for example, have argued that females drive out surplus males in order to minimise the amount of ecological competition from males. On the whole, this claim is faulted by the common observation that males and females of many species exploit different niches, with the females often being able to feed in sectors of the micro-habitat (e.g. terminal twigs) that the heavier males cannot exploit (e.g. macaques: Caldecott 1986a; colobus: Clutton-Brock 1973; guenons: Cords 1986).

A more plausible suggestion is that females will begin to take an active interest in the males once they start to suffer from serious reproductive sup-pression *if* in doing so they can alleviate their physiological problems. This might occur either because the number of females in the group is large enough to generate significant levels of stress (e.g. gelada: Dunbar 1980a) or because fighting among the males has the same effect (e.g. colobus: Dunbar in press). If the males themselves are the source of the problem, then the females have two options: they can either drive the males out or, at the very worst, they can leave themselves. Clearly, females will only be able to prevent additional males joining their group if the males are not too much larger than they are. While macaque females are often able to displace and drive off their relatively small males, for example, baboon females are rarely able to do so because their males are very much larger (Packer and Pusey 1979). Colobus females, on the other hand, seem to be more willing to leave their groups, either to establish a new offshoot group in a new territory or to transfer to another existing group (Marsh 1979b, Dunbar in press). Constraints on individual movement may limit a female's options in this respect, so that in some species females may be forced to wait until a second male has joined their group before they can precipitate group fission in order to emigrate (e.g. gelada: Dunbar 1984a).

Where the source of the problem is harassment by other females, males may themselves become valuable alliance partners that females can use to buffer themselves against other females. In such cases, there may be some pressure to encourage additional males to join the group. One way of achieving this would be for females to make it difficult for a single male to monopolise a group by entraining their reproductive cycles. Even if only some females synchronise their cycles, they may be able to attract enough males into the group to meet their needs. There is some evidence, for example, that the females of baboon groups may be in reproductive synchrony and that this synchrony is unrelated to any environmental factors because neighbouring groups are not in synchrony with each other (Kummer 1968, Altmann 1980, personal observation). There is also evidence from the gelada that the females who are under the greatest stress (and thus suffering most severely from reproductive suppression) are the ones that are most interested in encouraging other males to join their units (as indicated by their willingness to desert in favour of a new harem male when one tries to join their unit: Dunbar 1984a). Disguising ovulation and extending the period of sexual receptivity (oestrus) would obviously have the same effect.

What females are prepared to tolerate will, in the final analysis, depend on (1) the costs they incur from socially induced stress, (2) the extent to which large group size permits them to exploit the habitat more effectively (particularly with respect to predation risk), and (3) the extent to which the presence of additional males may be an advantage to them (either as additional protection against predators or as potential coalition partners within the group itself).

Solitary species.   Conventionally, the dispersal of females into their own individual ranging areas such as is found among the semi-solitary species is considered to be the limiting case. However, the only issue of importance from the male's point of view is whether or not he can defend a collection of females. Whether or not those females habitually live together in a group is of little significance, aside from the fact that it will be easier to keep an eye on five or six females if they are together in one place than if they are scattered over a wide area in dense forest.

Thus, exactly the same considerations apply when the females live solitarily: can the male defend a large enough area to enclose the ranges of several females? If he can, then he creates a polygynous mating system that, in terms of its evolutionary consequences, is essentially a one-male group. Providing (1) that the females have small ranging areas, (2) that the females' reproductive cycles are not too closely synchronised, and (3) that the inter-birth interval is not less than 12 months, then a male can pursue such a strategy. These conditions appear to be met in most of those cases where solitary males hold large territories that overlie the smaller territories of several solitary or semi-solitary females (galago: Bearder and Doyle 1974, Charles-Dominique 1977, C. Harcourt and Nash 1986; mouse lemur: Martin 1973). Owing to the difficulty of monitoring females when they are scattered in dense vegetation, however, we would not expect males in these systems to be able to monopolise such large harems as those in species where the females lived in groups. This is born out by the evidence that males in semi-solitary species generally seem to have regular access to only about two females on average (Charles-Dominique 1977, C. Harcourt and Nash 1986) whereas, in most species of small primates that form stable one-male groups, males are usually able to monopolise access to between four and ten females (e.g. forest guenons: data summarised by Andelman 1986). None the less, it is clear that a male in a semi-solitary species will do better by pursuing this strategy than by adopting the alternative strategy of staying with one female.

Where the three conditions listed above are not met, a single male will be unable to prevent other males gaining access to at least some of the females living within his territory. In this case, the male has two options. Either he can allow other males to join him in order to defend a cooperative territory while access to individual females is contested as they come into oestrus (a form of multimale polygyny adopted by chimpanzees: Wrangham and Smuts 1980) or he can concentrate on guarding a single female on a permanent basis in a form of monogamy (a solution that it has been suggested is adopted by a number of monogamous species, including gibbons, the Mentawi langur and some cebids: Kleiman 1977, Rutberg 1983).

It is worth emphasising that this last point underlines the distinction between mating system and social grouping. These different functional units often take different forms in the same species, while different species may have similar forms in one respect but not the other. This distinction was often glossed over in traditional views where the number of males in a group was regarded as a defining feature of the society.

★

So far, my primary concern has been to establish a set of general principles governing the grouping patterns of individual animals. In most cases, however, these general principles are insufficient on their own to explain (let alone predict) the social system of any given species. To be able to do this, we need to interpolate these principles into the particular ecological and reproductive context of the individuals concerned. Since these may in fact differ not just from one species to another but also, within species, from one population to another, we can expect significant differences in the forms of social system that we encounter in the field. I emphasise this because it seems to me that the traditional species-specific view of societies (that each species has the same form of social system throughout its range) still has an insidious hold on our thinking.

I shall postpone further discussion of these issues until I have had an opportunity to examine the dynamics of social systems in finer detail. Chapters 8 to 11 will be concerned with this. I shall return to the question of the evolution of social systems in Chapters 12 and 13.

Notes

1. Note that exactly the same set of curves will result if groups form for reasons of resource defence: the point is only that some factor promotes the formation of large groups because lifetime reproductive output is greater in large groups. Several candidate principles can fill this role. Thus, although the *interpretation* of the curves will differ, the shape will remain substantially the same. I prefer predation as the explanation because the weight of evidence favours this hypothesis. Note also that we could not use the results of this analysis to test the predation-defence hypothesis, since both the resource-defence and the predation-defence hypotheses would yield the same results.

2. For graphical convenience, I have represented the optimum group size as lying at the intersection of the relevant cost and predation graphs. In fact, with graphs of these particular shapes, the optimum will generally lie slightly to the right of their intersection, but this will not make any difference to the conclusions we draw.

3. Because of the paucity of data in the outer arms, a curvilinear regression is heavily biased by the mass of data in the mid-range and so generates a nearly linear relationship between group size and rainfall. The slope of this relationship is not, however, significant: linear regression set by least squares gives a slope of $b = -0.012$ ($r^2 = 0.039$, $t_{29} = -0.643$, $p > 0.30$ 2-tailed). But this clearly ignores a very significant negative relationship over the mid-section within the range 500-1200 mm rainfall: slope $b = -0.089$ ($r^2 = 0.435$, $t_{21} = -3.793$, $p < 0.01$ 2-tailed). An N-shaped curve of this kind is often the product of an interaction between two different curves: in this case, these could be an inverted-U-shaped curve and a positively increasing exponential curve, each reflecting the influence of a different environmental variable.

Chapter 8
Mating Strategies

A female mammal, once fertilised, gains little by repeated matings. Her interests are best served by concentrating on the business of gestating and rearing the fetus(es). Males, on the other hand, are less constrained by their reproductive biology and can, in principle at least, increase their reproductive output by continuing to mate with as many females as they can. With a gestation period in the order of six months in most primates, there is little that a male can do to help the female in the business of rearing until after the infant is born. Only in respect of the indirect help that a male can confer on a female carrying his offspring is he able to make any contribution at all. The extent to which he can contribute even in this way depends largely on the nature of the species' ecological strategies. I will consider the male's contributions to parental care in more detail in the following chapter. Here, I shall concentrate on the prior problem of decisions about mating.

Reproductive Behaviour in Primates

Among primates, the female undergoes a regular menstrual cycle, generally of about 4–5 weeks in length (Table 8.1). There may, however, be considerable variation both in the mean length of cycles for different females and in the range of cycle lengths experienced by individual females.

From a physiological point of view, the cycle is divided into two phases: a follicular phase roughly from the onset of menstruation to the following ovulation and a luteal phase from ovulation to the next menstruation. During the follicular phase, the ovaries are involved in the production of Graafian follicles (hence 'follicular'), each containing an ovum (or egg) under the control of the hormone oestrogen. Ovulation marks the point at which the ovum is released from the ovary into the Fallopian tubes where it is ready to be fertilised. Following its release, physiological activity shifts over to the preparation of the womb (or uterus) for implantation of the fertilised ovum after conception has occurred, much of this preparation being under the control of

151

Table 8.1: Length of menstrual cycle in selected Old World monkeys

Species	Cycle length (days) Mean/median	Range	Source
Yellow baboon	31.4	23–47	Hausfater 1975
Chacma baboon	35.6	29–42	Gillman and Gilbert 1956
Mandrill	33.5	20–244	Hadidian and Bernstein 1979
Drill	35.0	13–82	Hadidian and Bernstein 1979
Japanese macaque	26.5	8–52	Takahata 1980
Pigtail macaque	35.3	6–396	Hadidian and Bernstein 1979
Celebes black macaque	35.5	10–115	Hadidian and Bernstein 1979
Moor macaque	32.0	14–85	Hadidian and Bernstein 1979
Sooty mangabey	34.5	10–177	Hadidian and Bernstein 1979
Agile mangabey	29.8	22–39	Gautier-Hion and Gautier 1976
Greycheeked mangabey	29.8	25–42	Gautier-Hion and Gautier 1976
Whitecollared mangabey	32.5	26–35	Gautier-Hion and Gautier 1976
Spotnosed guenon	28.0	27–54	Gautier-Hion and Gautier 1976
Talapoin	36.0	29–53	Gautier-Hion and Gautier 1976

a hormone (progesterone) produced by corpora lutea (hence 'luteal') in the Graafian follicle from which the ovum was released.

The length of the luteal phase of the cycle is virtually constant for all individuals of a given species, but the follicular phase may vary considerably even within individuals from one cycle to the next. It is this variance in the follicular phase that is mainly responsible for the long tail to the right in the distribution of menstrual cycles. The variability in this part of the cycle is caused by factors as diverse as illness, injury and stress (Rowell 1972a).

In a number of primate species (including baboons, macaques, mangabeys, talapoins and chimpanzees), females possess a sexual skin on the perineum that undergoes a cyclic swelling under the control of the ovarian hormones (notably oestrogen). The sexual skin gradually swells over the follicular phase of the cycle, reaching maximum distension about one week before ovulation occurs. It undergoes a sudden collapse as the fluids are resorbed 1–4 days after ovulation. Deflation takes place over the course of a few days, following which the skin is 'flat' for about 10 days until menstruation occurs. Although not all species of primates have sex skins that tumesce in this way, most undergo cyclical changes that can be detected by their males. These changes include a detectable swelling of the lips of the vulva (Rowell 1972a, Dixson 1983) and pheromonal signals from vaginal secretions (Michael and Keverne 1968, Keverne 1982).

Dixson (1983) has reviewed the distribution and physiology of sexual swellings in primates. Their disjunct distribution among the Old World monkeys and apes suggests that they have evolved independently at least four times. In all cases, however, their anatomical features suggest that they are generally outgrowths of the cyclical vulvular swellings commonly found in all primates.

In one genus (the macaques), there is evidence that the character has been lost in some of the evolutionarily more recent lineages. In this genus, the degree to which females characteristically exhibit swellings seems to be inversely related to the complexity of the male penile morphology and the structure of the female's internal genitalia, though why this should be the case remains unclear. Those species of primates that possess swellings all live in multimale social systems where males compete with each other for access to individual females as they come into oestrus. This correlation suggests a possible function in terms of the incitement of male–male competition as a means of facilitating female choice. This hypothesis, however, is not wholly convincing since many multimale-grouping species lack swellings of any kind (e.g. lemurs, all New World monkeys, vervets) while some species that form one-male groups possess analogous features (the fluid-filled vesicles that develop around the various areas of bare skin on the chest and rump of the gelada). Moreover, given the demonstrated efficacy of olfactory cues in sig- nalling a female's receptivity, the need to have an additional visual cue would seem to be less pressing. It is possible, however, that the visual signal serves to communicate a female's state of receptivity over distances that are too great for olfactory signals to be effective. This might be important either because these species are largely terrestrial and thus live in larger more dispersed groups than most arboreal species or because females actively wish to attract males to join and stay in their groups.

Mating typically exhibits a clear peak around the time of ovulation in most (but not all) species (e.g. baboons: Bielert 1982; macaques: Michael and Zumpe 1970, Enomoto et al. 1979; patas: Rowell and Hartwell 1978). This closely parallels the relationship between the likelihood of ovulation and the day of the menstrual cycle (Figure 8.1), suggesting that males and females may be able to identify the most profitable time to mate (in terms of maxi- mising the likelihood of fertilisation) even if they cannot pinpoint the exact day of ovulation. There is some circumstantial evidence that matings and con- sortships are more likely to occur during the morning than the afternoon (baboons: Hausfater 1975; gelada: U. Mori 1979a, Dunbar 1984a; chim- panzees: Thompson-Handler et al. 1984), possibly reflecting the fact that ovu- lation is more likely to occur early in the day rather than later (cf. Balin and Wan 1968). Other species, however, exhibit a less marked cyclicity of sexual behaviour during the menstrual cycle: in gelada, mating is distributed more evenly across a longer portion of the cycle (Dunbar 1978), while van Noordwijk (1985) could find no correlation at all between the frequency of copulation and the degree of swelling of the female's sex skin in longtailed macaques. This might suggest a tendency towards the concealment of ovu- lation, possibly in order to persuade consorting males to remain longer with them.

In most species, males and females differ in the roles they adopt in sexual interactions. Because female mammals bear a disproportionate share of the

Figure 8.1. *Probability of ovulation occurring on any given day of the menstrual cycle
(with the first day of menstruation defined as day-1) for rhesus macaques*

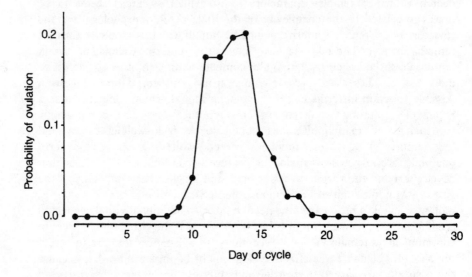

*Source: Data from timed matings and laparotomies for 104 pregnancies given by
Hartman (1932). Similar results were obtained by van Wagenen (1945)*

costs of reproduction, we can generally expect males to take the lead in initiating sexual interactions whereas females will be more selective and respond
only slowly in the light of the male's courtship. Typically, a male initiates
copulation by approaching an oestrous female and sniffing or touching her
vulva. If the female is willing to mate, she will stand and present her rear to
the male in a characteristically braced posture, thereby inviting copulation
(for further details, see Rowell 1972b).

There are two important exceptions to this generalisation. First, in many
species, this pattern is true only of older females. Nulliparous females (i.e.
those that have never given birth) tend to be more active in soliciting the
male, and a shift to male-initiated copulations occurs only after the female has
reproduced for the first time (baboons: K.R.L. Rasmussen 1983; gelada:
Dunbar and Dunbar 1975, Dunbar 1984a; macaques: van Noordwijk 1985;
orang utan: Galdikas 1981). This age-related difference in proceptive
behaviour seems to reflect the fact that younger females, and particularly
those that have not yet produced infants, have to work harder to rouse the
male's interest than older females do. Anderson (1986) has reviewed the literature from some 15 species (mostly, but not exclusively, Old World monkeys)
and found that males show a marked preference for older parous females in
all cases. One obvious functional explanation for this is that males are less

willing to risk wasting time on females whose fertility is in doubt. This applies not just to females that are chronically infertile (and will never reproduce) but also to young females because the first few menstrual cycles after a female has undergone puberty are normally infertile (Rowell 1972a). A male who formed a consortship with an adolescent female might be wasting time trying to impregnate her when he could have been consorting with a more fertile older female with greater certainty of success. A male's interest in younger females may thus hinge on the likelihood that a mature female will come into oestrus and be lost to another male.

The second exception arises from the fact that sexual interactions are two-way processes. Males do not impose themselves on females wholly uninvited. Rather, it is an interaction aimed at achieving a specific goal of equal interest to both parties. To achieve this, communication about willingness to mate is necessary and this is often multichannel and bidirectional. This has been illus-trated very elegantly by a series of experiments on rhesus macaques by Keverne (1982). When the male's sense of smell was neutralised by coating his nasal passages with a removable wax plug, there was a significant drop in the frequencies with which males attempted to initiate copulation, yet the fre-quency of matings remained at pre-experimental levels. Detailed behavioural observations revealed that the female was compensating for the male's apparent lack of interest by increasing her own rate of solicitation. In other words, finding that the usual (olfactory) channel by which the male acquired information on her reproductive state was not functioning effectively (i.e. it was not eliciting sexual behaviour from the male), the female stepped up her signalling through another channel (i.e. visual signals) in order to get her message across. What this implies is that the normally high frequency of male-initiated copulations is due less to the female's lack of interest than to the fact that it may be unnecessary for her to do anything when the male is already receiving all the information he needs. None the less, she continues to monitor his response in order to ensure that he does what she wants him to do. The incentive for each partner to initiate matings will obviously depend partly on the level of competition for access to members of the opposite sex. Hence, although competition between males will generally tend to be higher and force males into taking the active role, there will be species where the level of competition is lower and the active role will pass over to the females (e.g. gelada: Dunbar 1978; brown capuchin monkeys: Janson 1984).

In some species (notably baboons, macaques and chimpanzees), both the female and the male give loud vocalisations during and immediately after copulation. These calls are not simply a consequence of exertion during mating (as has sometimes been supposed), but are highly structured acoustically and are clearly intended to convey information. Hamilton and Arrowood (1978) have suggested that such calls may serve either to stimulate mutual arousal in the copulating pair (in order to facilitate sperm transfer) or to incite other males to compete for the female (as seems to be the case in, for example, elephant seals: Cox and LeBoeuf 1977). It is difficult to see how the latter explanation might apply in the case of most primates since in most instances the calls are given (or at least reach their peak of audibility) at or after the point at which the male ejaculates, by which time it is presumably too late to encourage competition among rival males. In any case, the male's vocalisations are invariably much louder than the female's. While it is certainly significant that subordinate males often suppress their copulatory vocalisations in order to avoid attracting the attention of high-ranking males (Hamilton and Arrowood 1978, de Waal 1982), it remains unclear just what males would gain from advertising their actions so conspicuously unless it has something to do with a declaration of 'ownership'. It is possible that these calls evolved originally to facilitate sperm transfer, but it may well be that intra-sexual selection has since enhanced these calls in males as a means of reinforcing the dominant males' monopoly over the females.

One last difference found among primates is the distinction between those species that engage in repeated mounts before ejaculation (e.g. many macaques) and those that have a single mount to ejaculation (e.g. most baboons). This difference is well documented and quite striking (Rowell 1972a), but its significance remains unclear since its distribution does not follow clear taxonomic divisions. Caldecott (1986b) has suggested that in macaques multiple mounting has been evolved essentially by females as a mechanism for inciting male–male competition, and he points to the fact that its distribution within this genus seems to correlate with other behavioural patterns that imply intense male competition (e.g. long consortships, antagonistic male–male relationships, high socionomic sex ratios in favour of females, and frequent male migration). Surprisingly, perhaps, it correlates rather poorly with sexual swellings, a feature which is also supposed to function in stimulating male competition. The problem clearly merits further research.

Gaining Access to Mates

In considering the male's problem of how to acquire mates, the most profitable strategy is to begin by examining the simplest possible situation. This involves identifying the constraint-free strategy (i.e. the strategy that, in the absence of any constraints other than the basic social system, would be the most effective means of gaining access to a given resource). We need to be careful not to be confused by the fact that a natural context is actually *never* free of such constraints. None the less, in order to understand how these constraints affect the decisions of animals, we need to know how the basic system works in their absence: this provides us with the essential baseline against which to evaluate the consequences of these constraints. In effect, then, we have to take the system apart, identify its key components, assess their behaviour in isolation and then reconstruct the system by adding complicating factors one variable at a time.

In this section, then, I shall concentrate on the male's constraint-free strategy for acquiring mates. I take this to be the exercise of brute power (i.e. aggression) in contests where females are essentially passive prizes for the victor. I shall use the predictions generated by models based on these assumptions to search for the points at which such simple explanations are inadequate to account for what we actually find in nature. Having identified these points, we can then use this information to direct our search for the most relevant complicating factors (i.e. the constraints) that need to be taken into account. I explore the consequences of these constraints in the following two sections.

General considerations

The male's problem, in a nutshell, is one of gaining access to breeding females. How he does this depends on the kind of groups the females live in. If they live in small groups over which an individual male can maintain control, then the male's problem is essentially one of gaining control over such a group. This will usually mean having to wrest control of a group away from another male, who will naturally be unwilling to accept his loss of breeding opportunity. Once a male has gained control over a group of females, however, his problem switches to one of keeping rivals at bay in order to prevent them ousting him in his turn. His access to the females when they come into oestrus is more or less unaffected by any other considerations (unless many of them happen to come into oestrus at the same time). In contrast, when females live in large groups and many males are able to gain access to them simultaneously, the male's problem is one of ensuring access to individual females during the period when ovulation is most likely to occur in order to ensure that he is able to fertilise the female. If he abandons a female in order to mate with another, he risks a rival fertilising her instead. This can happen either because fertilisation is only probabilistic on any given mating (i.e. males cannot predict exactly when ovulation occurs) or because successive males 'flush out' the previous male's semen from the female's reproductive tract. Although there is no detailed data on sperm competition in primates, sperm-flushing is known to occur widely among insects (Parker 1970), and in at least one bird species (the dunnock) females actively eject sperm from a previous mating when they mate with a second male (Davies 1983). The fact that primate seminal fluid contains a coagulating agent (van Wagenen 1936) suggests that the formation of vaginal plugs following ejaculation may represent an attempt to reduce the likelihood of rival males fertilising a female.

The contrast in reproductive context may have further implications for the animals' styles of sexual behaviour. This is illustrated by a comparison of sexual behaviour in *Papio* baboons and the gelada (Table 8.2). In baboons, males compete directly for access to individual females as they come into

Table 8.2. Comparison of mating behaviour in gelada and *Papio* baboons during the ten days of peak oestrus

Variable	Gelada[a]	Baboons[b]
Copulations initiated by male (%)	34.0	66.0[c]
Copulation rate (n/h)	0.55	1.19
Male inspects female (n/h)	0.21	1.53
Female presents to male (n/h)	1.10	0.72
Male herds female (n/h)	0.09	0.81

[a] From Dunbar (1978).
[b] From Hausfater (1975).
[c] From Saayman (1970).

oestrus, and the maintenance of a consort relationship is a key feature of a male's attempts to control access to females. Male baboons initiate the majority of copulations, mating is frequent and copulation is commonly followed by grooming. Aside from following (and some herding), grooming is probably the most important behavioural mechanism that a male uses to maintain a consort relationship with a female. Among gelada, on the other hand, males compete only for long-term control of groups of females. Once that control has been established, a harem-holder's access to individual females as they come into oestrus is not normally challenged. The general tenor of sexual relationships is therefore very different in this species. Females initiate a high proportion of the matings, copulation occurs at much lower rates, and pairs rarely groom afterwards. Harcourt (1981) summarises very similar contrasts in the sexual behaviour of multimale-group-living chimpanzees and one-male-group-living gorillas. In both sets of species, these behavioural differences are mirrored by corresponding differences in reproductive anatomy: gelada and gorillas have a much lower ratio of testis weight to body weight than *Papio* baboons and chimpanzees (Harcourt *et al.* 1981b). These anatomical contrasts seem to be a direct consequence of the differences in behaviour. For baboon and chimpanzee males, displacement by a rival is an ever-present threat, and one way of maximising the likelihood of fertilising the female is to ensure that she receives as much sperm as possible. This can be achieved both by copulating frequently and by producing large quantities of sperm on each occasion. Since male sperm production and storage capacities are limited, relatively large testes are required to produce sufficient quantities of sperm to avoid too rapid depletion of available sperm after only a few copulations (Harcourt *et al.* 1981b). Janson (1986) reports an analogous contrast in the male's role in initiating matings between brown and white-fronted capuchin monkeys: in this case, the contrast may have ecological roots in that dominant male brown capuchins are able to offer females exclusive access to preferred food sources and this may encourage them to be more active in soliciting the male.

Mate acquisition in multimale group contexts

A male in a multimale group faces a problem in trying to maximise his reproductive rate because his uncertainty as to exactly when the female will ovulate makes it difficult for him to know how long he should continue to mate with her. At what point should he desert her if a second female comes into oestrus? The longer he continues to consort with the first female, the more certain he can be that he will have mated with her when she finally ovulated. But because the probability of ovulation on any day of the cycle has a relatively sharp peak (Figure 8.1), the rate at which this certainty increases begins to decline once past the modal date of ovulation. The cumulative probability of ovulation is thus logistic (or sigmoid) in shape (Figure 8.2). Since the male gains diminishing returns by staying with the female after the peak, we can easily determine the point at which it will pay the male to shift to a second female in order to maximise his reproductive output. Figure 8.2 shows the cumulative ovulation probability curves for two females whose menstrual cycles are displaced in time (in this case, female-2 lags behind female-1 by three days). We assume that the male is mating with female-1 and has to decide between shifting to female-2 either at time t or at time $t+1$. If the male shifts at time t, his total reproductive output is:

$$R_t = \int_0^t P_1(t) + \int_s^\infty P_2(s) \tag{8.1}$$

where s is the time in female-2's cycle that corresponds in real time to t on female-1's cycle, and $P_1(t)$ and $P_2(s)$ are the respective cumulative probability density functions for the occurrence of ovulation for each of the two females. Similarly, if the male decides to wait until time $t+1$ before shifting to female-2, his total reproductive output will be:

$$R_{t+1} = \int_0^{t+1} P_1(t) + \int_{s+1}^\infty P_2(s)$$

Obviously, it will pay the male to shift to female-2 at time t only if:

$$R_t > R_{t+1}$$

Substituting the full expressions and doing a little algebra shows that the male should shift at time t if:

$$\int_t^{t+1} P_1(t) < \int_s^{s+1} P_2(s)$$

Figure 8.2. Cumulative probability that a male will have fertilised a given female if he mates with her continuously up to the tth day of her menstrual cycle (based on the data given in Figure 8.1). The probability curves for two females whose cycles are separated by a lag of l = 3 days are graphed. The male's problem is whether he will maximise his net reproductive output by switching from female-1 to female-2 at time t or at time t+1

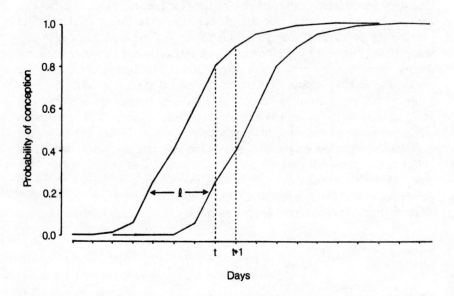

The conditions under which this criterion is met will, however, depend not only on the shape of the cumulative probability function, but also on the temporal separation (or lag) between the two females' cycles — in other words, on the degree of synchrony in their menstrual cycles (Figure 8.3). On the whole, the longer the lag between the two cycles, the later the male should shift from one to the other. It is worth noting, however, that while there is a great deal of tolerance in the timing of this shift for long and short lags, timing is much more critical at lags that correspond to the ovulation 'window' (i.e. the width of the peak in the distribution of the ovulation probability).

We have modelled the situation for a species whose females provide the males with moderately clear signs of the imminence of ovulation. If the male is given no such information or the actual occurrence of ovulation is more variable during the female's cycle, the cumulative ovulation probability curves will rise more slowly to their asymptotes than those in Figure 8.2. If we recalculate the optimal day on which to desert female-1, we find that the curves shown in Figure 8.3 are now flatter over a much wider range of lags. In other words, the longer the period during which ovulation might occur, the less worth the male's while it will be to desert one female in favour of another.

Figure 8.3. Optimal day on which to desert female-1 for female-2 for a range of lags, l, between the two females' menstrual cycles. The number of fertilisations obtained is calculated from the distributions in Figure 8.2 using equation 8.1. Note that, for short or long lags, there is wide tolerance for the timing of the optimal day for desertion, but that the peak in the graphs is more marked (and the tolerance correspondingly less) at imtermediate lags. Also note that as the lag between the two females' cycles increases, so it pays the male to delay deserting his current female. If the lag is very short (l=1 day), however, it may not be worth the male's while deserting at all

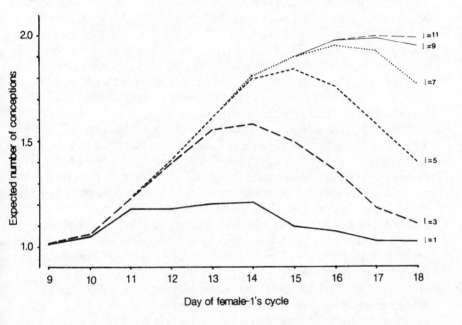

Concealed ovulation may thus be one means by which females can ensure a male's loyalty.

A male in such a nominally promiscuous mating system faces a second more serious problem, however. His access to oestrous females is determined in the first place by his ability to dominate and displace other competing males. (I discuss the conceptual difficulties that have beset the concept of dominance in Chapter 10, p. 206.) A number of 'priority-of-access' models have been developed that relate frequencies of consorting and/or conceptions to the dominance ranks of the males in a group and the number of females who are in oestrus at any given moment (see Altmann 1962, Suarez and Ackerman 1971). In general, these models assume that the dominant male can monopolise access to an oestrous female and that a lower ranking male can only mate providing there are enough females simultaneously in oestrus to preoccupy all the other males that rank higher than him in the hierarchy.

Despite persistent claims to the contrary, there is a considerable body of evidence to suggest that, in qualitative terms at least, dominance rank does determine a male's ability to mate with females (baboons: Hall and DeVore 1965, Hausfater 1975, Seyfarth 1978a, Packer 1979b; macaques: Kaufmann 1965, Lindburg 1971, Enomoto 1974, Smith 1980, Taub 1980, Chapais 1983a, van Noordwijk 1985; vervets: Struhsaker 1967c; howler monkeys: Jones 1985; capuchin monkeys: Janson 1984; spider monkeys: Milton 1985). This relationship, however, is far from universal, even within a given species. Some studies have found no correlation between male dominance ranks and various indices of reproductive success (baboons: Strum 1982, Smuts 1985; macaques: Loy 1971, Eaton 1974; guenons: Cords *et al.* 1986; capuchin monkeys: Oppenheimer 1968; chimpanzees: Tutin 1979), while others have reported a correlation during some years but not others within the same group (macaques: Duvall *et al.* 1976, Witt *et al.* 1981; vervets: Cheney *et al.* 1986; chimpanzees: de Waal 1982, Nishida 1983).

More problematic, however, is the fact that even in those cases where there is a correlation between dominance rank and mating success, *quantitative* tests of the models' predictions have proved disappointing (e.g. Hausfater 1975, Chapais 1983a). Dominant males invariably do less mating than they ought to, while many low-ranking males successfully manage to mate when they ought to have been denied access (Figure 8.4). Bearing in mind that our aim should be to predict *exactly* the behaviour of animals, we clearly need to look more closely at the assumptions of these models and at the way in which they are being tested in order to try to find out why the results are less good than we would wish.

Where priority-of-access models go wrong

There are at least three reasons why we might not get a close correlation between dominance ranks and reproductive performance in particular cases. One is methodological, the other two essentially biological.

The methodological problem arises from the fact that the measures of reproductive success commonly used to test these models may not always be the most appropriate. It is rarely possible to determine paternity in the wild, so most studies have resorted to indirect indices and assumed that these correlate with the actual fertilisation rate. These include variables like the amount of time spent consorting with females, the number of different females mated, the total number of copulations, etc. While there may well be an ordinal relationship between these variables and fertilisation, it is likely that the relationship is at best statistical. Hence, deviations from predicted values are inevitable. Worse still, there is really no good reason to suppose that the frequency of copulation, for example, should be directly related to the probability of fertilisation. Many other variables (including the timing of mating relative to the occurrence of ovulation) are involved that make it possible for the males that do the least mating to fertilise the most females

Figure 8.4. A test of the 'priority-of-access' model of male mating success (see text). The proportion of females fertilised (filled bars) by individual male rhesus macaques of the Cayo Santiago group F is compared with the proportion expected on the assumption that dominant males can monopolise females in oestrus only providing several females do not come into oestrus simultaneously (open bars)

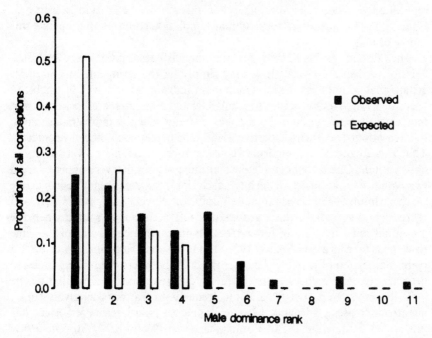

Source: Chapais (1983a).

simply because they can time their copulations more reliably in relation to the occurrence of ovulation. Stern and Smith (1984), for example, were able to use genetic paternity exclusion techniques on three groups of captive rhesus macaques to show that the correlations between true paternity and various behavioural measures of reproductive success can be very poor.

Additional problems can arise as a result of differential sampling of the various animals, especially under field conditions. Drickamer (1974b), for example, obtained a positive correlation between rank and reproductive success in a free-ranging group of rhesus macaques on Cayo Santiago, but found that this correlation vanished when he took into account the fact that males differed in the frequency with which he was able to sample their behaviour. Low-ranking animals tended to remain near the periphery of the troop where they were less conspicious. Several other studies have noted that low-ranking males often try to attract females to the periphery of the group where they are less visible to the dominant males (baboons: K.R. Rasmussen

1986; chimpanzees: Tutin 1979, de Waal 1982). Ruiz de Elvira and Herndon (1986) have demonstrated experimentally that subordinate male rhesus macaques tend to initiate the majority of their copulations when they are out of view of the higher ranking animals. However, this particular problem cannot account for all the positive correlations between dominance rank and reproductive success because many of the studies (e.g. Hausfater 1975, Chapais 1983a) have used focal animal sampling techniques that obviate this source of bias.

The second problem concerns the models' assumptions about males' abilities to displace rivals: these may simply be too stringent. In multimale groups that contain many males, dominant individuals may well be unable to prevent rivals gaining access to individual females on some occasions, but much will depend on both the number of rivals and the individual animals' relative power (i.e. their respective abilities to defeat each other in contests). That this might be an important consideration is indicated by two sets of observations. One is that completely contradictory results were obtained from two studies that used paternity exclusion to determine reproductive success. Smith (1980, 1981) found a significant correlation between rank and reproductive success in three groups of captive rhesus at a time when the groups all contained four or fewer breeding males. In a later study of the same three groups, Stern and Smith (1984) found no such relationship but by this point maturation of juveniles had increased the numbers of breeding males to between five and eight. Studies of many different taxa suggest that as the local density of males rises (and the frequency of attacks by rivals consequently increases), so males become unable to guard resources effectively (insects: Campanella 1974, Campanella and Wolf 1974, Borgia 1981; reptiles: Berry 1974, Brattstrom 1974; birds: Davis 1959; mammals: Davis 1958, LeBoeuf 1974). Similar effects have been noted in primates too (e.g. lemurs: Budnitz and Dainis 1975; guenons: Cords 1984, Tsingalia and Rowell 1984, Cords et al. 1986). The other line of evidence to support this suggestion is the fact that lower ranking males are often able to gain access to cycling females by forming coalitions among themselves (baboons: Packer 1977, Smuts 1985; macaques: Witt et al. 1981; chimpanzees: de Waal 1982, Nishida 1983). Coalitions may allow individuals to defend a key resource like an oestrous female temporarily against a more dominant individual without influencing their long-term dominance relationships. I shall have more to say about coalitions in the final section of this chapter.

The third possibility as to why the priority-of-access models are less successful than we might expect is that they ignore the role that the females might play. The models invariably assume that a male's access to a female is determined purely by his ability to displace other males. Though this is true to some extent, there is considerable evidence suggesting that the females do exert a great deal of influence over what the males can do in certain cases. Bachmann and Kummer (1980), for example, have shown experimentally that a hamadryas baboon male's willingness to intervene in a relationship between a female and another male depends on quantifiable aspects of the strength of the bond between the pair in question. A number of other studies have reported that females often form preferential consortships with particular males (macaques: Michael and Saayman 1967, Kaufmann 1965; baboons: Seyfarth 1978a, Strum 1982, Smuts 1985, K.R. Rasmussen 1986; chimpanzees: Tutin 1979, de Waal 1982, Hasegawa and Hiraiwa-Hasegawa 1983). In addition, Packer and Pusey (1979) found that macaque females were significantly more aggressive towards males than baboon females were, probably because sexual dimorphism is much less marked in macaques. They suggested that macaque females were consequently able to exert much greater selectively in mating partners than baboon females were able to do (see also Caldecott 1986b).

Faced with the problem of a female's expressed preference for a lower ranking male, a dominant male has two options. Either he can try to drive the rival away from the female or he can try to persuade the female to change her preference in his favour. The difficulty with the first of these is that even if the male succeeds in driving the rival away, this will not necessarily mean that the female will then turn her attention to him. Experimental studies of macaques have amply demonstrated that females often prefer the less aggressive of two males (Herbert 1968, Michael et al. 1978). Similarly, de Waal (1982) noted that female chimpanzees would often go off to a quiet secluded corner of their compound to mate with preferred low-ranking males out of sight of the dominant males who would otherwise have interfered (see also Tutin 1979). This suggests that a strategy aimed at inducing the female to change her mind might be more profitable. Chapais (1983a) in fact found that the majority of attempts made by rhesus macaque males to interfere in the consortships of other males involved actions directed at the female rather than the male. Such tactics apparently proved quite effective in persuading females to change allegiance to the more dominant male. In those cases where the male did try aggression but failed to gain the female, he often subsequently tried a combination of friendly and submissive gestures (e.g. lip-smacking and fear grimaces) directed at the female.

Some authors (e.g. Bernstein 1976, Fedigan 1983) have assumed that the existence of these confounding factors necessarily negates the hypothesised relationship between dominance rank and reproductive success and its evolutionary interpretation. Such a view implies a very simplistic relationship between causes and their effects in biological systems. Indeed, it implicitly assumes that animals are incapable of making any attempts to overcome the adverse circumstances in which they happen to find themselves in real life. There is, however, considerable evidence showing that not only do animals attempt to escape from the consequences of low dominance rank but also that such behaviour is directly caused by the initial competitive edge that the ability to dominate others gives an individual in any contest for access to a limited resource. If we are to avoid fruitless arguments about hypotheses of such transparent simplicity that they are never likely to be valid, we must take a more sophisticated view of social behaviour and the evolutionary processes that underlie it. I shall have more to say about this in the following section. First, however, we need to consider the mate acquisition strategies open to males in one-male-group social systems.

Mate acquisition in one male-group contexts

In order to be able to breed, a male in a social system based on formal one-male groups has to gain control over a group of females. The general and perhaps most obvious strategy is to challenge an incumbent male and, having defeated him, take the group over. A male's ability to effect a takeover depends on several factors. First, takeovers invariably involve very severe fights. An incumbent harem-holder will be reluctant to surrender his harem too early, for defeat may signal the end of his reproductive career. A male attempting a takeover must, therefore, be physically strong enough to risk a once-and-for-all fight. Secondly, a male's success in any takeover must also depend on the females' willingness to change allegiance to a new male. If the females have no good reason to prefer a new male, they may be reluctant to abandon a known (and possibly reliable) male for an unknown one. Conversely, of course, females who want to change males may tip the balance in favour of an incoming male even in those cases where the incoming male is not as intrinsically powerful as the incumbent male. The females' preferences in these cases are obviously determined by factors that may be quite irrelevant to the males' own interests, though the males' respective personal qualities may be one consideration.

There are a number of documented examples in which the outcome of a takeover contest was clearly influenced by the females. Hall (1967) noted that in patas monkeys the females determine whether or not a maturing male offspring can stay on in his natal group to replace the incumbent male as harem-holder at a later stage. In the gelada, the females effectively vote on whether or not to accept a new male when one challenges their harem-holder. Each female signals her willingness to change allegiance to the new male by responding to his invitations to groom: once a majority of females have signalled their decision in this way, the incumbent harem-holder simply gives up trying to drive the intruder away and accepts his loss of status as inevitable (Dunbar and Dunbar 1975, Dunbar 1984a). During the process of this 'voting' procedure, the females are involved in a great deal of fighting among themselves as those who do not want to change males attempt to prevent those that do from interacting with the new male. U. Mori (1979b) and Dunbar (1984a) recorded several instances in which an incumbent gelada male was clearly overpowered by an intruder, but retained control of his female(s) because the female(s) concerned actively refused to desert. In each of these cases it was the female rather than the male that drove off the intruder who had recently taken over the rest of the unit. In addition, the fact that the probability of a successful takeover is an approximately linear function of harem size in gelada can be attributed to the females' reproductive decisions. As harem size increases, more and more of the females begin to suffer from reproductive suppression and are consequently more willing to desert their harem male (see also Kummer 1975). Tsingalia and Rowell (1984) noted that when intruding male blue monkeys try to mate with the group's females, the females concerned will sometimes betray their presence so that the group's resident male comes and drives them off.

The precise mode of attack used by males to acquire harems of females depends on the nature of the social relationships that characterise the society in question. Although fighting is always involved, the way in which that fighting is carried out depends on what the incoming male is actually trying to do. It is often assumed that, in contests over the control of groups of breeding females, the males are fighting to determine who has the right of access. This undoubtedly happens, but it need not always be the case. Detailed analysis of takeover fights in the gelada, for example, has shown that the intruding male is in fact attempting to solicit interactions with the group's females. Fights between him and the harem-holder occur when the incumbent male attempts to drive the intruder away so as to prevent him from interacting with the females (Dunbar 1984a). The intruder's response to the incumbent male's harassment determines the severity of the fighting, but not necessarily the outcome: that is determined solely by the females casting their 'votes' for or against the new male by actually interacting with him. In this case, the incumbent male has only two options: either to drive the intruder away so that he cannot get close enough to solicit the females or to reinforce his own bonds with his females by grooming with them. Gelada males alternate frenetically between both strategies.

This contrasts strikingly with the situation reported for some of the other species where intruders concentrate directly on attacking the incumbent harem-holder and driving him out of the group (e.g. langurs: Sugiyama 1967; colobus: Oates 1977b, Marsh 1979a: guenons: Struhsaker 1977). The contrast in approach generates significant differences in the subsequent behaviour of the defeated harem-holder. In most species, he is forcibly ejected from the group by the successful intruder; he then rejoins an all-male group, perhaps to attempt another takeover of his own at a later date. In the gelada, on the other hand, the defeated male stays on in his unit after undergoing a ritualised 'reconciliation' with the new male (see Dunbar 1984a). Precisely why these differences should exist remains unclear.

Male Lifetime Reproductive Success

So far, we have considered only the instantaneous reproductive success of males at a given moment in time (or during a particular breeding season). Evolution, however, operates in terms of *lifetime* reproductive success, and this is a function of the interaction between age-specific reproductive success and survivorship (see p. 57). What we really need to know is:

$$R_0 = \sum_x l_x m_x$$

Or, if we are considering a dominance hierarchy, it may be more convenient to determine:

$$R_0 = \sum_i t_i m_i \tag{8.1}$$

where m_i is the rank-specific fecundity rate for a male occupying rank i, and t_i is the rank-specific tenure (i.e. the length of time a male can expect to remain in rank i). Where males hold harems, this second equation reduces to a simple two-state situation in which the male is either a harem-holder (with a reproductive rate determined by the size of his harem) or not (with a reproductive rate of zero). I shall refer to an individual's reproductive success at any particular time as its *instantaneous reproductive rate* in order to distinguish age- or rank-specific reproductive *rates* from lifetime reproductive *output.*

The important point to remember is that a male's rank does not remain constant throughout his life. The constraint-free strategy (see p. 156) in most cases rests on the use of intrinsic power (sometimes also known as *resource-holding potential* or RHP: Parker 1974a). In other words, a male's rank (or his ability to hold a territory or a harem of females) depends only on the extent to which he can displace and defeat other males. Power is, of course, a composite function of a male's physical size, fighting ability and psychological skills. In the first instance, sheer physical aggression will determine a male's ability to drive off other individuals, but a male may be able to offset any inadequacies in physical attributes by greater fighting skill or psychological cunning. For present purposes, however, I shall focus on physical power for the sake of analytical simplicity, while noting that these other factors might well be involved.

Physical power is largely a function of an animal's size and weight. This shows a characteristic inverted-J-shaped trajectory when plotted against age: initially, it rises rapidly as the male completes his physical growth, achieves a peak during the period of his physical prime and then goes into a slow and prolonged decline as the effects of old age set in (Figure 8.5). Although a male's rank at any given moment actually depends on his power relative to that of all the other individuals in his group, all other things being equal the average male's expectation of rank over the course of his lifetime will have the kind of J-shaped trajectory depicted in Figure 8.5. Packer (1979a) showed that this was in fact the case for males in three troops of olive baboons. If we integrate this trajectory into the graph for rank-specific reproductive rate

Figure 8.5. An animal's power trajectory over age. The vertical broken line indicates the approximate timing of puberty. The curve is notional but reflects an animal's changing physical powers over its lifetime and the effect that this has on its ability to defeat other members of its group

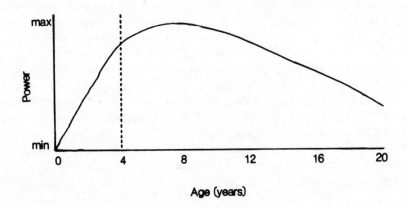

using equation 8.1, it is clear that both the highest rank that a male attains and the rate at which he loses rank as he ages will have a dramatic impact on his lifetime reproductive output.

There are two significant consequences of this. One is that if all males have the same lifetime rank trajectory, then they will all end up with the same R_0 at the end of their lives even though at any given moment males may differ very considerably in instantaneous reproductive rate. It is important to note that this does *not* mean that there will be no selection for characters that confer high dominance rank. Even though all males achieve the same R_0, they do so *only* because they have the same lifetime rank trajectories as a direct consequence of the competition for high rank. If any male opted out of this competition, it would automatically be forced to the bottom of the hierarchy where the rank-specific reproductive rate is low. This would inevitably mean that it ended up with a low R_0, with consequent heavy selection against such behaviour. At the same time, of course, every other male's mean rank would have increased slightly, with a consequent improvement in each one's R_0. Thus, one consequence of intense competition for high rank can be that all individuals end up with the same lifetime reproductive output.

The second consequence is that achieving high rank during his prime may impose particularly severe costs on a male and cause him to lose rank faster once past his prime (either because he dies sooner or because the injuries sustained in fighting for high rank detract from his ability to compete later in life). If this happens, the male's R_0 will depend on the relationship between the maximum rank achieved and the rate of decline in rank later on. The crucial factor here is likely to be the level of competition for access to females (itself partly a function of the adult sex ratio and partly a function of the way in which females are packaged in the environment — i.e. on harem size).

As the level of competition rises, the costs of defending a given resource or status increase, thereby inevitably reducing the length of tenure. An inverse relationship between the level of competition and the length of tenure as a harem-holder or territorial male (or, more generally, the expected length of reproductive lifespan) has been documented at many taxonomic levels (frogs: Wells 1978; dasyurid marsupials: Lee and Cockburn 1985; antelope: Gosling 1974; deer: Mitchell et al. 1976; seals: Anderson and Fedak 1985). In some cases, this is due to the fact that animals engaged in defending territories or harems during an intense breeding season are unable to feed and therefore lose weight rapidly. Studies of a number of temperate-zone species have shown that males who tax their reserves too heavily by fighting and not feeding during an autumn (fall) rut risk entering the winter in poor condition and therefore suffer high winter mortality (deer: Gibson and Guinness 1980a; feral goats: Gordon et al. 1987). Few primates have mating seasons as intense as this (ring-tailed lemurs may be an exception: A. Jolly 1966a), but, over the long term, defence of a territory or consortship can be just as disruptive of time and energy budgets (see p. 96). In addition, any form of active defence necessarily carries with it the risk of injury and death (e.g. macaques: Wilson and Boelkins 1970; baboons: Hausfater 1975, Smuts 1985; gelada: Dunbar 1984a; lemurs: A. Jolly 1966a).

Evidence to show that tenure or life expectancy is reduced by increased competition is not widely available for any primate species, partly because few field studies have had the opportunity to study more than one group over a limited period of time. Two sets of observations, however, suggest that a male's ability to control access to a resource is determined by the pressure exerted on him by competitors. One is the observation by Stern and Smith (1984) that demographic variables can affect the linearity of the relationship between male dominance rank and the proportion of offspring sired in captive groups of rhesus macaques. They noted that when the number of males in the group was less than five, there was a simple positive correlation between these two variables, but that the correlation vanished once the number of males exceeded this level. This suggests that dominant males are less effective at maintaining their priority of access when the number of rivals (and hence the amount of fighting they have to do) exceeds a certain level. The second piece of evidence is the finding that in gelada the rate of harem takeover is genuinely 'elastic' in the sense that pressure exerted by competition itself distorts the success rate (Figure 8.6). What these data imply is that a male's ability to maintain control over a group of breeding females depends on the pressure being exerted on him by rivals waiting for the opportunity to own a breeding group. Confirmation of this is provided by more detailed observations which suggest that males are more likely to lose their harems if they have to fight several different males in quick succession (Dunbar 1984a).

Figure 8.6. Probability of takeover per unit per year (standardised for a common mean harem size) in four bands of gelada, plotted against the proportion of males attempting entry to reproductive units that did so by way of takeover (as opposed to becoming a follower). The data suggest that as the proportion of males opting to try a takeover increases, so the likelihood of success increases, probably because incumbent harem-holders are unable to resist repeated attempts at takeover. Increased likelihood of successful takeover clearly implies that tenures as harem-holder will decline proportionately

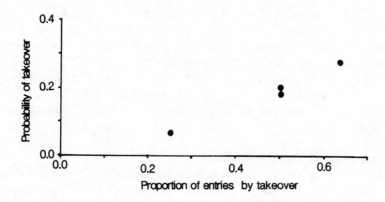

Source: reproduced from Dunbar (1984a), with the permission of the publisher.

The effect of these factors is to reduce a male's expected lifetime reproductive output. Once the costs of pursuing a given strategy become heavy enough, the gains from that strategy will be wiped out and the animals may in fact find it advantageous to opt for alternative strategies that do not incur such severe costs (Gadgil 1972, Maynard Smith and Price 1973, Maynard Smith 1982). Providing the instantaneous reproductive rate is not zero, the increased longevity gained by opting out of the competitive arena will begin to offset the loss on instantaneous reproductive rate that a male incurs by not reaching the highest rank. When the level of competition is sufficiently high, the gain in life expectancy (or tenure) will neutralise the losses incurred and the profitabilities of the two strategies will equilibrate. At this point, two alternative strategies will emerge that will remain in stable equilibrium so long as the demographic conditions do not change. These two strategies will usually be the constraint-free strategy and its complement. In other words, they will usually consist of a high-gain/high-risk strategy and a low-gain/low-risk strategy that offer an individual the option of trading a high instantaneous reproductive rate against a long life expectancy. Whereas an animal on a low-risk strategy can expect to gain a comfortable average R_0 at the end of its life, one on a high-risk strategy may be able to do very much better but at a significant risk that it might end up doing very much worse. Thus, on average, high- and low-risk animals will do equally well, but only because the variance in R_0 is significantly greater for high-risk animals than for low-risk ones. Alternative strategy sets of this kind have been widely documented at all levels of the Animal Kingdom (for a recent review, see Dunbar 1982a).

Alternative Strategies of Mate Acquisition

Although the constraint-free strategy is invariably based on aggressive competition, the alternative strategies adopted in opposition will depend crucially on the nature of the social/mating systems within which the animals actually live. In fact, in order to appreciate the nature of the alternative strategy sets in any given case, we have to be able to identify precisely what it is that males are competing for in proximate terms. Since the situation is generally more clear-cut in one-male-group mating systems, I shall deal with these first.

Options in single male systems

In one-male-group social systems, males conventionally acquire control over groups of breeding females by fighting and displacing incumbent harem-holders. Some males, however, do not contest access to the females in this way but instead join (or mature into) a group as a subordinate 'follower' (hamadryas baboons: Kummer 1968; gelada: Dunbar and Dunbar 1975; gorilla: Harcourt and Stewart 1981). Followers are normally careful to avoid trespassing on the harem-holder's hegemony over the reproductive females and in particular his sexual access to them.

In the gelada, the harem-holder's increasing tolerance of his presence with time eventually allows the follower to build up social relationships with one or two of the more peripheral females. In due course, these develop into full breeding relationships so that the follower establishes what amounts to a nuclear unit within the parent unit. About two years after he joined the unit, the follower will lead his females out of their parent unit to establish a new independent unit of his own (Dunbar 1984a).

In the hamadryas and the gorilla, a harem-holder typically 'passes on' his harem intact to his follower who may well be his son. A male still has the option, of course, of leaving his natal unit to search for females that he can acquire elsewhere. In these two species, however, this is only possible by 'kidnapping' individual females from another male's units (see Kummer 1968, Harcourt and Stewart 1981). This is probably because in both species the reproductive units are rather small (they typically contain only two breeding females: see Kummer 1968, Harcourt et al. 1981c) and thus do not have the surplus females that a harem-holder can afford to relinquish to a follower in the way that occurs in the larger breeding units of the gelada. (I use the term 'kidnapping' loosely here — in gorillas, such males acquire females by attracting them out from other units, though in the hamadryas they may coerce females into following them: see Kummer 1968, Harcourt 1978.)

In the gelada, males pursuing the takeover and follower strategies incur different costs and benefits (Table 8.4). Briefly, these amount to the fact that while takeover strategists gain many more females at the outset of their breeding careers than followers do because they can take over entire units, this in itself leaves them more vulnerable to attack in their turn; consequently, their tenure as harem-holder is much shorter. In addition, while followers are always successful in entering a reproductive unit, takeover strategists often fail and require more attempts before succeeding (all of which takes time).

Table 8.3. Costs and benefits incurred by male gelada acquiring reproductive units by two different methods

	Follower	Takeover
Number of females at start (mean)	1.89	4.71
Probability of wounding per entry attempt	0.00	0.50
Probability of successful entry per attempt	1.00	0.70
Mean age at start of tenure (years)	6.2	7.7
Mean expectation of tenure (years)	6.1	3.6
Estimated mean number of offspring in a lifetime	4.72	4.94

Source: Dunbar (1984a).

This, combined with the fact that a male must wait until he has completed physical growth in order to be able to compete effectively in a fight with a harem-holder, means that takeover strategists cannot embark on their reproductive careers as early as followers can. When these differences are taken into account and their effects on lifetime reproductive output are evaluated, it turns out that the takeover and follower strategies are equally profitable (Figure 8.7).

In the hamadryas and gorilla, in contrast, it is clear that males should do very much better by waiting to take over their fathers' harems. That way a male gains the whole group of females (perhaps as many as five in a large group), whereas the alternative (kidnapping) yields at best a single mate. Moreover, in the hamadryas it seems that males are only able to kidnap immature females and must wait until these have matured before they can begin to breed (a delay of 2 to 3 years). The cost of pursuing an inheritance strategy in both cases clearly lies in the delay that is incurred before a male can begin to reproduce: the male must wait until his father dies or reaches an age at which he is willing to relinquish control over some or all of his females.

This being so, the timing of a male's birth in relation to his father's age will

Figure 8.7. Results of simulation of the lifetime reproductive output of a gelada male. The gross number of offspring produced by a male who starts his breeding career at three different ages is plotted against the size of his harem at the start of his breeding career. All males are assumed to die at the same age (12 years), and size-specific probabilities of takeover and harem fission, along with changes in the age-specific fertility of the females, are assumed to be the main factors influencing a given male's lifetime reproductive output. The mean lifetime reproductive expectations of males who acquire their harems by becoming a follower (A) or by takeover (B) correspond to initial harem sizes of 1.9 and 4.7 females for males aged 6 and 8 years, respectively (dotted lines)

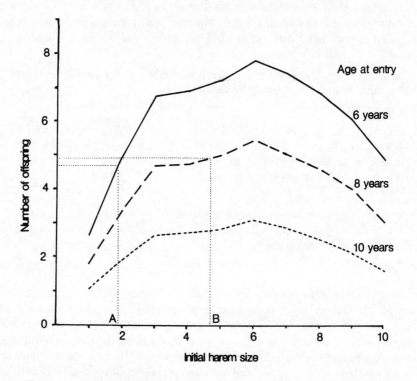

Source: reproduced from Dunbar (1984a), with the permission of the publisher.

be an important source of asymmetry in the potential gains that individual males will make from the inheritance strategy. Figure 8.8 shows the relationship between the lifespans of a father and three sons born to him at different stages in his tenure as harem-holder. A son born early on (Son A in the figure) may have to wait so long for his father to die that he himself would be left with too short a reproductive lifespan to yield an appreciable return. Instead, he might be better off leaving the group as soon as he can in order to form an initial unit by kidnapping a female. Conversely, a son born further on in his father's tenure (Son B) will mature at just about the time his father is likely to

Figure 8.8. Influence of birth order on the probability of a son being able to take over his father's harem on the father's death. Vertical bars mark each son's birth; the broad bar marks the period of adulthood when a male is capable of holding a harem. Son A wastes up to a third of his potential reproductive career if he waits to inherit his father's harem, and Son C will be too young to hold a harem at his father's death. Only Son B is likely to be in a position to inherit his father's females. Son B should therefore stay in his natal group to await his father's demise, but Sons A and C should leave in order to search for harems that they might be able to acquire in other ways (e.g. by takeover or by kidnapping individual females)

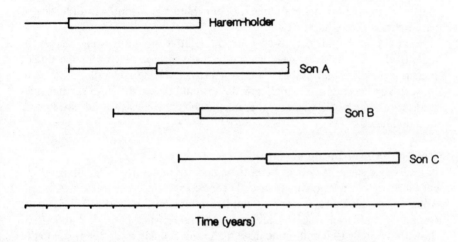

die. Such a male will be in an optimum position to inherit the group and he should stay on in the group. An offspring born late in his father's life (Son C) is likely to be still immature when the father dies. He is unlikely to inherit the group because another male (an older sibling or an intruder from outside) will be in a better position to take it over. Since this new male will not die for some years and is likely to pass on the unit to his own son when he does so, Son C has no real option but to leave. Harcourt and Stewart (1981) were able to show that whether or not a gorilla male inherits his natal unit depends on whether he was born before or after the incumbent male acquired the group (i.e. on whether or not he is the incumbent's son).

Species that habitually form monogamous pairs may face a similar decision. In gibbons, for example, an offspring has the choice between remaining on or near its parents' territory in the hope that the parent of its own sex will die in the near future (so that it can then take the parent's place) or leaving to set up its own territory elsewhere (Tilson 1981). Since the likelihood of a parent dying is in most cases negligible, the majority opt to leave. These then have a choice between staying near the parents (in which case the parents may help it to acquire a territory nearby) or moving away in search of an unmated individual with its own established territory. In the first case, it may have to wait some time before a potential mate turns up, and in the second it will not have the advantage of parental help in setting up a territory should it need to do so. In marmosets and tamarins, on the other hand, the inheritance of a territory (and a mate) by sons may be more common (Neyman 1980, McGrew and McLuckie 1986). In this case, however, offspring may gain additionally through kin selection by remaining on the parental territory in order to help rear their younger siblings rather than by dispersing in search of their own territories (McGrew 1987). The difference in this case may lie in the callitrichid's unusual reproductive strategy, which makes it possible for older siblings to contribute significantly to the costs of parental care (see Chapter 12).

It would not be difficult to cost out the relative advantages and disadvantages of the various strategies for each of these species, given a knowledge of their reproductive and lifehistory characteristics. Doing so would allow us to identify key questions on which new data should be sought from further field studies. As a heuristic exercise, therefore, such an analysis would pay significant dividends.

Options in multimale systems

In multimale social systems, the options open to males are of a rather different kind. Here, males compete directly for access to individual females as and when these come into oestrus. Saunders and Hausfater (1978) have used computer simulations on the lifehistory characteristics of male baboons to show that a male that cannot achieve high rank is likely to do relatively badly in terms of lifetime reproductive output. One solution to this problem is for a male to opt out of the competitive arena altogther and instead concentrate on forming an intense long-term relationship with a single female which would at least guarantee him sexual access to one female. Packer (1979b) has suggested that male baboons can be differentiated into two classes: those that opt for high rank (often achieved by moving repeatedly from one group to another and competing for access to individual females in oestrus in each group) and those that migrate only once during their lifetime (and then concentrate on building a long-term pair bond with a particular female). In other baboon populations, Smuts (1985) and Altmann et al. (1986) found that males pursue these two options at different stages of their lives, opting for a high-gain/high-risk dominance-based strategy while young and physically powerful and a low-gain/low-risk 'special relationship' strategy when older. Special relationships of this kind have been noted in virtually every population of *Papio* baboons that has been studied intensively; in addition to the three studies quoted above, Ransom and Ransom (1971), Seyfarth (1978b), Busse and Hamilton (1981), Anderson (1983) and Collins (1986) have all presented unequivocal evidence for long-term relationships between individual males and females.

Essentially similar alternatives have been reported for chimpanzees (Tutin

1979, de Waal 1982, Hasegawa and Hiraiwa-Hasegawa 1983) and macaques (Sugiyama 1976, Colvin 1983a, van Noordwijk and van Schaik 1985). In a particularly detailed study of longtailed macaques, van Noordwijk and van Schaik (1985) were able to demonstrate that males pursued two contrasting strategies of migration: one ('bluff immigration') was highly aggressive and aimed at achieving high rank within the group, while the other ('unobtrusive immigration') involved submissive behaviours aimed at making the immigrant's entry as inconspicuous as possible. Whereas unobtrusive immigrants were drawn from all age classes, only young males in their physical prime attempted bluff immigrations. Even so, attempts by bluff immigrants to acquire high rank in their host groups were not always successful. The most successful males in these terms were males who returned to their natal groups after a period as solitary males. These males were also able to take over the dominant position in much larger groups than other bluff immigrants could. Unfortunately, van Noordwijk and van Schaik (1985) were unable to determine the reproductive success of males pursuing these various strategies, so that we can say little about the functional significance of these options.

Because a male's power declines with age, males in multimale groups may also be able to maximise their reproductive outputs by minimising the rate at which their rank in the hierarchy declines with age. Males might be able to do this in one of two ways, namely either by adopting a generally lower profile so as to have a longer reproductive lifespan or by forming coalitions.

In theory at least, a male can choose to maximise the number of offspring he produces over a lifetime, R_0, by maximising either the length of his active reproductive lifespan or by maximising his age-specific reproductive rate over a shorter period of time. For males in a dominance hierarchy, this will generally mean either opting for a shallower lifetime rank trajectory or opting to compete for the highest possible dominance rank. Figure 8.9 shows the lifetime rank trajectories of three males pursuing rather different strategies. With rank-specific reproductive rates obtained by smoothing Chapais's (1983a) estimates for rhesus macaques (given in Figure 8.4), all three males achieve virtually identical lifetime reproductive outputs. Hausfater (1975) suggested that some baboon males might pursue this strategy, but he had no real evidence to support this possibility. In fact, it seems probable that the most important factors determining a male's lifetime reproductive output in this particular population of baboons are the rank a male manages to achieve early in his reproductive career and the length of his reproductive lifespan (Altmann et al. 1986). This may be because rank trajectories need to be very finely balanced for lifetime reproductive outputs to be equilibrated when males pursue alternative strategies based on different lifetime rank trajectories. In a mating system as fluid as a priority-of-access dominance hierarchy, the uncertainties may be too great: going for bust may simply be the best strategy. Moreover, in the relatively impoverished Amboseli habitat, the high mortality risks make a low-trajectory strategy too risky because of the

Figure 8.9. Three different lifetime rank trajectories for males that yield exactly the same lifetime reproductive output when rank determines a male's reproductive success at any given moment. With smoothed rank-specific reproductive rates based on the data in Figure 8.4, the number of conceptions that each rank trajectory yields over a lifetime are 13.8 (solid line), 13.8 (dashed line) and 13.2 (dotted line)

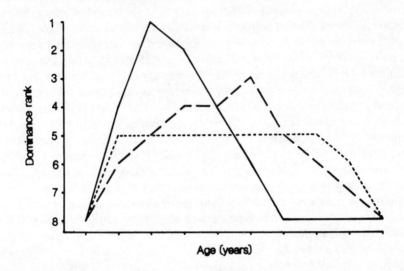

high likelihood that a male will die before he has lived long enough for such a strategy to pay off.

The alternative way of minimising the rate of rank decline is to form a coalition with one or more other males. Coalitions may not only help a male to gain a higher rank than he would otherwise have held, but they can also be used to slow down the rate of rank decline once the male is past his prime. Strum (1983), for example, noted that whereas young male baboons tended to rely more on their own intrinsic power in acquiring oestrous females from other males, older males made more use of coalitions in which allies provided each other with reciprocated aid in defence of consortships (see also Smuts 1985; Noë 1986). Packer (1977) also found that male baboons formed alliances based on reciprocal support in agonistic encounters during consortships. Rather similar use of coalitions has been described in macaques (Witt *et al.* 1981) and chimpanzees (de Waal 1982, Nishida 1983).

Evolutionary considerations

In certain cases, the lifetime reproductive outputs of the various alternative strategies may equilibrate (i.e. be equal), and the set of strategies may then be held in evolutionary balance. Rubenstein (1980) has modelled the conditions under which alternative strategy sets will be in evolutionary equilibrium. In general, a frequency-dependent balance between costs and/or benefits is required, with density-dependent effects playing an important role in establishing the conditions under which this can occur. Alternative strategies need not always equilibrate, however, and those strategy sets that are not equilibrated provide an important class of evolutionary phenomena.

There are two quite different reasons why one strategy in a set of alternatives may in fact yield fewer genes contributed to the next generation (see Dunbar 1982b). First, the sub-optimal strategy may not be a genuine strategy at all but rather a *tactic* within the normal constraint-free strategy. Studies of many species have shown that animals often pursue a variety of alternative tactics on the side that add marginally to their overall fitness. Some of these may be 'side payment' tactics pursued while the animals are in fact pursuing the constraint-free strategy. Examples include the occasional stealing of matings on other males' territories by normal territorial males (fishes: Keenleyside 1972, Fernald and Hirata 1977). Others may be developmental stages which most males pursue at some point during their lifetime while waiting for an opportunity to pursue the conventional constraint-free strategy. Examples include young non-territorial males that sneak matings on the territories of breeding males at a time when they are too young to hold territories in their own right (deer: Gibson and Guinness 1980)

Galdikas (1985) has reported that young male orang utans sometimes attempt to steal matings by violently attacking females. Whereas older males prefer to form extended consortships with the females that live within their territories, young males (being smaller and more agile) are able to pursue females into the trees where the older males are too ponderous to follow to prevent them from doing so.

These examples underline the need to make a clear distinction between strategies and tactics. By assuming that the two kinds of behaviour are of equal biological and evolutionary status, we would be misled into thinking that a 'sneak' strategy is equivalent to harem-holding as a reproductive strategy, and we would then be puzzled as to why it continued to survive when it was so obviously unsuccessful. In fact, of course, it is not equivalent but is a subsidiary tactic to which males resort in order to gain marginal improvements in their fitness while pursuing the conventional constraint-free strategy. Such tactics persist in the population precisely because they allow males to make such marginal gains. Where all males achieve roughly the same lifetime reproductive output by conventional means, such marginal gains play an important evolutionary role in weighting a male's contribution to the species' gene pool (Dunbar 1982b).

Not all such cases are of this kind, however. A second type of non-equilibrated strategy set is that in which some males genuinely do pursue a sub-optimal strategy because they are unable (through small physical size, injury received early in life, etc.) to compete effectively on the constraint-free strategy. Sub-optimal strategies of this kind are usually known as 'best of a bad job' options (or 'Hobson's Choice'). Males may pursue such strategies if their competitive abilities are so poor that they would gain even lower values of R_0 by trying to compete in the conventional arena. By opting for a less competitive strategy, such a male may gain fewer genes contributed to the next generation, but at least it manages to make some contribution. Its offspring may then be better placed to rectify the balance. Competitive disadvantages (or 'phenotypic asymmetries') that place an animal in this situation will often be historical (i.e. developmental) in origin and can arise in a number of ways (Dunbar 1982b). Small physical size relative to other members of the same birth cohort, for example, can arise as a result of the animal being forced on to a low growth plane at a critical stage during development either through a temporary deterioration in local habitat conditions or because of the mother's inability to provide sufficient food for her offspring (itself due either to accidental injury at a critical time or to low dominance rank).

Although there is considerable circumstantial evidence to suggest that males of a wide range of primate species pursue alternative strategies of mate acquisition, we have almost no quantitative data on lifetime strategy-specific reproductive output that can be used to determine the functional significance of such behaviour. In a number of cases, it is clear that the alternatives are tactical rather than strategic, suggesting that animals may try to maximise lifetime reproductive output by making the best of the circumstances in which they happen to find themselves at any given moment. There are, none the less, a few instances in which different behaviour patterns constitute genuine strategic alternatives, and in at least one case (the gelada) these alternatives can be shown to yield similar lifetime reproductive outputs. It is worth pointing out that in this case the relative frequencies of the two strategies vary from one population to another (and even within a given population over time) in response to variations in the demographic structure of the population (see Dunbar 1984a). The crucial point to bear in mind, however, is that the functional status of a given alternative strategy cannot be assessed unless we know how such behaviour maps on to the animal's lifetime reproductive history. High instantaneous rates of reproduction do not necessarily mean high lifetime reproductive output. We thus require detailed information on lifetime power trajectories as well as data on instantaneous (or status-specific) reproductive rates.

★

I have concentrated throughout this chapter on male mating strategies. This does not mean, of course, that females do not have mating strategies as such. Indeed, we have dealt specifically with female mating strategies at various points in the discussion. However, female mating strategies are less conspicuous than those of males, so that rather little work has been done on this aspect of their behaviour. Moreover, females' concerns are generally of a longer term nature than those of males: they are usually more interested in long-term relationships (which may involve preferential mating with particular males) and these relationships are usually concerned with coalitionary support and rearing patterns rather than short-term interests in mating *per se.* Thus, although females undoubtedly have mating strategies, it is more appropriate to consider their behaviour in the more general context of their longer term rearing strategies. These I consider in more detail in the next two chapters.

Chapter 9

Rearing Strategies

Once she has been fertilised, a female primate's primary concern shifts from the business of mating to that of ensuring the survival of her fetus and its development into a functional adult. Because primates are born helpless, this is a major long-term investment on her part and its costs far outweigh the minor consideration of finding a suitable mate. Males, of course, also have an interest in ensuring that their offspring mature. But the constraints of mammalian reproductive biology (something that primates are saddled with) make it more difficult for them to play a direct part. None the less, in some cases, males do contribute significantly to the rearing programme, radically altering the social system in consequence.

In this chapter, I begin by outlining the general features of infant development and rearing patterns among primates, then examine the costs that females incur in rearing young before going on to consider ways in which females seek to maximise their chances of successfully rearing offspring. Finally, in the concluding section, I briefly consider the roles that males can play in the rearing process.

Primate Rearing Patterns

There are two important stages to the development of an infant primate. The first is the period during which it is totally dependent on its parent(s) for nourishment, care and transport. In Old World monkeys, this generally covers the first 6–12 months of life. The second period begins at the point where the infant is no longer dependent on its parents for food and transport, and ends when it can effectively fend for itself in adult society. In Old World monkeys, this lasts from about the end of the first year up to the age of about four years. During this period, the animal is largely dependent on its mother and close relatives for social support against other group members, as well as for protection against predators. This period is also one in which it learns its essential survival techniques — what foods to eat and where to find them, where to find water and safe sleeping sites, how to move through the environment so as to minimise exposure to predation, etc. Its mother plays an important part in transmitting this information, but, in the more social species, peers and other group members have important roles to play as well. I shall be concerned with some of the more social aspects of rearing in Chapter 10. Here, I focus primarily on parental care during the first year of life when the parents' input is both intensive and extensive.

With the exception of a few prosimians that use nests, all primates carry their newborn infants with them, usually clinging to the fur on the mother's belly. In some species, this may require some assistance from the mother because the infant cannot hold its own weight until a few days after birth. Initially, the infant is carried and nursed by the mother at all times, but after the first few weeks it begins to explore its surroundings, moving a few metres away from its mother whenever she has settled to feed or rest. As it grows, the time spent away from the mother increases and its contacts with other animals multiply. Figure 9.1 illustrates this gradual shift in the relationship between mother and infant in a variety of New and Old World monkey species. Most other taxa exhibit similar trends, but the time course will

Figure 9.1. Percent of time for which infants are in physical contact with their mothers as a function of the infant's age. The plotted values are monthly medians for all sampled infants

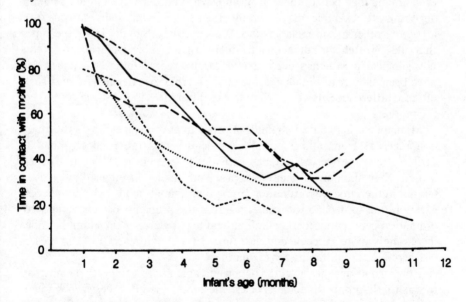

Sources: yellow baboons (solid line) from Altmann (1980); gelada (long dashed line) from P. Dunbar (unpublished data); rhesus macaques (dotted line) from Berman (1980); vervet monkeys (short dashed line) from Struhsaker (1971); red howler monkeys (long and short dashed line) from Mack (1979).

obviously vary in relation to body size. Independence usually occurs earlier, at about 3–4 months of age, in 0.6 kg callitrichids (Ingram 1977, Wright 1986) and very much later, at 1–2 years of age, in 30 kg apes (e.g. chimpanzees: Goodall 1967).

The apparent uniformity of these patterns, especially between populations of the same species living in different habitats, has been interpreted as implying that they reflect underlying developmental sequences that are biological universals (e.g. Johnson and Southwick 1984). While it is true that there are general patterns, the uniformity is much less conspicuous at a quantitative level. Thus, Simpson *et al.* (1986) have commented that free-ranging rhesus macaque infants spend less time away from the mother and the mothers are responsible for maintaining contact with the infants for much longer than is the case in captive animals. This makes some sense given the fact that the dangers (from both ecological and social sources) are likely to be significantly greater in the wild than in captivity. Similarly, Hendy (1986) found that infant baboons at Gombe were more independent and were both given less protection and rejected/punished more often than infants in the Mikumi population: one possible reason for this difference is that infants were relatively safer in the forested habitat at Gombe where day journeys were short and refuges readily available. More is thus likely to depend on the danger as perceived by the mother, and significant differences can be expected even among members of the same group. Within groups, the mother's dominance rank, for example, may radically affect the risk that an infant runs from attack by other group members (see Wasser 1983). Suomi (1981) has noted that there is often considerable individual variance in behavioural and pyscho-physiological parameters even among captive-reared infant macaques that have been subject to similar rearing and maternal experiences. At least some of these differences may be attributable to differences in the social contexts of the mothers.

Monkey mothers have two kinds of problem at this stage. One is the attention received from other individuals; the other is the need to maintain sufficient control over the infant's activities to prevent it getting lost or left behind when she moves. These two problems change in intensity over time, but not in the same way. Attention from other animals tends to be most severe early on and to decline with time, whereas the problems of behavioural co-ordination between mother and infant tend to get worse as the infant becomes increasingly more independent with time. In addition, each problem creates different kinds of difficulty for the mother.

Newborn infants are a source of considerable interest to other members of the mother's group. In some species, the mother is often besieged by individuals attempting to interact with the infant. Such interest generally has to be channelled through the mother, largely because the mother will react aggressively towards individuals that appear to her to threaten the safety of her infant. As a result, individuals wanting to interact with an infant will often approach the mother and groom with her first. Once the prospective inter-actee's 'good faith' has been established, it will then turn its attention to the infant. Species differ considerably in the extent to which mothers tolerate other individuals interacting with their young infants, however. Baboons and macaques are relatively intolerant, whereas colobus and langur monkeys tend to be very tolerant (even to the extent of allowing others to take newborn infants away for considerable periods of time: see McKenna 1979, 1981).

Among non-colobines, the attention of other individuals may be a source of considerable distress for monkey mothers (see Seyfarth 1978b, Altmann 1980), and may lead to considerable restrictiveness on the part of the mother in an attempt to reduce the risk to her infant (Ransom and Rowell 1972, Altmann 1980, Wasser 1983). High-ranking females are generally less restric-tive and less protective of their infants than low-ranking ones. This is mainly because low-ranking females are more vulnerable to other individuals inter-fering with their infants. Wasser (1983) has pointed out that high-ranking females could in principle enhance their own fitness by maltreating the infants of unrelated females in order to reduce those individuals' reproductive rates. There is considerable evidence to suggest that females do harass other animals' offspring (see below), though whether evolutionary 'spite' is the reason they do so remains unclear. Primiparous females also tend to be more restrictive than multipares, but this is probably due to inexperience (see Dolhinow et al. 1979).

Because their mothers are less willing to allow them to stray, the offspring of restrictive mothers usually take longer to develop independence and have less intense social experiences as a result. Hooley and Simpson (1981) were able to detect consistent differences in the personalities of infant rhesus macaques born to primiparous and multiparous mothers that could be attri-buted to differences in mothering style. At 16 weeks of age, for example, the male infants of primiparous mothers scored significantly lower on the person-ality dimension *Excitability* than did those of multiparous mothers.

So far, few attempts have been made to assess the fitness consequences of maternal behaviour in primates, partly because the prolonged period of development makes it impossible to do so without very long study periods. In addition, few studies have been able to sample more than a handful of infants, so that sample sizes are often very small indeed once data-sets have been partitioned for detailed analysis. None the less, Altmann (1980) obtained evidence to suggest that contrasting mothering styles did influence infant survival in different ways. Interestingly, in this case restrictive and non-restrictive mothering styles seemed to have complementary effects. In her baboon troop, restrictive mothers lost fewer infants during the first few months post-partum than non-restrictive mothers did, but their infants gained their independence much later, and, as a result, were less likely to survive if they were orphaned.

The mother's other problem lies in ensuring that the infant coordinates its behaviour with her own. A poorly meshed relationship imposes a heavy cost on the female and puts the infant at risk (especially if sudden flight is necessary following the appearance of a predator). A female who, for example, is constantly having to rescue her infant from several metres away will often have to interrupt feeding or grooming bouts to do so. Not only may she lose her place to another animal, but she will have to expend unnecessary energy in moving, first to retrieve the infant and then in finding a new feeding place or social partner. Two contexts illustrate the problems of meshing particularly well. One concerns the timing of the infant's access to the nipple; the other is when the mother wants to move from one feeding place to another. From the mother's point of view, it may often be important that the infant learns to mesh its behaviour with hers as early as possible in order to minimise her own energetic costs. When ecological conditions are on the margin, both the infant's and the mother's survival may depend on how well meshed their relationship is.

Weaning behaviour, as it is often mistakenly called, has attracted a good deal of attention, partly because it is usually assumed to be a dispute between the mother and the infant over the amount of milk that the mother should provide. In fact, the dispute is often about the *timing* of the infant's access to milk rather than about the *quantity* that it should have. The problem arises because, as the infant grows, its presence on the nipple interferes with the mother's ability to do anything else. (For the theoretical background to parent–offspring conflict, see Trivers 1974.)

This problem is especially acute for ground-feeding primates where the mother commonly sits to feed. Once the infant is more than a few months old, she can neither see over its head nor get her arms comfortably round it to reach food on the ground immediately in front of her if the infant is on the nipple. Gelada females, for example, endeavour to train their infants to go on the nipple only when they are resting or engaged in social activity, during neither of which is the infant's presence an encumbrance (P. Dunbar, unpublished data; see also Altmann 1980). In the gelada, there is some evidence to suggest that female infants learn to mesh their behaviour with the mother's more quickly and with fewer mistakes than males do, probably because they spend more time near the mother than do males. Rowell and Chism (1986b) found a similar sex difference in the adventurousness of infant patas monkeys, though there is no reason to expect that all primate species should exhibit this difference. Altmann (1980), for example, found no significant sex differences in any behavioural measure of infant independence, but she did find that the offspring of non-restrictive (*laissez-faire*) mothers learned to mesh their behaviour with the mother's significantly more quickly than the infants of restrictive mothers. The slowness of restricted infants in this respect seems to reflect the fact that the infant's ability to learn was inhibited by the frustration it experienced in having its behaviour constantly restricted by the mother. It is well known that experimentally induced frustration inhibits animals' abilities to learn conventional maze and discrimination problems (see Dickinson 1980).

At a later stage, once the infant spends a significant proportion of its time away from her, the mother has the converse problem of ensuring that the infant does not get left behind whenever she moves. Initially, of course, she carries the infant with her, even though this may mean she has first to collect it from a peer group some distance away. Once the infant is more mobile, the mother simply signals her intention to move to it and expects the infant to jump into the normal carrying position of its own accord. She may signal her intention to leave by presenting her lowered dorsum to the infant, tapping it on the head or contact calling to it. While the infant is dependent on its mother for most of its nourishment, it may in fact be more economical for the mother to carry it even short distances than to let it walk by itself because, either way, she has to provide the energy. Given the low conversion coefficient of lactation (see next section) and the high energy consumption of an active infant (p. 37), she will use less energy carrying the infant than she will in providing the milk for it to run along beside her. Once the infant is beginning to take an appreciable proportion of its energy requirement by feeding for itself, the mother begins to train it to follow to her signal rather than expecting to be carried.

At first, this produces a great deal of tantrum behaviour from the infant, but eventually it learns to follow automatically (see Nash 1978). Once this has become firmly established, the mother may no longer notify the infant but expect it to monitor her movements and follow as necessary (though she may still continue to check visually that the infant is following her).

Hinde and his co-workers (Hinde 1969, Hinde and Atkinson 1970, Hinde and Simpson 1975) have, in particular, emphasised the fact that infant development is a two-way process in which the infant's behaviour also influences the mother's responses rather than being something that is purely a consequence of the mother's actions (see also Harper 1981). This is reflected in the way that the responsibility for maintenance of contact between mother and infant changes over time. A number of studies of both captive and wild monkeys have shown that initially it is the infant that tends to break contact with the mother and the mother that re-establishes contact, but that the reverse tends to be the case after the third month of life as the mother begins to respond less to the infant's behaviour and to expect it to become more independent (Figure 9.2).

The interactive nature of the relationship between mother and infant is further illustrated by the way in which each may compensate for the absence of an expected response from the other. This has been demonstrated in experiments in which infants and their mothers have been separated temporarily from each other (Spencer-Booth and Hinde 1971b) or in which the mother or infant was anaesthetised (Rosenblum and Youngstein 1974). Generally speaking, the loss of the partner (either physically or in terms of its

Figure 9.2. Responsibility for maintaining contact between mother and infant as a function of the infant's age. The plotted variable is the Hinde and Atkinson (1970) index of responsibility: values below 0 indicate greater responsibility by the mother in maintaining contact with the infant, values above 0 indicate greater responsibility on the infant's part. Plotted points are monthly medians for all sampled infants from studies of free-ranging primates

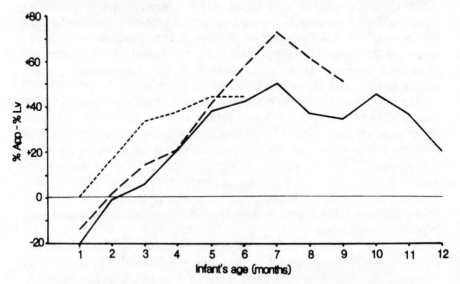

Sources: yellow baboons (solid line) from Altmann (1980); gelada (long dashed line) from P. Dunbar (unpublished data); rhesus macaques (short dashed line) from Berman (1980).

responsiveness) leads to a rebound effect in which the other partner intensifies its efforts to maintain the relationship between them (often by maintaining close physical proximity).

The ultimate crisis in the mother–infant relationship comes should the mother die, so leaving her infant to fend for itself. Although the adoption of orphans by other group members has been reported (for a recent review, see Thierry and Anderson 1986), the loss of the mother generally has severe consequences for the infant. In all studies, infants less than a year old at the time of the mother's death have invariably died shortly afterwards. Only infants that have achieved nutritional and locomotory independence have a significant chance of survival. Most of these will attempt to form relationships with other individuals so as to regain the social support normally provided by the mother. Although older siblings are the typical substitutes (e.g. baboons: Hamilton *et al.* 1982; gelada: Dunbar 1979a; vervets: Lee 1983b), unrelated individuals do sometimes become foster parents (macaques: Berman 1983). Lee (1983b) was able to show that, although the amount of time spent feeding by orphans was not affected in any way, none the less most orphans (but particularly males) were involved in fewer friendly interactions, suffered more harassment and, of course, received less coalitionary support when under attack. Thus, even if orphans do survive, their expectations of contributing to the species' gene pool are likely to be significantly reduced. From the mother's point of view, then, it is particularly important that she get the balance in her reproductive strategy just right so as not to overtax herself by conceiving again too soon, so forcing her to divide her time and energy between too many competing interests.

Ecology of Motherhood

Energy costs of parental care

Altmann (1980) developed a formal model of maternal time budgets in baboons in order to study the ways in which the mother's behaviour is affected by the energy demands of the growing infant as it matures. Her model relates the proportion of time a female must spend feeding on any given day, F_t, to her own basal metabolic requirement and the energy that the infant requires both for BMR and growth during the immediate future. Assuming that the infant's energy requirement for growth is a constant function of its age (i.e. that it grows at a constant rate), the mother's feeding time requirement is:

$$F_t = Am^{0.75} + \frac{A(i_0 + t\Delta i)^{0.75}}{E} \qquad (9.1)$$

where A is a constant that converts energy requirement into time spent feeding, m is the mother's body weight (in kg), i_0 is the infant's weight at birth (also in kg), Δi is the increment in the infant's weight due to growth (in kg/day), t is the infant's age (in days) and E is the conversion coefficient of lactation (i.e. the efficiency with which ingested energy is converted into milk energy). From data on the weights of laboratory and wild baboons, Altmann estimated m to be 11 kg and i_0 to be 0.775 kg. The energy requirement can then be determined from Kleiber's (1961) relationship between metabolic requirement and body weight (see p. 33). Studies of captive baboons suggest that Δi has to lie between 0.005 and 0.01 kg/day to maintain normal growth (see p. 37). Finally, data from mammals suggest that E lies between 0.6 and 0.8 (see p. 37). By interpolating these values into equation 9.1 for a time (late pregnancy) when the feeding time requirement was known (43% as determined by observation), Altmann estimated that $A = 7.08$. Note, here, that A incorporates a scaling factor for both ingestion rate (the rate at which an animal can harvest energy) and for the additional energy required over and above BMR to fuel normal life processes (see p. 34).

With these values, it is possible to calculate the increment in feeding time requirement needed to maintain the mother's body weight and allow her infant to grow at the levels required for normal development. From general time-budget data, Altmann calculated that at the end of pregnancy a female spends 43% of her time feeding, 23% moving, 17% resting and 17% in social activity. For the purposes of simplification, Altmann assumed that moving and resting time were constants and allowed the balance of time left after the feeding-time requirement had been taken out to be available for social activity, additional feeding time or more rest. In fact, as we saw in Chapter 6, monkeys in general, and baboons in particular, probably try to conserve social time and instead use resting time as a reserve on which to draw for additional feeding time when this is required. Evidence from the gelada suggests that mothers only give up social time when the additional requirement for feeding time exceeds the resting time available (Dunbar and Dunbar in press).

By standardising the increase in time spent feeding against the pre-natal requirement, we can plot data for both the Amboseli baboons and the gelada on the same graph since the animals are of approximately the same size (Figure 9.3). In both species, the time spent feeding increases as predicted, reaching a peak in the fifth month, after which it begins to fall. However, neither set of data matches the predicted values at a quantitative level within the first few months, the one being consistently below and the other consistently above expectation. In the gelada, those months in which the observed points exceed the predicted values all occur during the dry season when the quality of the forage is declining steadily with time. Because of this, females will need to eat proportionately more food to extract the same quantity of nutrients: in other words, the value of A will increase as the dry season progresses. Sensitivity analyses of the various parameters in equation 9.1 showed that only quite small changes in A were required to bring the observed values in line with those predicted by the model, whereas E would have to change substantially to have such an effect (Dunbar and Dunbar in press). The sudden fall in the sixth month in the gelada data can be attributed to the fact that in this species the infants begin to feed for themselves from the third month onwards and this rises to an asymptote in the sixth month (see Dunbar and Dunbar in press).

The important point to note about the two sets of data, however, is that both reach their peak at the point at which females exhaust the resting time that they have available for conversion into feeding time. The baboons manage to do this without forgoing any social time, but the gelada significantly reduce their social time allocation during critical months when the feeding-time requirement is especially high. They can manage this because their social time allocation is in fact substantially higher than that for the Amboseli baboons: even after giving up a substantial quantity of social time (equivalent to about 10% of the total time budget), they still manage to spend an average of 15% of their time in social interaction, a figure that is virtually identical to the 17% allocated to social time under normal conditions by the

*Figure 9.3. Maternal time budgets in baboons. The increment in feeding time required
to sustain lactation is plotted against the infant's age, taking the mother's pre-natal
feeding time as a baseline for comparison (horizontal dashed line at 0). The change in
feeding time predicted by Altmann's (1980) model of maternal time budgets is indicated
by the solid line. Actual time spent feeding in each month for yellow baboon mothers
(dotted line: from Altmann 1980) and gelada mothers (short dashed line: from Dunbar
and Dunbar in press) are plotted against a common baseline. Also shown as fine dotted
and dashed lines across the graph are the amounts of time that baboon and gelada
mothers have available for additional feeding time in the form of resting time. That part
of the time budget above each of these lines represents the time committed to social
interaction and travel. As the infant's demand for milk increases with time as it grows,
so the mother's time budget is placed under increasing stress as she uses up all her
available surplus time. When her feeding time requirement exceeds her available resting
time, a mother has to reduce her social activity in order to make the necessary extra time
available*

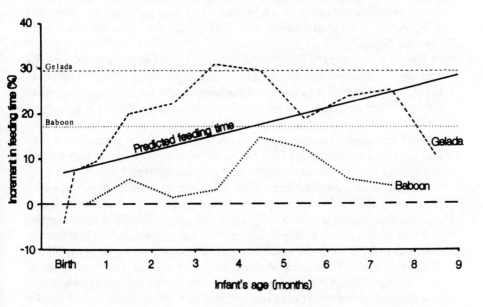

Amboseli baboons. Significantly, when the gelada females do surrender social
time in order to spend more time feeding, it is casual interactions that they
give up: relationships with female associates and dependent offspring (their
'key' relationships that are important in the long term) are preserved throughout (Dunbar and Dunbar in press).

Optimising parental investment decisions
Although we conventionally measure a female's reproductive output in terms
of R_0 or v_x, these only provide an approximation to fitness because they
consider only the numbers of infants born to a female, not the number that
survive to reproduce in their turn. From the female's point of view, any

decision about the amount of parental care to invest in a given offspring has to depend on how that parental care affects the likelihood that the offspring will survive to sexual maturity. We thus need to add a term to the equation for net reproductive rate (equation 4.2, p. 58) that adjusts fecundity to take account of offspring survival:

$$R'_x = \sum_{y=x} l_y m_{y(i)} c_i \qquad (9.2)$$

where R'_x is the expected number of offspring that a female aged x can expect to rear successfully during the rest of her lifetime, c_i is the probability of an infant surviving to maturity given that i units of parental care are invested in it, $m_{y(i)}$ is the female's age-specific fecundity given the expenditure of i units of parental care, and l_y is the female's survivorship (for further discussion of these terms, see Chapter 4). These three variables represent the three key components of survival, mating and rearing that determine an individual's personal fitness (see p. 21).

Ignoring the survivorship component for the moment, the relationship between the quantity of parental care, i, and both c and m_y will be approximately as in Figure 9.4 (a and b). A minimum of parental care is required to give an infant any chance of surviving to maturity, so that the benefit from small quantities of parental care is likely to be negligible; but thereafter the infant's chances of reaching maturity increase steadily in proportion to the amount of care invested until it reaches an asymptote beyond which further increments in parental care will have little or no influence. Hence, in Figure 9.4(a), the relationship between c and i is sigmoid (S-shaped). Conversely, since continued investment in parental care prevents a female returning to reproductive condition (see p. 60), fecundity, m_x, will be a simple negative function of the amount of parental care given, i (Figure 9.4b). What the female is looking for is the value of parental investment, i_{opt}, that maximises R'_x, the number of offspring she can expect to rear to maturity. Ignoring survivorship still, this is simply the product of $m_{y(i)}$ and c_i summed across all the remaining years of the female's life. If we plot the resulting values of R'_x against i (Figure 9.4c), we can determine the optimum investment, i_{opt}, by finding the point at which the tangent drawn through the origin just cuts the slope of R'_x: this gives us the level of parental investment that optimises the rate of return (in terms of numbers of offspring reared to maturity) on the investment. (This point is usually known as a *Nash equilibrium*.)

Having established the basic principle, let us now introduce some complicating factors. We can begin by considering what happens to R'_x if either of the relationships for fecundity or offspring survival are altered by changing environmental conditions. Figure 9.5 shows what happens if the slope of the $m_{y(i)}$ relationship increases (i.e. the interbirth interval lengthens) or the c_i curve reaches an asymptote at a lower level (i.e. infants have a lower probability of surviving to maturity for a given level of parental care). In both cases,

Figure 9.4. Relationship between the amount of parental care invested in an infant and (a) the infant's chances of achieving reproductive maturity, c, (b) the mother's age-specific birth rate, m_x, and (c) the mother's lifetime reproductive output, R'_x (this being the product of c and m_x). In (a) the curve is presumed to be sigmoid on the grounds that a minimum investment in parental care is necessary to give an infant any chance at all of surviving but that there will be a point beyond which further investment does not substantially improve the infant's chances of surviving. In (b), the curve is presumed to be linear on the grounds that the delay incurred before a female can conceive a new infant is directly proportional to the amount of time and energy invested in parental care of the current infant, so that the birth rate per unit time will decline proportionately. The optimal amount of parental care, i_{opt}, that maximises lifetime reproductive output is found by locating the point at which a tangent drawn through the origin just touches the R'_x graph.

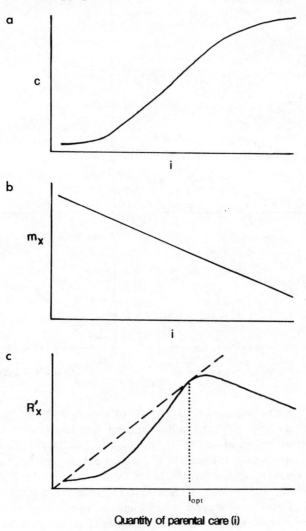

Quantity of parental care (i)

Figure 9.5. Two ways in which declining habitat conditions might affect a female's decision on how much parental care to invest in her current offspring. (a) The mother's ability to conceive again declines more steeply with increasing parental care. (b) The infant's chances of reaching sexual maturity are reduced so that the c graph reaches its asymptote at a lower level. The effect in both cases is to move the peak in R'ₓ to the left, thereby reducing the optimal level of parental care, Iₒₚₜ. Thus, the female should terminate lactation and other expensive forms of parental care earlier as habitat quality declines

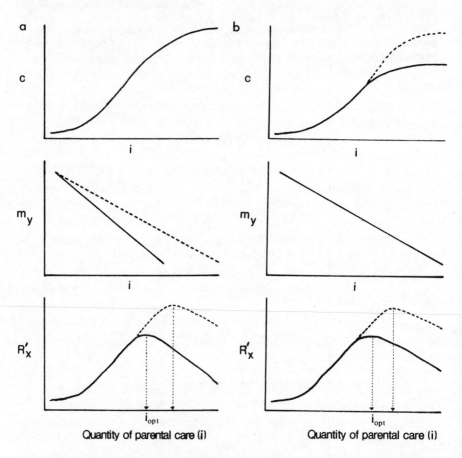

the effect of worsening conditions is to shift the peak in the R'_x curve to the left. Conversely, improvements in habitat conditions that make it easier for a female to conceive again sooner or make it more likely that an infant will survive will tend to push the peak in R'_x to the right.

So far, rather few studies have looked at this kind of investment problem in primates. However, two sets of data tend to confirm the predictions from Figure 9.5. First, the data for gelada and baboons given in Figure 9.3 demonstrate that, in the poorer quality habitat at Amboseli (it receives less than one-third of the rainfall of the gelada habitat), the mother's feeding time allocation has returned to approximately normal levels by the time the infant reaches 7 months of age, whereas as late as the ninth month gelada females are still feeding at significantly higher levels than prior to the infant's birth. If feeding time requirement is determined largely by investment in lactation, as seems likely, then the Amboseli baboons are terminating lactation earlier than are gelada females. Second, Lee (1983a, 1986) found that female vervets living in poor quality territories weaned their infants significantly earlier than those living in better quality territories. In addition, within territories, females whose previous infant had been born the year before (and who were thus presumably under some energetic stress through lack of time to recuperate between successive pregnancies) began to wean their infants up to three months earlier than those whose previous infant had been born two years before. Note that in both cases there is no intrinsic reason why the females in poorer conditions could not go into energy deficit in order to invest the same amount in their offspring as those in better habitats. We know, for example, that women can lose weight during lactation for this reason (Hytten and Leitch 1964, Naismith and Ritchie 1975). Moreover, there is some evidence to suggest that captive rhesus macaque mothers may go into protein deficit during pregnancy in order to ensure an adequate supply for the fetus (Riopelle *et al.* 1976). Rather, Lee's observations suggest that the females opt not to do so because the long-term costs are too great.

In marginal ecological conditions, the balance between sufficient investment in lactation to ensure the infant's survival and the mother's risk of overtaxing herself (and so reducing her own survival prospects) may be very fine. Altmann (1983) estimated that mortality rates for adult females during lactation are approximately double what they are during other stages of the reproductive cycle. Given that investment in parental care, i, will influence the female's own survivorship, l_x, negatively, the effect on R'_x will be to shift the peak still further to the left.

This will also mean that, irrespective of the effect of i on l_x, the female's age at the time of her investment decision will be an important consideration. Once past a certain age, the increasing mortality rate to which she is subject (Table 4.1) will mean that the probability of being able to rear another offspring in addition to the current one drops dramatically. If she terminates parental care too early in the hope of squeezing one more infant in before the end of her natural life, she may end up losing both infants because she has insufficient time to bring her next infant up to a point where it can survive on its own if she dies, while under-investment on the first infant will have placed its life at risk too. At this stage of her life, she would do better to concentrate on her current infant and make sure that it can get through to adulthood. Hence, the length of the period of parental care should increase as the female approaches the end of her life. This prediction is clearly supported by the fact that interbirth intervals invariably get longer as a female ages in all species (see Table 4.1). However, other factors (e.g. increasing physiological inefficiency with age) might also cause this effect, and we really need to be able to show that older females lactate for longer than do females in the prime of life.

A further complicating factor is that the demand curve of infants may differ in relation to their sex. There are three considerations that might prompt a mother to invest more parental care in one sex of offspring. First, a male's reproductive success is more heavily dependent on his own abilities than is a female's because males commonly migrate to other groups; a female, on the other hand, can continue to depend on the support of her mother and other relatives long after puberty. The need to ensure that sons achieve as high a growth plane as possible early on may induce mothers to invest more heavily in their sons than in their daughters. The second consideration stems from the fact that, in some species, females depend heavily on coalitionary support from their daughters during the later part of their lives (e.g. gelada: Dunbar 1980a, 1984a; vervets: Horrocks and Hunte 1983a). Investing more heavily in daughters to ensure that they achieve high growth planes may therefore have important consequences for their value as allies later on. Thirdly, there is considerable evidence to suggest that female offspring suffer much more from harassment by other group members than male offspring (macaques: Dittus 1977, Silk et al. 1981a, Simpson and Simpson 1985; baboons: Wasser 1983; vervets: Horrocks and Hunte 1983a). Sackett (1982) even found that in pigtail macaques mothers carrying female fetuses were attacked significantly more often than those carrying male fetuses. There are two reasons why other animals might harass daughters more than sons. Dittus (1979, 1980) argued that this is part of the population's self-regulating mechanism: individuals gain more in terms of personal fitness by limiting the reproduction of other group members and this is most effectively done by selective female infanticide, especially when ecological conditions are on the margin. Alternatively, Simpson and Simpson (1985) have suggested that when a female's rank in the hierarchy is determined by the size of her kin-based coalition, it pays individuals to try to reduce the size of coalitions that other individuals will have in the future by creating high mortality among their daughters. Either way, mothers will be inclined to invest more heavily in daughters in order to reduce the risks that they run.

So far, no attempt has been made to evaluate the relative importance of these three factors in order to determine the optimal decision from the mother's point of view. Since all three factors are relevant, the problem cannot be solved *a priori* but will require careful empirical evaluation. In some captive macaque populations mothers do have shorter interbirth intervals after producing sons than after daughters (Simpson et al. 1981, Simpson and Simpson 1985), suggesting that they regard the future value of daughters as allies as being more important. This is by no means the case in all populations, however. Small and Smith (1984) found no significant differences between male and female infants in birth weight, growth rates during the first year or subsequent interbirth intervals in another population of captive rhesus, while Rowell and Chism (1986b) found that sons were nursed significantly more in later months than were daughters in captive patas monkeys. Since the weighting given to the three factors will differ depending on whether females form kin-based coalitions and on whether females migrate regularly, we can expect significant differences between species and perhaps even between populations of the same species. In addition, since daughters born late in a female's life will not mature in time to act as an ally for the mother, we might expect a shift in favour of male offspring as the mother ages even in those species where females commonly invest more in daughters.

We can extend the logic of this argument one step further by pointing out that the offspring's own reproductive prospects should be an important consideration from the mother's point of view. Although the *average* lifetime reproductive success for each sex has to be the same, the *variance* in reproductive success is very different in the two cases. This is due (a) to differences in mortality rates (males are more likely to die before maturity and a higher proportion will therefore make no contribution to their mother's fitness), and (b) to the effects of the polygyny skew (i.e. of those males that do survive, some will be *very* much more successful than others, whereas female reproductive rates are more constrained: see Vehrencamp 1983). This difference in the variance may induce low-ranking females (whose own reproductive prospects are poor) to invest more heavily in individual offspring of the less risky sex (i.e. daughters) in order to play safe and ensure a reasonable average fitness. Because high-ranking females have more offspring anyway, they can afford to risk a little more in order to gamble on gaining an additional advantage on the very high reproductive rates that some males can achieve.

Data on which to test these predictions are not available for primates, and those for mammals in general are somewhat contradictory (see review by Clutton-Brock and Albon 1982). This almost certainly means that a species' biological context and socio-ecological strategies are crucially important in determining the choice of option that any individual makes, thus highlighting once again the dangers of searching for predictions that are universally true of all species. The decision is also likely to depend on the size of the reproductive differential between high- and low-ranking females in any given context. Where the difference is small, high-ranking females will not have the extra 'capital' with which to take risks, and there should be no difference between females of different rank in the way they distribute their parental investment. Harcourt (in press) has suggested that female dominance hierachies are more conspicuous in habitats where food resources are patchily distributed, irrespective of the actual richness of the habitat as a whole. Differences in rearing strategies of the kind discussed here should thus be more conspicuous in such habitats.

Varying the sex ratio of offspring

An alternative option open to a female is to vary the sex ratio of her offspring so as to take advantage of these differences in the reproductive value of the two sexes (for the background theory, see Trivers and Willard 1973, Clutton-Brock 1982). Precisely the same set of factors will affect the neonatal sex ratio as affect parental investment, so that similar predictions can be expected.

Once again, the available data on neonatal sex ratios are highly variable. Some studies have found that dominant females tend to produce more sons than daughters, with subordinate females tending towards the reverse (macaques: Paul and Thommen 1984, Meikle *et al.* 1984), but others have found the reverse relationship (macaques: Simpson and Simpson 1982, Silk 1983; baboons: Altmann *et al.* 1986) and yet others have reported no significant differences (vervets: Cheney *et al.* 1986a; gelada: Dunbar 1984a). This strongly suggests that the decision is dependent on the biological and social context of the population concerned and that, once again, we should be careful to avoid looking for superficial universal rules that apply to all species.

Because the value of a daughter as a future coalition partner will vary with her position in the mother's birth order, we can again expect the sex ratio of a female's offspring to vary over her lifetime. As yet, there are few or no data to assess whether or not this does occur, though data for gelada do suggest a (non-significant) tendency for females to produce more daughters than sons on their first births, as we would expect given the importance of a first-born daughter as a coalition partner (Dunbar 1984a).

Although it is quite clear that neonatal sex ratios can be varied from the standard 50:50 ratio, we have no idea at present how females achieve this nor at what stage in the process of reproduction it is brought about. The options seems to be either in terms of the likelihood of conception (or perhaps implantation) or in terms of selective abortion of fetuses of the less desirable sex at a later point during pregnancy. Gosling (1986) has presented evidence to show that coypu do selectively abort litters of one sex. What emerges from this analysis is that a great deal more research needs to be done on the problem before any firm conclusions can be reached. We do not yet adequately understand how the various components relate to each other at a theoretical level, nor do we have enough data to determine this empirically. By the same token, we cannot dismiss the phenomenon as a figment on these grounds alone, as some have been tempted to do. Where we do not understand the biology, we are not at liberty to make categorical assertions.

Social Aspects of Rearing

The social milieu in which a young monkey grows up plays a crucial formative role in its adult behaviour in two important respects. First, the network of relationships centred on its mother provides the infant with a source of allies and protectors who can buffer it against harassment by other members of the group. Secondly, interactions with peers allow the infant to learn the social techniques that will enable it to form effective relationships as an adult (including the ability to exploit other individuals to its own advantage: see Kummer 1978).

An infant's small size makes it particularly vulnerable to both exploitation and maltreatment by the adult and near-adult members of its group. The risk of harassment by higher ranking group-members may therefore prompt low-ranking mothers to invest more in their offspring (irrespective of their sex) than high-ranking individuals in order to provide them with more protection. Among gelada, for example, low-ranking mothers allowed their offspring to spend more time on the nipple than high-ranking mothers did (Dunbar 1984a). Such effects are likely to depend on the demographic state of the population, however. Berman (1980), for example, noted that the Madingley rhesus colony mothers became increasingly more relaxed with time as the groups were allowed to grow naturally and thus develop kinship structures. Eventually, the behaviour of the mothers was virtually identical to that observed in the large free-ranging groups on Cayo Santiago in the Caribbean. Berman (1980) suggested that mothers may be more relaxed and less restrictive in their mothering when they know that they can count on the support of close relatives. Over and above these considerations, the most important factor influencing a female's decision about investment in her offspring is likely to be the extent to which any such additional investment improves the infant's chances of survival.

Berman (1982) has found significant differences between matrilines within rhesus monkey groups in the social experiences of infants. She found that high-ranking matrilines were socially more cohesive and that infants born into them interacted with many more kin than infants in low-ranking families. Although there was no significant difference in the quality and frequency of their interactions with those relatives that they did interact with, infants in high-ranking matrilines could call on much more support in threatening situations (see also Datta 1983b). Bernstein and Ehardt (1986) have also stressed the important role that relatives can play in the socialisation of infants by punishing behaviour that is 'anti-social'. Providing we avoid the group-selectionist pitfalls lurking behind this explanation by interpreting 'anti-social' to mean 'not in the interests of the animals or kin group concerned', it is clear that relatives can exert an important influence in this respect. (The functional significance of such behaviour is not clear in this case since Bernstein and Ehardt provide no information on the *context* of aggression towards kin; moreover, the risks involved in threatening close kin are much lower than those incurred by threatening non-kin, so it is far from clear what we can learn from these observations alone.)

The social aspects of rearing play a particularly important role in those contexts where an older sibling provides both protection and substitute parental care. Caretaking behaviour of this kind, particularly by juvenile females, is quite common in some species of primates, notably those like the colobines where females are more tolerant of the attention of others (for reviews, see Lancaster 1971, Hrdy 1976, McKenna 1979).

From the mother's point of view, this arrangement may be beneficial in that it allows her some respite in which to feed or socialise without needing to be constantly aware of her infant's whereabouts. There are, however, serious costs both in terms of ill-treatment of infants by spiteful or inexperienced handlers and in terms of prolonged separation if the infant is unable to nurse as a result. In hot climates, dehydration is a serious risk to infants separated from their mothers for any length of time (Rhine *et al.* 1980). Females will consequently be reluctant to allow very young infants to be taken from them, and most caretaking does in fact involve infants that have been weaned (McKenna 1979). The caretaker, on the other hand, may benefit from the opportunity to learn 'how to mother' in a relatively non-critical environment where the mother is close enough to rescue the infant if the caretaker gets into difficulties. While this might explain a mother's tolerance of older daughters, 'learning to mother' is not itself an adequate explanation for tolerance towards less closely related helpers because it assumes a group-selectionist view. Presumably there is a trading of costs and benefits in all such cases, but data are not available that could be used to estimate these with any degree of confidence.

This behaviour is, however, taken to an extreme in the callitrichids: offspring who remain on their parental territories are reproductively suppressed (p. 67) and contribute a significant amount of help to the parents by carrying and caring for the offspring (Ingram 1977, Garber *et al.* 1984). Studies of both canids (Moehlman 1979) and primates (Garber *et al.* 1984) have shown that parents who have help from older offspring are more successful in rearing their infants than those who do not. This may be especially important in cases where both parents are needed to rear young but one of them dies prior to weaning: a helper can then take over the duties of the dead parent and so save the litter from what would otherwise have been certain death (Garber *et al.* 1984). However, this is only likely to be a serious consideration in those cases where more than one offspring is born at a time.

Such altruism on the part of the helpers none the less requires an explanation, for it is seldom worth an individual's while giving up all prospect of personal reproduction even in order to help its parents reproduce more effectively (p. 21). Two circumstances might, however, militate in favour of such behaviour. One is when the individual's prospects of finding a mate or territory of its own (and hence of reproducing) are very low, at least within the immediate future (Emlen 1982); the other is when the helper is at least as closely related to its siblings as it would be to its own offspring (Hamilton 1964). This second condition, however, can only occur in mammals under strict monogamy where successive litters have the same father as well as the same mother. It is certainly significant that reproductive suppression and active caretaking are only known to occur among primates in the predominantly monogamous callitrichids. Whether the animals are also prevented from breeding themselves by a shortage of suitable territories or mates, however, remains unknown. (For a more detailed review of these issues, see Emlen 1984.)

In more general terms, a network of social contacts on which she can draw for assistance is so important a part of a female's reproductive strategy that an individual lacking such support often faces serious reproductive disadvantages. Female gelada who lack allies, for example, occupy lower ranks in the dominance hierarchy and, as a direct result, have lower reproductive outputs than females who do have allies (Dunbar 1984a). In such circumstances, a female's options may be very limited. If she lacks the power (or attractiveness) to acquire coalition partners, her only alternative to sitting it out is to emigrate to another group.

Although female migration is relatively rare in primates (see Chapter 5), it is significant that in those cases where it has been documented, reproductive performance has often been a key factor differentiating females who migrated from those who did not. Harcourt *et al.* (1976), for example, noted that a failure to breed successfully often seemed to be the factor precipitating emigration by female gorillas. Marsh (1979b) found that most red colobus females who transferred between groups did so following the takeover of their own group by another male: he suggested that they did so in order to avoid losing their young infants through infanticide by the new male. Finally, studies of Japanese macaques have demonstrated that females are more likely to leave their natal groups if they do not have ready access to a member of the dominant clique of central males (Grewal 1980, Yamagiwa 1985) or had no living female relatives in the troop (Sugiyama and Ohsawa 1983b).

In most cases, however, migration is likely to be a last resort. Whenever a female's status in society depends on the formation of coalitions and such coalitions are formed predominantly among close relatives, transfer to another group may well reduce the likelihood of joining a coalition because an immigrating female will be even less closely related to the members of her new group than she was in her old group. Indeed, in species where coalitions of closely related females are an important social component, females may be very unwilling to allow unrelated strangers to join them. In the gelada, for example, female immigration is extremely rare and is actively resisted by females, even though harem-holders may try to encourage it (Dunbar 1984a, Ohsawa and Dunbar 1984). Among *Papio* baboons, on the other hand, coalitions formed with an adult male seem to be at least as important as those formed with female relatives (Smuts 1985): female transfer between groups is not only easier in these species (e.g. Kummer 1968), but is also far from rare (e.g. D.R. Rasmussen 1981a).

The social experiences of an infant can also be expected to have important implications for its behaviour as an adult, particularly when the ability to function effectively within a complex social setting determines its reproductive success. This has been emphasised not only by the numerous studies of social deprivation carried out in the 1960s by Harlow and others (e.g. Seay *et al.* 1964, Arling and Harlow 1967, Sackett 1968), but also by a number of more recent studies that have assayed the behaviour of deprived animals in much finer detail. In a series of studies of rhesus macaques, Anderson and Mason (1974, 1978) found that infants reared in social isolation developed less complex sets of relationships as juveniles and interacted with fewer individuals than animals reared in peer groups. The deprived animals were socially less adroit and, unlike their peer-raised conspecifics, were unable to exploit the power relationships between other individuals to compensate for their own physical disadvantages. Similarly, Young and Hawkins (1979) found that infant baboons reared alone with their mothers (and hence lacking experience of peers) had higher scores for aggressive behaviour than infants reared in social groups (several mother–infant pairs plus an adult male), but had lower scores on such variables as locomotion, non-agonistic social behaviour, exploration and rough-and-tumble play. Socially experienced individuals are more effective in their relationships and can integrate and maintain many more relationships simultaneously. These differences may have significant implications. Smuts (1985), for example, noted that adult males who were socially more adept were more successful at entering groups and achieving high status and, in consequence, had higher reproductive success than less skilful individuals. Experimental studies of macaques by Herbert (1968) and Michael *et al.* (1978) have demonstrated that females often prefer the less aggressive males as mating partners. Berry (1974) described a similar phenomenon in lizards.

Male Parental Investment

Although parental care is largely the responsibility of females, the contribution made by male primates is not always negligible (for a general review, see Kleiman and Malcolm 1981). Males can contribute to the costs of rearing offspring in three general respects. First, males can provide direct assistance to the female by helping to transport and/or care for infants. This frees the female from some of the costs of reproduction and allows her to spend more time feeding without having to concern herself with the infant's whereabouts. Secondly, males can provide indirect help by actively providing the female and her offspring with protection, both against predators and against conspecifics. This may have important consequences for the survival rates of both the infants and their mothers. Thirdly, it has been suggested that males may contribute towards the costs of parental care by defending a territory that provides the female and her offspring with exclusive access to essential resources. I shall argue (in Chapter 12) that this seems to be the least likely reason for the evolution of territoriality in primates. Indeed, there is no unequivocal evidence to suggest that female primates either need or benefit from exclusive access to resources other than those that are highly localised (where territorial defence is often impossible). I shall therefore not consider territorial behaviour as an important form of paternal investment, though I recognise that, given that males are territorial for other reasons, territoriality may have some secondary advantages in this respect.

Unlike the males of many monogamous birds and mammalian carnivores (see Kleiman 1977), male primates are generally precluded from helping the female with the business of feeding her young. The most that can be done is to help in transporting infants and in caring for them (e.g. grooming them and keeping an eye on them while they are playing away from the mother). This kind of behaviour is not particularly common among primates, but it is by no means unknown. Males of many different species have been observed playing with infants, carrying them or helping them across particularly difficult gaps in the canopy (see reviews by Mitchell and Brandt 1972, Parke and Suomi 1981). Regular transport of the young, however, has been documented only in the siamang (Chivers 1974) and among the callitrichids (see Kleiman 1977, Kleiman and Malcolm 1981). In both marmosets and titis, for example, the male is responsible for more than 85% of the carrying of the infants throughout the period of dependency (Ingram 1977, Fragaszy et al. 1982). The female's responsibility in these cases is limited to lactation and she has the infants only long enough to nurse them at the appropriate intervals. So assiduous is the male marmoset in this respect that he will forcibly remove the infants from the female if she delays in returning them to him once they have finished nursing (Ingram 1977).

In polygamous species, males devote very much less attention to infants. None the less, they may provide both mother and infant with important sources of protection (macaques: Berman 1982, Taub 1984, Vessey and Meikle 1984; baboons: Ransom and Ransom 1971, Altmann 1980, Packer 1980, Stein and Stacey 1981, Smuts 1985, Collins 1986; capuchin monkeys: Izawa 1980). In most cases, males who take a close interest in infants are usually the known or probable father of the infant concerned. Studies of baboons have shown that infants gain considerable benefit from proximity to a large protector who is willing to come to their aid when a situation gets out of hand, particularly during the first few months of life when they are especially vulnerable (Altmann 1980, Stein and Stacey 1981, Smuts 1985). In most of these cases, the male concerned also had a very close relationship with the infant's mother. Stein and Stacey (1981) found that contact interactions between such males and the infants were relatively rare even though the males were often in close spatial proximity to the infants (mainly because of the male's continuing association with the mother). This relationship with the mother seems to be instrumental in promoting concern for the infant. In other words, the male's concern may be less for the infant itself than for his relationship with the mother. High-ranking males that opt to compete for access to all cycling females rather than forming a long-term pair bond with a single female rarely take such a close interest in their (putative) offspring.

Busse and Hamilton (1981) have also pointed out that, in chacma baboons, males may be instrumental in preventing new males entering the troop from killing off the young infants present at the time, most of whom will of course be the offspring of the resident males. Packer (1980) reported instances in which male olive baboons actively protected infants from predation by chimpanzees. Although active grooming of infants by males does occur in these species, this often seems to be related more to pacifying the infant than to any functional benefit that the infant might incur in terms of hygiene (Packer 1980).

Altmann (1980) has pointed out that a special relationship between an infant and its (putative) father may continue beyond the infant stage, so that the infant continues to derive protection as a juvenile or even a young adult. In some cases, this support may extend to promoting the offspring's reproductive prospects. Tilson (1981), for example, found that male gibbons sometimes helped their offspring to set up their territories when these were adjacent to the parent's territory.

Although it is quite clear that fathers can offer a great deal in the way of protection to their offspring, what remains to be ascertained is just what difference this makes to the survival of the infants. So far, the evolutionary value has been based on the unsupported assumption that defending an infant must be beneficial (e.g. Smuts 1985). In the long run, however, we will need data on the survivorship of both infants and their mothers for those that do have protectors to compare with the survivorships of those that do not.

Chapter 10
Conflicts and Coalitions

So far, I have focused on the primary biological considerations that relate to mating and the rearing of offspring, together with the strategies that animals use to resolve the problems they encounter at a tactical level. The core of these decisions lies in the fact that primates use social relationships to further their reproductive ends. I now take this one stage further by asking how considerations of maximising fitness might influence the decisions animals make in developing relationships with the other members of their social group. In this chapter, I focus mainly on dominance relationships. In the following chapter, I shall be concerned with the behavioural processes of negotiation that are involved in establishing these relationships.

The Constraint-free Strategy

In order to understand why the animals behave as they do, it is necessary to ascertain what they would do in an ideal world where there are no constraints on their choice of strategy. In other words, we need to determine the constraint-free strategy. Once we know what this is, we can then look at the context in which the animals are embedded and ask how this ideal strategy is affected by having to live in social groups.

In this section, I shall again argue that dominance achieved through outright aggression is the natural constraint-free strategy both where there is competition for limited resources and where animals live in large groups that impose serious levels of stress on them. In the first place, high rank confers priority of access to those resources on which successful reproduction depends; these include not only mates and food, but also access to social partners with whom to form alliances. There is extensive evidence from field studies to suggest that high-ranking animals gain priority of access to preferred food sources (vervets: Wrangham and Waterman 1981, Whitten 1983; see review by Harcourt, in press), have higher food intakes (capuchins: Janson 1985) or weigh more (macaques: A. Mori 1979a, Sugiyama and

205

Ohsawa 1982a; vervets: Whitten 1983). In the second case, high rank buffers an animal against the stresses of group-living by reducing both the frequency of interruptions to activity and the absolute amount of aggression received from other individuals. In the constraint-free condition, dominance is solely a consequence of an individual's own *intrinsic* power (in the sense of Datta 1983b). Having established that this is so, I shall go on in the following sections to show how the social context allows low-ranking animals to use alternative strategies to buffer themselves against the disadvantages of low rank. The most important of these is the formation of alliances with other individuals, thus adding a new dimension to an individual animal's power by allowing it to call on support from third parties. Datta (1983b) refers to this 'external' support as *extrinsic* power. First, however, it is necessary to resolve a number of difficulties surrounding the concept of social dominance.

The concept of dominance

Although the concept of dominance has been criticised on a number of different grounds (see Gartlan 1968, Rowell 1974, Bernstein 1976, 1981, Hinde 1978), many of these criticisms arise from misconceptions (see Wade 1978) or are tangential to the way in which dominance is used here. Seven misconceptions are worth discussing in order to avoid problems of interpretation at a later stage.

1. It is important to be clear that dominance as used here refers to a *relationship* between two animals (Hinde 1978) and is not an inheritable property as such. One animal is said to be dominant over another because it is able to displace that individual, though it may well achieve this in part by virtue of attributes such as physical size or aggressiveness that are genetically inherited. An animal holds a particular dominance *rank* in a hierarchy at a given time because of the sum of its dominance relationships with all the other members of its group. Dominance rank is not, therefore, a property of an individual that can be inherited genetically. An individual's rank depends on how its intrinsic power relates to those of the other individuals in its group at the time and the way in which 'natural' ranks are modified by the extrinsic power on which an individual can call. In general, I follow Hinde (1978) in maintaining a distinction between dominance (as a characteristic of a *relationship*) and high/low rank (as a characteristic of an individual's *status* within a particular group of animals). In principle, it is perfectly possible for the lowest ranking animal in a group to transfer to a new group and become its highest ranking animal without changing its power characteristics in *any* way.

2. Dominance is not an end in itself, but rather a means to an end (namely the acquisition of resources). Although animals are sometimes said to compete for high rank as an end in itself (de Waal 1982, Fedigan 1983),

this needs to be interpreted against the long-term objectives of animals: dominance is a means to an end, a proximate goal that provides an animal with priority of access to resources that are of crucial long-term importance in terms of reproduction. If primates do appear to compete for high rank for its own sake, it is almost certainly because they appreciate the relationship between the proximate and ultimate goals involved (where by 'ultimate' in this context I mean access to resources). That a baboon or chimpanzee should *understand* that being the dominant animal will allow it to gain priority of access to resources should it need these at any future time can hardly be said to make excessive intellectual demands on animals as intelligent as these.

3. A lack of correlation between the frequencies with which animals gain access to different commodities cannot of itself invalidate the concept of dominance. Behavioural indices of access to particular goals only measure a dominant animal's *willingness* to exercise its potential monopoly over a given commodity. Only the very naivest interpretations of dominance would suggest that animals have to exercise their 'rights' of access *all the time*. By the same token, of course, the value of the concept as an explanation of behaviour requires that it should actually predict priority of access at least sometimes (Richards 1974, Deag 1978), otherwise it cannot help us to understand why animals behave as they do. We do need to be careful to distinguish between the operational use of the term 'dominance' (defined in terms of access to a given resource) and its use as an explanatory concept (that summarises certain salient characteristics about an individual). An animal does not gain access to a resource because it gains access to that resource, as an operational definition would imply: defined thus, it merely describes what happens without offering an explanation. Rather, an animal is dominant because it can bring greater intrinsic power to bear in keeping competitors away. Ideally, our assessment of dominance ranks should be independent of the resources for which the animals are competing: that is to say, our estimates of dominance rank should reflect the animals' respective *capacities* to defeat other individuals, not simply a statement of whether or not it did so. Where this capacity (or power) is a simple function of some biological attribute like size, this obviously poses few problems. But the ability to dominate another individual often depends on factors besides sheer physical power: in species as intelligent as primates, psychological attributes may be just as important, particularly those that allow the animals to draw on extrinsic sources of power (see, for example, Richards 1974). Though these difficulties do not make the concept of dominance any less valid, they should caution us against using them in a slipshod way.

4. We should beware of a tendency to dismiss dominance as irrelevant simply because the highest ranking animals do not always gain priority of

access to a given resource. The fact that low-ranking individuals evolve alternative strategies to circumvent the immediate disadvantages of low rank (thereby gaining a share in a resource: see p. 163) does not mean that dominance as such plays no role. On the contrary, alternative strategies evolve precisely because of the effectiveness of dominance as a strategy for gaining access to desirable resources. If there were no effects due to dominance, there would be no need for alternative strategies.

5. Dominance hierarchies do not necessarily depend on high rates of aggression and are not, therefore, simply an artefact of the high frequencies of fighting commonly found under captive conditions (e.g. Rowell 1967). Despite occasional claims to the contrary, dominance hierarchies have been documented in a wide range of species at all taxonomic levels. Low frequencies of aggression originally led me to conclude that dominance was not an important factor in the social behaviour of the gelada (see Dunbar and Dunbar 1975). In fact, dominance hierarchies turned out on more detailed analysis to be a characteristic of all gelada reproductive units, and their consequences proved to be the fundamental driving force behind many of the characteristic features of the gelada's social system (Dunbar 1984a, 1986).

7. One final criticism of hierarchies has been that they force a complex set of relationships between individuals into an artificial linear structure: in many cases, it is argued, dominance relationships take complex non-linear forms which can involve lower-ranking individuals dominating higher-ranking ones (e.g. *non-transitive* hierarchies in which A dominates B and B dominates C, but C can dominate A) or that they may consist of sets of individuals of equal rank. Moreover, Landau (1951) pointed out that, as the number of animals in the hierarchy decreases, the likelihood of obtaining a strictly linear hierarchy by chance alone increases (see also Appleby 1983a). While this is certainly true, it is also the case that it is significantly easier for animals to form themselves into a strictly linear hierarchy when the set of relationships that has to be integrated is smaller. In the gelada, all hierarchies are linear, including the largest ones, but the hierarchies do become increasingly unstable (as reflected in the frequency of statistical as opposed to deterministic dominance relationships) as the size of the hierarchy increases (Dunbar 1984a). The essential point here is not whether linear hierarchies can be generated at random, but whether that linearity affects the access the animals concerned have to resources. In the long run, it is the functional *consequences* of linear dominance hierarchies that are important, not their *causes*.

Economics of contests

That dominance should be an important factor should not be surprising. As a first option, a show of aggression to frighten rivals away is invariably a worthwhile strategy. It works because not every individual will be willing to contest the issue, so that an initial display of aggression, even in bluff, may discourage competitors from attempting to gain access to a resource.

In principle, whether two animals are willing to fight over a resource item will depend on (1) the value of the resource to each of them, (2) each individual's *own* estimate of its likelihood of winning the fight, and (3) the cost that each will incur in escalating the conflict into a fight. An animal should only risk a fight if the profit it can expect to gain from winning the fight (when devalued by the probability of a successful outcome) is greater than the costs involved when the probability of losing is taken into account. In formal terms, it is worth fighting for a resource only if:

$$p_{ij}(G_i - C_{ij}) > (1 - p_{ij})C_{ij} \qquad (10.1)$$

where p_{ij} is individual i's estimate of the probability that it will win a fight against individual j, $(1 - p_{ij})$ is the probability of losing, G_i is the value of the resource to i (i.e. its 'gain'), and C_{ij} is the cost to i of fighting j, with the benefit and cost being measured in the same units (usually either energy expended and gained or numbers of future offspring). Note that p_{ij} is a composite function of several factors. Obviously, one important component is likely to be the animal's size relative to its opponent's but this may be weighted by experience from previous fights with this individual in other contexts or by information gained from watching that individual in contests with third parties. In addition, where the resource being contested is another animal (a mate or a potential ally or even simply a grooming partner), then that individual's intimated preferences for one or other of the rivals will be an important factor.

By multiplying out all the brackets in equation 10.1 and adding terms, we obtain the simple rule: Escalate a fight if and only if

$$p_{ij}G_i > C_{ij} \qquad (10.2)$$

In other words, it is only worth an animal's while fighting for a resource when the value of the resource devalued by the probability of winning is greater than the costs expended in fighting.

Equation 10.2 suggests that an individual should only fight if either (1) the probability of winning is high or (2) the gain from winning is very high, or (3) the costs of fighting are very low. (For further discussion of the role of asymmetries in animal contests, see Maynard Smith and Parker 1976, Maynard Smith 1982.) If none of these conditions holds, then it is unlikely that there will be sufficient profit to be gained from fighting. Despite the fact that it is difficult to quantify all three of the component variables in equation 10.2, a number of observations confirm that asymmetries in any of them will play an important role in an animal's decision on whether or not to contest access to a given resource (see also Popp and DeVore 1979).

Two animals should *both* be willing to fight, for example, only when they are fairly evenly matched. An animal that is at a significant physical disadvantage will have a value of p_{ij} that is so low that fighting would be a serious waste of energy. That fights tend to occur predominantly between individuals of similar size and/or weight has been widely documented in mammals (sheep/goats: Geist 1971, Dunbar and Dunbar 1981; deer: Appleby 1983b; macaques: Datta 1983a). This is to be expected because the punch an animal can deliver depends almost entirely on its weight. Hence, in the first instance, we can expect size and weight to be the most important factors influencing the probability of winning a fight.

In contests over access to another animal, that animal's expressed preference for one of the contestants provides an important source of asymmetry that will alter each animal's estimate of winning. Clear evidence of this is provided by a series of experiments on hamadryas baboons. Kummer (1968) noticed that, in the wild, hamadryas males seem to respect each other's hegemony over the females in their respective units. To determine the proximate basis for this inhibition, Kummer *et al.* (1974) allowed a male (the 'rival') to observe another male and a female interacting in a compound. The rival was then released into the compound with the pair. Instead of attacking the paired male or attempting to kidnap the female, rivals invariably avoided all contact with the pair and would sit staring out through the mesh of the cage, often nervously fiddling with grass or stones. When the males' roles were reversed with a new female, exactly the same results were obtained: the former owner-turned-rival now avoided all contact with the pair and made no attempt to take the female from his rival-turned-owner. The males' physical abilities to dominate each other in fights (as determined from independent tests over food items) played no part at all.

In a later series of experiments, however, Bachmann and Kummer (1980) did find that the degree of asymmetry in relative dominance between the two males was important. If the rival was *very* much more dominant than the owner (and hence had a very good chance of forcing the owner away from the female), then the rival was more likely to attempt to fight for the female. On the other hand, if the two males were more evenly matched, the rival would attempt a takeover only if the bond between the female and the owner appeared to be weak (as indicated by the female's willingness to interact with the owner and to follow him around the compound). Thus, the rival seemed to be weighing up the likelihood of winning a fight in terms of (a) his own power relative to that of the owner *and* (b) the likelihood and the female would be prepared to desert the owner even if he did win. Smuts (1985) observed rather similar behaviour in olive baboons, a species that does not form harem-based reproductive units, although females may form special long-term relationships with individual males. Smuts found that in 14 of the 17 instances where a consorting male was displaced by a rival, the behaviour of the females concerned (in particular, their willingness to follow the male when he moved) suggested that she had lost interest in her current male consort.

Equation 10.2 also suggests that the value of the contested resource to an individual, G_i, is an important consideration. An animal that has just fed to satiation is not likely to contest access to another pile of food. A physically more powerful animal may therefore be willing to concede access to a resource that it does not particularly need. Lee (1983c) has shown that among juvenile vervet monkeys 'wins against the hierarchy' (i.e. encounters won by the animal that is normally the subordinate) are significantly more frequent in certain contexts than in others. In three different groups, the mean proportion of such reversals was 34% over access to social partners, 20% over access to resting places (e.g. in the shade), but only 6% over access to feeding sites. Clearly, in the semi-arid environment of the Amboseli habitat, access to food sources is likely to be significantly more important in terms of immediate survival than access to shade, so that animals are less willing to concede control over them. Similarly, at certain times of day, shade is probably more important than access to social partners.

Finally, a number of studies have shown that the amount of competition for access to a resource directly affects an animal's willingness to risk a fight for it. Janson (1985) found that, in a wild group of capuchin monkeys, dominance rank determined access to food resources when the amount of competition was high, but not when competition was low (in which case, the amount of aggression received tended to be more important). This suggests that low-ranking animals are more willing to contest access to a given resource when the high-ranking animals are less likely to insist on their priority of access because alternative sources of food are readily available nearby. Furuishi (1983) obtained similar results from a study of Japanese macaques: subordinate animals were more likely to infringe the personal space of high-ranking individuals when food sources were highly clumped in space. Finally, in a captive study of rhesus macaques, Belzung and Anderson (1986) found that the tendency for low-ranking animals to hang back (as measured by the latency before feeding) was significantly lower and the number of fights they became involved in significantly higher (a) when food was dispersed rather than clumped and (b) when they were competing for preferred foods (Figure 10.1). In other words, low-ranking animals were more willing to risk a fight if higher ranking animals had other food sources to go to (i.e. less asymmetry in p_{ij}) or if the food was a highly prized one (i.e. high G_i).

Determinants of RHP
Asymmetries in the probability of winning a dyadic interaction (i.e. in its Resource Holding Potential or RHP) arise from two main sources. First, an animal's intrinsic power is primarily related, in the first instance, to its physical size. Power consequently tends to follow an inverted-J-shaped curve when plotted against age, its peak corresponding to the peak in physical

Figure 10.1. The willingness of high- and low-ranking female rhesus monkeys to contest access to a food source (as measured by their latency to approach to the food when it becomes available) is strongly influenced both by the extent to which the food is clumped (squares) or dispersed (circles) and by their preference for the actual food items

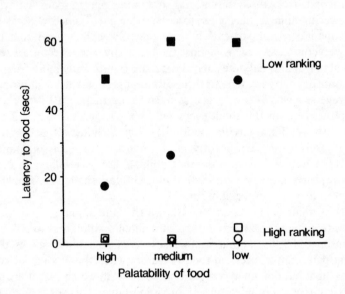

Source: Belzung and Anderson (1986, Table 1).

condition and fighting ability that occurs during an animal's prime. Animals, however, differ significantly in their growth planes and physical size, so that each individual will have its own power curve. An individual's growth plane is determined partly by its genotype (in humans, 5–40% of the variance in adult body weight has been attributed to genotypic effects: Falconer 1960, Cavalli-Sforza and Bodmer 1971) and partly by environmental conditions experienced during critical periods during development (Dobbing 1976). An animal that is caught during such a sensitive period in a situation where its mother is unable to provide it with adequate quantities of food will be forced into a lower growth plane from which it cannot later escape even if food is superabundant.

The second source of asymmetries in RHP is an animal's past experiences. An animal that spends all its time at the bottom of the hierarchy in a group of particularly aggressive individuals may remain at the bottom of the hierarchy if it transfers to a new group, even though the members of its new group are actually less aggressive. Ginsburg and Allee (1942), for example, demonstrated experimentally that dominant rats which then experienced defeat subsequently behaved submissively towards those same individuals over whom they had previously been dominant even though the behaviour of those individuals was unchanged. Similarly, although hamadryas baboon females normally compete for proximity to their harem-holder, Sigg (1980) was able to use mild electric shock to train the dominant female to allow the subordinate female to stay between her and the male. This kind of effect probably explains Rowell's (1966b, 1974) observation that it is subordinate monkeys that often seem to be responsible for perpetuating dominance relationships by their submissive behaviour towards higher ranking individuals. Presumably, they do so only because they have experienced the dominant animal's ability to win fights or are able to infer the likely consequences on the basis of cues which they have learned are indicative of high power in an animal.

An individual's rank within a group does not depend only on its intrinsic age-specific power, of course: it also depends on how its own power relates to that of other individuals in its group. This will depend both on the power trajectories of the particular individuals in the group and on their pitch (or elevation) relative to each other. In principle, individuals will rise and fall in the hierarchy as their position on their power trajectory crosses those of other individuals in the group (Figure 10.2). This usually means that an individual's rank will rise and then fall over the course of its lifetime such that it occupies higher ranks during the years of its physical prime and lower ranks during the

Figure 10.2. Hypothetical example of the way in which the power trajectories of individual animals change with respect to each other over time. At different times, the rank orders of the animals in a group are determined by each individual's relative power, not by its absolute position on its own power trajectory. Not only may individuals differ in the relative levels of their power planes (e.g. individuals A and B), but an individual's own status may change as a result of the maturation of other individuals (e.g. individual A at times t₁ and t₂)

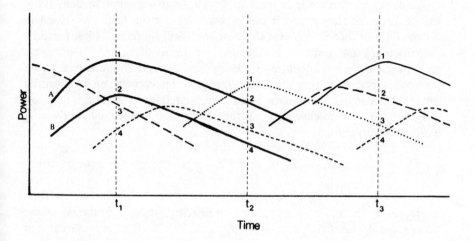

early and late parts of its reproductive career. Its problem thus amounts to finding strategies that will (1) allow it to achieve the highest possible rank during its peak in power, and (2) minimise the rate of decline as it moves into old age. These strategies are the subject of the next section.

Economics of Coalition Formation

Given that low-ranking individuals face severe disadvantages in terms of access to resources essential for survival and reproduction, we can expect such individuals to attempt to circumvent these costs in some way. If it cannot emigrate (see p. 81), then what options does it have? Among primates, the most profitable (and most common) strategy for solving the problems of low dominance rank is to form coalitions with other individuals. By teaming up, two individuals may be able to overpower a higher ranking animal (or at least minimise the loss of priority caused by low rank). The functional value of coalitions in this respect has been amply confirmed in a wide variety of species (baboons: Hall and DeVore 1965, Cheney 1977, Smuts 1985; vervets: Fairbanks 1980, Seyfarth and Cheney 1984, Fairbanks and McGuire 1986; gelada: Dunbar 1984a; macaques: Kawai 1958, Silk 1982, Datta 1983b). The important consideration in the present context is the decision an animal makes about whom to form an alliance with. Choosing the right ally can obviously make the difference between a successful coalition and one that provides little benefit.

The value of any coalition to a given individual depends on its own rank since the payoff gained from forming a coalition will depend on the number of ranks by which the coalition improves the animal's status. A coalition with a high-ranking partner will have less effect on the rank of an individual that is itself high-ranking than on one that is of relatively low rank.

We can define the value or *utility* of a coalition to a given individual as the gain in inclusive fitness that it can expect from a given coalition. This will depend on four factors: the cost to the animal itself (in terms of lost personal reproduction), the gain to the recipient (in terms of additional offspring gained), the genetic relatedness of the ally, and the gain in the animal's own personal reproduction resulting from subsequent intervention on its behalf by the ally. We can use the concept of future reproductive expectation (equation 4.2, p. 58) to write an expression for the utility, U_{ij}, to individual i of forming a coalition with individual j:

$$U_{ij} = ({}_{ij}R_x - {}_{i\cdot}R_x) + \sum_{j \neq i} r_{ij}({}_{ji}R_y - {}_{j\cdot}R_y) \tag{10.3}$$

where ${}_{i\cdot}R_x$ is i's current age-specific future reproductive expectation at the time of the decision (i.e. at age x), ${}_{ij}R_x$ is its reproductive expectation if it forms a coalition with j (with j's reproductive expectations being defined similarly, y being j's age at the time of the decision) and r_{ij} is the coefficient of relationship between i and j. Note that the term $_{ij}R_x$ includes both the gains and losses incurred by i when forming an alliance with j. The second element on the right-hand side of equation 10.3 is summed across all other individuals in the group because when i forms a coalition with j it cannot necessarily also form one with k, l, m, etc., and these individual's reproductive expectations may suffer in consequence: if i's support allows j to occupy a higher rank than it would otherwise have done, for example, k, l and m will all necessarily occupy lower ones. Equation 10.3 essentially says that the value of a coalition with j depends on the gain (or loss) that i incurs in its own personal reproduction plus the summed relatedness-weighted gains and losses incurred by all of i's relatives if it forms a coalition with j. Since the coefficient of relatedness, r_{ij}, decreases exponentially with each successive remove in a pedigree (Table 2.1), individuals who are more distantly related than second cousins will contribute negligibly to each other's inclusive fitness and can, to all intents and purposes, be ignored.

Given the definition of U_{ij}, an animal should form a coalition with another individual j only if:

$$U_{ij} > 0$$

Note that if:

$$(_{ij}R_x - {_i}.R_x) = \Delta_i.R_x < 0$$

(where $\Delta_i.R_x$ signifies the change in i's future reproductive expectations brought about by forming a coalition), then i should form a coalition with j only if:

$$r_{ij}(\Delta_j.R_y) > \Delta_i.R_x$$

thus confirming Hamilton's Rule for the spread of altruistic behaviour by kin selection (equation 2.1, p. 18). Although this looks a complicated way of expressing Hamilton's Rule, it has the advantage that the kinds of mistakes pointed out by Grafen (1984) are less likely if we conceive it in this way.

More generally, i should prefer a coalition with j over one with k only if:

$$U_{ij} > U_{ik}$$

Since all individuals other than i, j and k are equally affected by i's decision, it follows from equation 10.3 that i should prefer j over k if:

$$_{ij}R_x - {_{ik}}R_x > r_{ij}(_j.R_y - {_{ji}}R_y) - r_{ik}(_k.R_y - {_{ki}}R_y) \qquad (10.4)$$

Inequality 10.4 generates a number of different predictions. Some of the more obvious are:

1. when i is related to neither j nor k (i.e. $r_{ij} = r_{ik} \simeq 0$), i should choose whichever individual has the greatest effect on its own reproductive expectations;
2. when i is equally related to j and k (i.e. $r_{ij} = r_{ik} > 0$), i should choose:
    (i) whichever provides it with the greatest gain in its own personal reproductive output, or
    (ii) if $_{ij}R_x = {}_{ik}R_{x}$, whichever individual gains most from i's support;
3. when I is more closely related to one of them (i.e. $r_{ij} \neq r_{ik}$), i should choose:
    (i) whichever yields the greatest gain in its own personal reproductive expectations, or
    (ii) if $_{ij}R_x = {}_{ik}R_x$ and $\Delta(_{j}.R_y) = \Delta(_{k}.R_y)$, whichever has the highest coefficient of relationship to i.

These predictions emphasise the fact that we need to know how each of the component variables is affected in a given case, and that attempts to test Hamilton's Rule or any other similar predictions using data on only a single variable are unlikely to be meaningful no matter how clearcut they are. Although this makes it difficult to find data adequate to test the predictions listed above, none the less we can adduce some evidence in some or most of them.

A number of studies summarised by Seyfarth (1983) confirm prediction (1): individuals do prefer the higher ranking of two unrelated individuals (see below, p. 224). Data from gelada confirm prediction (2i): given a choice of two equally related individuals (daughters), females show a significant preference for forming alliances with the one that has the greatest impact on their own reproductive outputs (in this case, the older of two daughters: Dunbar 1984a). Datta (1981) also found that, in rhesus macaques, animals exhibited a marked preference for the more dominant (and therefore more valuable ally) of two equally related individuals. (Note that species may differ in the effect that a daughter can have on the rank of the mother: in the gelada, a mother's rank, once she is past her prime, depends on the coalitionary support of her daughters whereas in macaques the mother's rank is apparently unaffected by anything the daughters can do: see below.) Evidence in support of prediction (2ii) is provided by the many studies which show that a mother always supports the daughter that gains the most in terms of reproductive output, this often being the younger one (macaques: Sade 1967, Missakian 1972, Koyama 1967; gorilla: Harcourt and Stewart 1987; vervets: Horrocks and Hunte 1983b). Finally, Datta (1983b) found that low-ranking rhesus macaques were more likely to 'betray' their close relatives by supporting a less closely related opponent than were higher ranking individuals: given that coalitionary support from a high-ranking individual will have a greater impact on the status of a low-ranking individual than on that of a higher ranking one, this can be interpreted as evidence in support of prediction (3i).

It has been suggested that an animal should behave altruistically (e.g. by giving coalitionary aid) towards its various relatives in direct proportion to their coefficients of relatedness to it (e.g. Kurland 1977, Massey 1977, Kaplan 1978). As Altmann (1979) has pointed out, this assumption is false because it commits the common gamblers' fallacy of distributing stakes in proportion to the odds. Rather, if the benefits of an altruistic act are a constant function of investment, then an animal should invest all its available support in its nearest relative. However, if the return is a diminishing function of investment, as will often be the case, then it should invest all its effort in that individual which gives the best return after being devalued by the coefficient of relatedness. Reiss (1984) has shown that this must be so mathematically, and the same point emerges from equation 10.4.

The age of a prospective coalition partner may be an important consideration in any such decision. Because of the way fecundity varies with age, a rise in rank of the same magnitude can have a very different effect on the reproductive expectations of females of different age (Figure 10.3). If rank has less influence on fecundity as a female ages, then a change from, say, rank five to

Figure 10.3. Hypothetical example of the effect of rank on lifetime fecundity schedules. The schedules for individuals of high, medium and low rank are graphed against age. Because rank affects fecundity in different ways at different stages of the female's lifespan, an increase in rank will have a much greater effect (and thus be more valuable and hence more attractive) at certain ages (e.g. t_2) than at others (e.g. t_1 and t_3)

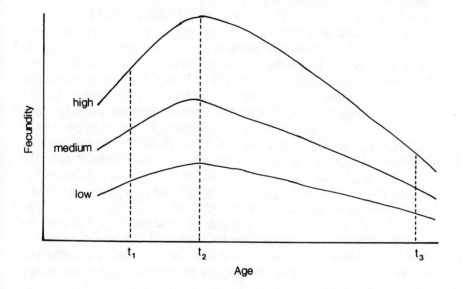

rank two will have a much greater impact on the reproductive expectations of a young female than on those of an old one. Hence, with a choice between two females of equal relatedness (say, older and younger daughters or sisters), a female should prefer to form a coalition with the younger whenever her choice has little impact on her own reproductive output. The tendency for mothers to support their younger daughters against their older ones (macaques: Schulman and Chapais 1980; vervets: Horrocks and Hunte 1983b) is some evidence in support of this prediction. Note, however, that the converse may also be the case: in some species, a coalition can have a greater impact on the reproductive expectations of an *older* female (gelada are one example), so that equation 10.4 would predict the converse behaviour (as in fact is the case: Dunbar 1984a). This simply emphasises the fact that we have to understand the detailed dynamics of a particular system before we can make any predictions about what animals should do.

Another important asymmetry arises from the difference in ranks between two prospective coalition partners. A coalition will have a relatively small effect on the rank of a high-ranking individual (thus yielding only a small gain in reproductive performance), whereas a low-ranking ally will gain a much larger change in rank. Consequently, a low-ranking animal will have a much greater interest in a given coalition than its high-ranking partner will. High-ranking animals can therefore expect to gain primarily through improvements in the fitness of relatives. Low-ranking individuals, in contrast, will gain most by forming coalitions with high-ranking individuals (irrespective of their degree of relatedness) in order to boost their own ranks and will have least interest in coalitions with relatives since these will also tend to be low-ranking. Cheney (1983) found precisely such a contrast in the strategies of high- and low-ranking vervet monkeys.

So far, rather few attempts have been made to cost out the lifetime reproductive consequences of forming coalitions with different individuals. In an analysis of coalition formation and grooming relationships among female gelada, however, it was possible to show that females did select their partners in relation to the effect that forming a coalition with a given individual had on the female's inclusive fitness (Dunbar 1984a). Both the female's age and her birth rank significantly influenced the value of forming a coalition with her own mother and eldest daughter. Coalitions formed with a first-born daughter are very much more profitable for the mother because a first-born daughter is just reaching her peak in power at the critical moment when the mother's is beginning to decline, whereas later-born daughters reach their peaks too late to have a significant impact on the mother's rank (Figure 10.4). A coalition formed between a mother and daughter also has the advantage that the female gains additionally through kin selection. Approximately half of the daughter's estimated gain in inclusive fitness is due to the increase in the fitness of collateral relatives for coalitions formed between a mother and her first-born daughter, but the proportion declines significantly as the

Figure 10.4. Hypothetical power trajectories for a female (solid line) and her three successive daughters, these being her first-born, third-born and fifth-born offspring. Only the first-born daughter has sufficient power to have a significant effect on her mother's rate of decline of power in her old age. The later in life that a female produces her first daughter, the less value that daughter will be to her as a ally

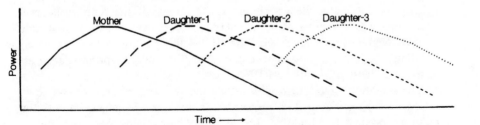

daughter's birth rank declines (Dunbar 1984a).

One of the most serious difficulties in attempting to cost out the reproductive value of different strategies lies in estimating the strategy-specific costs. Most studies have simply overlooked these, making instead a direct comparison of frequencies of coalitionary support with the ally's genetic relatedness. Sometimes, the costs of intervening on an ally's behalf can be indirect. Cheney and Seyfarth (1986), for example, found that vervet monkeys are significantly more likely to threaten a relative of an individual with whom they (or one of their own relatives) has recently been involved in a fight than they are likely to if no fight had occurred. In other words, being an ally may expose you to retaliatory action in the future, even in cases where support was not actually given during the fight that provoked the retaliation. Mere knowledge of past coalitionary associations may be enough to make the ally a legitimate target. Such 'hidden' costs would be very difficult to quantify in the field.

It is quite clear that animals do take these costs into consideration when deciding whether or not to go to another individual's aid. Datta (1983a), for example, found that the likelihood of a female rhesus macaque supporting a relative depended on the identity of the opponent. If the opponent was high-ranking (hence, a serious risk if the encounter escalated into a real fight, not to mention the prospect of intervention by the opponent's powerful family), coalition partners were reluctant to become involved. Similarly, a member of a low-ranking family was less likely to aid a relative if the opponent's mother was within 2 m of it (and hence close enough to intervene on its behalf) than if she was further away. Members of high-ranking families, in contrast, showed no such bias, partly because they were likely to be physically more powerful than their opponents and partly because they could call on larger numbers of more powerful relatives for support. Chapais (1983b) also found that among adult female rhesus an ally's relatedness could be a crucial factor determining the risk an individual was prepared to take in going to another animal's assistance in a fight. In all cases where a female supported an unrelated individual, the opponent was always lower ranking than at least one of the two allies. But when females gave support to a relative against an unrelated individual, they did so against both lower and higher ranking opponents. In other words, a female was prepared to accept a much higher risk of retaliation by the opponent (or its family) if the female she was supporting was a relative of hers. There is similar evidence to suggest that female longtailed macaques also weigh up the risk of retaliation when deciding whether or not to intervene on behalf of their offspring against an aggressor (Netto and van Hooff 1986).

The most serious problem with estimating the costs of a given course of action lies in translating the observed cost into units of genetic fitness (or offspring produced). In many cases, agonistic interactions may not result in significant observable injury to the animals involved. In the gelada, for example, a female was threatened, on average, only about 3.5 times a day: most of these involved passive displacements or a single threat given from a distance, and only 6% (one every five days) involved physical contact or a prolonged exchange of threats. Furuishi (1983) also found that the average number of agonistic encounters per female per day in wild (i.e. unprovisioned) Japanese macaques was only about 3.5 (see also A. Mori 1977b). Such low frequencies of interaction are hardly likely to generate serious direct costs in terms of injury. However, they may well impose significant costs in terms of stress, both to the aggressor and to the aggressee. Bowman et al. (1978) demonstrated experimentally in talapoins that being at the bottom of the hierarchy was enough to stress females so severely that the quantities of endogenous opiates they produced resulted in reproductive suppression. While persistent reproductive suppression is likely to be rare in wild animals, animals involved in a significant number of encounters, even if only as an ally, may incur a cost in terms of an increased risk of anovulatory oestrous cycles.

In some cases, it may be possible to estimate the gains from a given strategy in such a way that the costs are taken into account. Grafen (1982), for example, recommends that we compare the net reproductive outputs of individuals who behave in the way we are interested in with those of individuals who do not, since net lifetime reproductive output will automatically take both costs and benefits into account. Thus, in costing out the value of different potential allies in terms of the influence on the ranks held by a female gelada (and hence on her lifetime reproductive success), Dunbar (1984a) used the observed rank-specific birth rates since these already incorporated the costs of coalitionary support.

One last phenomenon needs a comment. Animals sometimes support an individual in a fight when that individual is already dominant to its opponent: why should they bother to do so when their intervention is unlikely to alter the outcome? There are three possible reasons why they might do so, none of which is mutually exclusive. The first, suggested by Datta (1983b), is that in some cases at least the dominant animal's allies will support it in order to reinforce its status, especially if there is any risk that the lower ranking opponent might be attempting a challenge for higher rank. Examples of such behaviour were noted in the Cayo Santiago rhesus population (Datta 1983b). A second possibility, suggested by Cheney (1977), is that the supporter's intentions are more directly selfish: by supporting a dominant animal against its opponent, a lower ranking animal itself records a 'win' against that opponent and this may help it in future contests with that same opponent. It may also gain a psychological advantage ('prestige'?) by being seen to be an ally of the dominant animal, which will presumably imply that the dominant animal may come to its support on a later occasion. Evidence that vervets may be able to recognise the existence of coalitionary relationships between third parties (Cheney and Seyfarth 1986) provides some justification for this claim. Making use of opportunities like this may be a particularly important strategy during the process of rank reversal between the ally and opponent. This certainly seemed to have been an important consideration in the gelada: a follower that made a habit of joining the females once they had forced their harem-holder to give way did in fact later defeat the harem-holder and take the unit over (Dunbar and Dunbar 1975). The third possible explanation, suggested by Silk and Boyd (1983), is that by supporting the aggressor in a fight, a female might be gaining a reproductive advantage by stressing the opponent even more and so increasing the likelihood that it will suffer reproductive suppression (see also Wasser 1983).

Demographic Considerations

In practice, an individual has two primary options when looking for suitable allies. One is to form coalitions with a relative; the other is to form a coalition with the highest ranking individual available. Obviously, in certain cases both criteria will refer to the same individual, but most animals will not be related to the dominant animal and will have to choose between the options. Each option has its own advantages and disadvantages. Relatives, for example, are often valuable because they are readily available, and familiarity provides an important basis on which to build the trust that is essential for a coalition to function effectively. Their primary disadvantage is that they may not themselves be especially high-ranking, so that any gain obtained by allying with them may be marginal. High-ranking individuals, on the other hand, are obviously premium allies in terms of their effectiveness, but this very fact makes them an object of competition among all the other members of the group. Hence, a vicious circle can develop in which the highest ranking animals are monopolised as allies by the very individuals that a low-ranking animal is seeking an ally against.

The relative value of each of these two options will depend on (1) the availability of relatives, (2) the age and rank of the individual concerned, and (3) the lifehistory characteristics of the species and their effect on lifetime rank trajectories. The availability of close relatives has one obvious effect on an individual's choice: if it has no living relatives, it will not have this option open to it. Fluctuations in the mortality and birth rates make this predicament particularly likely in populations living in poor to moderate habitats (Chapter 4). In addition, if 'relative' means a relative of the same sex, then the vagaries of statistical sampling effects will often result in an individual having no living relatives (see Table 6.1). An animal with no suitable relatives may have no option but to compete for an alliance with a high-ranking animal.

For females, males may constitute an additional source of allies, particularly in those species where males are normally significantly larger than females. The male's value to a female may, however, depend critically on his size. Packer and Pusey (1979) found that female Japanese macaques were significantly more aggressive towards the males of their groups than were baboon females, and they attributed this to the fact that macaques are less sexually dimorphic in body weight than *Papio* baboons (Table 10.1). Significantly, female gelada also behave very aggressively towards their males and commonly drive them away (Dunbar and Dunbar 1975): gelada fall within the macaque range on sexual dimorphism, well below the range for *Papio* baboons.

Table 10.1. Sexual dimorphism in body weight in baboons and macaques

Baboon species[a]	Male/female weight	Macaque species[b]	Male/female weight
Olive baboon	1.87	Bonnet macaque	1.59
Yellow baboon	1.86	Toque macaque	1.59
Hamadryas baboon	1.75	Longtailed macaque[c]	1.57
Chacma baboon	1.74	Rhesus macaque	1.39
		Pigtail macaque	1.33
		Japanese macaque	1.19
Gelada	1.58	Barbary macaque	1.12

[a] Data from various sources collated by Dunbar and Sharman (unpublished): sample sizes are 6,3,3,6 and 2 populations for each species, respectively.
[b] Source: Clutton-Brock and Harvey (1977).
[c] Source: van Shaik *et al.* (1983b).

This raises the possibility that female macaques (and gelada) may view male allies rather differently from female baboons. Since male baboons are very much bigger and more powerful than the females, female baboons may find males more valuable as allies than females are likely to be; conversely, macaque females may find males no better than other females and may therefore prefer to have female allies since these have a longer lifespan and can therefore form longer-lasting coalitions. In fact, both gelada and macaque females show a marked preference for coalitions with other (related) females, whereas 'special friendships' with individual males are a conspicuous feature of all *Papio* populations (see p. 176). It may also be the case that the ranks of female baboons may sometimes depend on support received from males (see Hall and DeVore 1965). This is never true in the gelada (Dunbar 1983d) and seems not to be the case in rhesus macaques (Ehardt and Bernstein 1986). A similar contrast seems to occur in colobus monkeys. Struhsaker and Leland (1985) noted that when a new dominant male red colobus began to behave infanticidally towards the group's infants, individual females developed close relationships with particular males, apparently in order to gain their protection against the dominant male. Female black-and-white colobus, on the other hand, do not form significant relationships with the males of their group even when they are placed under considerable stress by high levels of competition among the males (Dunbar and Dunbar 1976); they prefer to maintain close relationships with other females. Whereas red colobus are highly sexually dimorphic in body weight (males average 1.81 times the weight of females — within the range for *Papio* baboons), black-and-white colobus are significantly less so (males average only 1.19 times the weight of females, within the range for macaques: see Clutton-Brock and Harvey 1977).

This does not, of course, mean that all female baboons will prefer male allies, but rather that, relative to macaques, male baboons have a higher valency in these terms than females do. Consequently, the value of female relatives (say, in terms of the size of the matriline) has to be that much greater to outweigh the advantages of a coalition with a male. Much will therefore depend on the demographic structure of the population and on the species' reproductive characteristics. In populations with low growth rates where kinship groups are small, females will tend to prefer alliances with males, whereas in populations with high growth rates and large kinship groups they will prefer female relatives as allies.

So far, we have considered an individual's decisions on whom to form coalitions with as though they were in fact unconstrained by the prospective ally's own preferences. In fact, an individual living in a social group encounters two key constraints: (1) there will be conflicts of priority with other members of the group over access to the best allies, and (2) those preferred allies will themselves have preferences as to whom to form alliances with. These are largely a consequence of the fact that an animal's time budget is not infinitely flexible and that, while having only a limited amount of time available for developing and servicing relationships, each relationship requires the investment of a certain amount of time to ensure that it functions as an effective coalition. In primates, social grooming is the main process used in servicing relationships (see p. 251) and it is the time that has to be invested in grooming that creates these problems.

Seyfarth (1977, 1983) has used simple models to explore the consequences of these constraints. He assumed (1) that animals compete to form coalitions with the highest ranking individuals, (2) that a specific amount of grooming has to be invested in such a relationship in order for it to function as an effective coalition, and (3) that an animal has sufficient time available to groom with only 1–2 individuals to this criterion. He was able to show that while the freedom of choice of high-ranking animals is more or less unconstrained, that of low-ranking individuals is severely limited by the preferences of the higher ranking animals. High-ranking individuals are able to exploit their priority of access to the best partners and will thus tend to monopolise other high-ranking individuals. Because of this, animals will tend to end up forming coalitions (and hence grooming) predominantly with their immediate neighbours in the hierarchy. A second consequence is that grooming will tend to be predominantly up the hierarchy because individuals will always attempt to improve their alliance options by grooming high-ranking individuals when these are free. Seyfarth was able to show that there was a close fit between the patterns of grooming in a number of different species and the predictions of the models.

Seyfarth originally formulated his model to show that animals could spend most of their time grooming with kin, not because of kin selection (as had commonly been supposed) but because competition for access to the most powerful allies left them with no alternative. Unfortunately, of course, this begs the question as to why relatives should occupy adjacent ranks in the hierarchy. All other things being equal, animals will rank on the basis of their intrinsic power and, since this is age-dependent, relatives (being normally of different ages) will not be of similar rank. In fact, all the evidence generally points to the fact that relatives occupy adjacent ranks because animals prefer to give coalitionary support to their relatives rather than to unrelated individuals (see, for example, Kurland 1977, Massey 1977, Silk 1982, Dunbar 1984a).

To try to distinguish between a genuine preference for relatives and an apparent preference due to lack of choice is more difficult than one might suppose, for the two hypotheses yield essentially indistinguishable predictions. It is only in their assumptions and the mechanics of their processes that they differ. None the less, two lines of evidence suggest that when animals do have a choice, they prefer relatives to non-relatives. The first is that while some studies have provided evidence in support of Seyfarth's model, others have failed to do so (see Seyfarth 1983). It is conspicuous that most of the groups that support the model's predictions lack well developed kinship lineages either because they had recently been convened in captivity (gelada: Bramblett 1970, Kummer 1975; vervets: Fairbanks 1980; hamadryas baboons: Stammbach 1979) or were natural populations living under marginal conditions where high mortality and low birth rates are likely to reduce lineages to the minimum size (baboons: Seyfarth 1976; vervets:

Seyfarth 1980). In contrast, those studies that have failed to confirm the model's predictions all derive from captive or wild populations with high growth rates and extended kinship structures (rhesus macaques: Sade 1972; bonnet macaques: Silk *et al.* 1981b; gelada: Dunbar 1982c, 1984a). The second source of evidence concerns the contrasting preferences made by different individual animals within groups. In groups where kinship relationships are known, those individuals who have close relatives in the group invariably prefer to groom with them, whereas those that have no living relatives generally compete for access to the most dominant members of the group (e.g. gelada: Dunbar 1984a).

Seyfarth's rule thus seems to apply primarily to groups of unrelated animals; in groups that have well-developed kinship structures, it applies only to those individuals who have no close relatives. As suggested by the analyses on p. 100, individuals have a hierarchy of choices available to them. So while Seyfarth's rule is both correct and universally true of all species, like all such rules it is context-specific and we would not expect it to find expression in every conceivable case.

This raises a key question: why should animals prefer relatives when they have them available? There are probably two main reasons. One is that, as we have seen (p. 214), an animal forming a coalition with a relative gains an additional advantage through kin selection that it does not gain from a non-relative. This is an important security to fall back on as a way of minimising the lost investment that will occur if the ally dies or reneges before repaying the 'debt'. (But note the distinction here between the reasons why animals form coalitions and the reasons why, given that animals decide to form coalitions, they will prefer to form them with relatives. Kin selection has often mistakenly been interpreted as an explanation for the formation of coalitions: it almost certainly is not, though it may well be a reason why animals choose to form coalitions with relatives when they decide to form coalitions. There are two separate decisions being made here and it is important that we do not confuse them.) The second consideration that might prompt animals to prefer relatives is that, if a coalition has to be a long-term investment in order to show a significant return, then an animal has to be very confident of its ally's behaviour in the future. The familiarity generated between animals that are reared together is probably the most effective (and perhaps the only) basis on which such knowledge of another animal's behaviour can be obtained.

Colvin (1983b) has developed this idea[1] in an alternative to Seyfarth's model. He argued that individuals of similar rank (1) share a wider variety of needs in common (e.g. harassment by the same high-ranking individuals) and hence are more likely to be able to perform a greater variety of services for each other, (2) are likely to incur similar costs in providing those services (hence an asymmetry in costs is less likely to deter one member of a coalition from action at a critical moment), and (3) are likely to be more familiar with each other because individuals of the same family are more likely to occupy adjacent ranks (so that they will be more effective at predicting each other's future behaviour). As Colvin points out, different groups of monkey (or even cohorts within a group) may exhibit preferences for different rules: two cohorts of juveniles from the Cayo Santiago rhesus macaque colony, for example, showed a marked preference for partners of high rank (Seyfarth's model), while two other cohorts preferred partners of similar rank (Colvin's model). Unfortunately, Colvin's model also suffers from the fundamental flaw that it begs the question as to why relatives should hold adjacent ranks. His model was originally developed with juveniles in mind, and this may be a less serious difficulty in this case since juveniles' ranks are imposed on them by their families. But where adults are concerned, we cannot make this assumption since rank here is partly a *consequence* of expressed preferences. This difficulty is itself instructive because it underlines the importance of identifying the constraint-free strategy at the outset and then asking how animals that are forced into low ranks might try to overcome the resulting disadvantages.

One final consideration of more general relevance is that animals form coalitions for reasons of reciprocal altruism in which the debt is paid back more or less immediately (Trivers 1971). In this sense, coalition partners only aid each other so long as it is in their interests to do so and so long as their partner continues to do the same for them. Short-term coalitions based on reciprocal altruism in which each individual expects to be repaid in kind within a limited period have been reported for baboons (Packer 1977) and vervet monkeys (Seyfarth and Cheney 1984). Such coalitions need not involve the reciprocal exchange of the same commodity, however. Providing the fitness consequences are positive for both partners, animals may trade different commodities that have different values to each of them. Smuts (1985), for example, has suggested that special friendships formed between a male and a female baboon are based on an exchange of different benefits: the female gains protection for herself and her infant, while the male gains priority of sexual access to the female. In contrast, coalitions based solely on kin selection in which one partner gives aid without any expectation of repayment are likely to be rare except when the costs of doing so are small.

Acquisition of Rank

In this final section, I want to consider briefly some developmental aspects of dominance rank. In particular, I want to focus on the process of rank acquisition in females. Rank acquisition in males is fairly straightforward and depends primarily on the individual male's intrinsic power, especially once it has left its natal group (e.g. baboons: Lee and Oliver 1979, Johnson in press; macaques: Sade 1967). In contrast, the females of most primate species tend to remain throughout their lives in close association with their immediate female relatives and are consequently involved in a very much more complex social process in which the influence of relatives often plays a vital part.

Studies of many species have documented the fact that a female's initial rank as a juvenile and young adult depends primarily on the rank of her mother (macaques: Kawai 1958, Sade 1967, Missakian 1972; baboons: Cheney 1977, Lee and Oliver 1979, Walters 1980, Johnson in press; gelada: Dunbar 1980a; vervets: Horrocks and Hunte 1983a). By no means all species exhibit this tendency, however. Harcourt and Stewart (1987) found that although higher ranking relatives support young gorillas, a gorilla's rank is not determined by its mother's. They suggest that this is because the long interbirth intervals and small group size result in too great an imbalance in intrinsic power in any given pair-wise relationship for additional aid to make a significant difference. Similarly, in a small provisioned troop of Barbary macaques, the females ranked in strict order of age, irrespective of their matrilineal affinities (Fa 1986). Clarke and Glander (1984) found that young female immigrants into a group of howler monkeys were able to rely on their higher intrinsic power to dominate the older resident females.

In those cases where mothers do exert a significant effect on the ranks of their daughters, it has been customary to distinguish between species in which daughters can outrank their mothers once they are old enough (langurs: Hrdy and Hrdy 1976; gelada: Dunbar 1980a; baboons: Sigg 1980, Hausfater *et al.* 1982a; howler monkeys: Jones 1980) and those in which the mother continues to dominate her daughters throughout her lifetime (macaques: Koyama 1967, Sade 1967, Missakian 1972). The latter species are characterised by what is generally known as the 'inverse rule': daughters always rank in inverse order of age with the youngest highest.

There are several aspects of the 'macaque' system that are puzzling. Thus, no explanation has ever been offered as to why the matrilines should rank in the order they do, most observers simply being content to note that their current ranks are still those of the original matriarchs when the troop was first observed two decades or more before. There are clearly two different questions here: (1) why did the original matriarchs hold the respective ranks they did? and (2) why do their descendants still occupy the same ranks long after the deaths of these matriarchs? A second puzzle concerns the apparent willingness of older (and hence intrinsically more powerful) daughters to accept a rank below that of a younger sister when in doing so they must lose fitness.

Determinants of matriline rank

It is clear that in small groups matriline rank is dependent on the relative intrinsic power of the most dominant member, with matriline size being unimportant (e.g. Dunbar 1980a, Fa 1986). In the large groups of the provisioned macaque populations, on the other hand, there is a clear tendency for matriline size to correlate with matriline rank (e.g. A. Mori 1975, Sade *et al.* 1976). A possible explanation for the 'macaque' system might thus lie in the sheer size of their kinship groups. In small natural groups, including those of the macaques, kinship groups are typically small (1–3 adult females). In such cases, as Harcourt and Stewart (1987) point out, the asymmetries in intrinsic power are so great that a female can probably do little more than buffer her relatives against harassment by individuals that rank below her in the hierarchy. Only rarely will her kinship group (or family) be large enough for its size to outweigh the intrinsic power of a more dominant female. Consequently, matrilines will tend to be ranked in accordance with the relative intrinsic powers of their most dominant members, with each individual's relatives ranking immediately below her as a result of the rank support that she gives them. As the size of these kinship groups has grown following the *ad libitum* provisioning of the Japanese and rhesus macaque populations, the size of coalition that a family can muster has become more important. But since high-ranking matrilines have been able to grow faster, so they have been able to reinforce their dominance by sheer weight of numbers. Moreover, as groups grow in size, the power asymmetries between individual females will decrease because there will be more females of about the same age (and hence at similar points on their lifetime power trajectories), leaving coalition size as the only significant asymmetry between two matriarchs.

The relevance of family size in this respect has been stressed by Johnson (in press). She found that, in her Gilgil (Kenya) group of baboons, daughters invariably fell in rank along with their mother when the mother was no longer able to sustain her own rank in the group, whereas this was not always so in the Amboseli baboons studied by Hausfater *et al.* (1982a). Johnson points out that the higher growth rates at Gilgil produced relatively larger kinship groups than at Amboseli (Table 6.2, p. 100): consequently, Gilgil females could call on support from other family members in order to suppress a female's rank once the mother was no longer able to support her, whereas at Amboseli they were less often able to do so and instead had to rely more on one-on-one interactions.

One obvious consequence of this will be that as family size becomes increasingly important, hierarchies should become more stable: the dominant family will remain dominant because the fact of its dominance allows it to grow faster and thus maintain its advantage. In addition, its very size will make it less susceptible to the kinds of small-sample statistical effects that can unexpectedly reduce the size of a small kinship group following the death of a member or the birth of a succession of sons. This would explain the relative stability of matriline ranks in the large provisioned groups of macaques (A. Mori 1975, Sade *et al.* 1976) when compared with groups with lower reproductive rates and smaller kin group sizes (e.g. Hausfater *et al.* 1982a, Fa 1986, Johnson in press). There is evidence both from the field and the laboratory that large numbers of juvenile females maturing simultaneously can cause major disruption in the adult female hierarchy and precipitate reversals of matriline ranks (Altmann and Altmann 1979, Samuels and Henrickson 1983, Ehardt and Bernstein 1986[2]).

Ranks within families

The inverse rule itself poses a more serious problem since the apparent acquiescence of older daughters in the preferment of their younger sisters seems to imply genetic altruism. There are at least three possible explanations for this phenomenon: (1) the costs incurred by the older sister are sufficiently low that the loss of rank does not make a significant difference to her inclusive fitness (i.e. her acquiescence is a consequence of kin selection); (2) the mother manipulates the ranks of her daughters in order to maximise her own fitness by aiding a younger daughter against an older one because a younger daughter has a higher reproductive value (and hence will contribute more to the mother's fitness); and (3) the suppression of an older daughter is a consequence of a coalition formed between the mother and her younger daughter in order to prevent the older daughter dominating the mother (and, of course, the younger sister).

Since all populations in which the inverse rule has been recorded are heavily provisioned and growing rapidly, it is just possible that the consequences of a loss of one or two rank places has a negligible effect on the reproductive output of an older daughter. If so, some explanation is necessary to explain why mothers then bother to aid either of their daughters since, if the loss to the older daughter is small, then the loss to the younger daughter of remaining in a lower rank would also be small. One possible explanation is that younger daughters are more vulnerable than older daughters (Horrocks and Hunte 1983b; see also Fairbanks and McGuire 1986, Harcourt and Stewart 1987). While this is a plausible explanation as to why a mother should support a juvenile daughter against an older daughter, it is not clear why she should then continue to do so even after the younger daughter is herself physically mature. Perhaps the most telling point against this hypothesis, however, is the fact that the older daughter's altruism is more apparent than real: detailed observations of rank reversals between siblings show that older sisters strongly resist their younger sibling's rise in rank (macaques: Datta 1981; baboons: Walters 1980).

The second possibility has been advanced by Schulman and Chapais (1980) with particular reference to the Cayo Santiago rhesus macaques. Their argument is based on the assumption that a mother will support whichever daughter has the highest reproductive value (see p. 58) at any given moment (see also Chapais and Schulman 1980). By conferring the higher rank on the daughter with the higher reproductive value, a mother will be able to maximise her own fitness. This can be seen graphically in Figure 10.5. From an older sister's point of view, it will be in her interests also to accept this reversal of rank with a younger sister providing:

$$V_i / V_j > r_{ij}$$

where V_i is the reproductive value of the older sister, V_j that of the younger sister, and r_{ij} is the coefficient of relationship between them. The logic of this argument, however, implies that a mother ought to allow her youngest daughter to surpass her in rank once her own reproductive value passes its peak (at about the age of 11 years). Since this does not happen, Chapais and Schulman (1980) introduce a further modification that allows a mother to include in her assessment the effect of her rank support on all her descendants. This yields a more complex equation which, given the lifehistory characteristics of the Cayo Santiago rhesus, does make it worth the mother's while remaining dominant over all her daughters throughout her life. Part of the reason for this is that by remaining dominant the mother has a higher probability of producing more offspring herself, and, since these are more

Figure 10.5. The Schulman–Chapais model of reproductive value and its influence on the relative ranks of macaque sisters. Because of the shape of the reproductive value curve, two sisters will always differ in their relative reproductive value. Initially (e.g. at time t_1), the older sister (curve 1) has a higher reproductive value than her younger sister (curve 2), but once past the peak in her own reproductive value she is overtaken by her younger sister so that the younger sister will thereafter always have a higher reproductive value (e.g. at time t_2). The Schulman-Chapais model suggests that the mother should favour whichever daughter has the higher reproductive value at any given moment, and that the older daughter should tolerate being outranked by her younger sibling providing the gain ratio gained from the younger sister's higher reproductive value is greater than the coefficient of relationship between them (see text for further details)

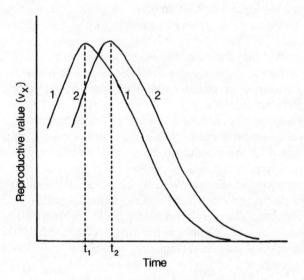

closely related to her than the offspring of her daughters (Table 2.1), it pays her to remain dominant.

The main difficulty with this explanation is that it begs the question as to how the mother is able to remain dominant over her daughters for so long despite her own declining physical powers. As the mother's power is eroded by increasing age, it hardly seems plausible to fall back on longstanding habit as the reason why a daughter in her physical prime should continue to defer to her mother when it must be in her interests to reverse ranks with her. For some reason, this curious paradox has remained unchallenged throughout the literature on this topic.

The third candidate explanation attempts to overcome this difficulty to some extent. Horrocks and Hunte (1983b) suggested that it is in the mother's interests to remain dominant over her daughters for as long as she can in order to produce more offspring of her own. Her primary concern should therefore be to form coalitions that enable her to prevent any of her daughters displacing her. To see how this might work out, we need to know just how the power trajectories of the mother and her successive daughters relate to each other over the course of the mother's life. Dunbar (1984a) gives a general formula for calculating the mean age at which a mother produces her first daughter, B_f, namely:

$$B_f = B_0 + \sum_{i=1}^{k} 0.5^i m(i-1)$$

where B_0 is the age at which she produces her first offspring of either sex (4 years old in this case), i is the offspring's rank in the mother's birth order (with a maximum of k infants born during her lifetime) and m is the interbirth interval (here 1.0 year). On this basis, the average Cayo Santiago female can expect to produce her first daughter at the age of $B_f = 5.0$ years. Successive daughters will then be produced at intervals of:

$$M_f = \sum_{i=1}^{k} 0.5^i mi$$

In this case, successive daughters will be born, on average, at $M_f = 2.0$ year intervals.

Figure 10.6 plots the respective power trajectories of the mother and each of her successive daughters, based on the power trajectories of the gelada (see Dunbar 1984a). If power is additive (such that the power of a coalition is essentially the sum of the individual power values of its members), then it is clear that a mother will always be able to outrank her older daughters by forming a coalition with the youngest one, at least until quite late in life. Obviously, it will be in the youngest daughter's interests to encourage her mother in this because she gains at the expense of her older sister(s). As with the mother, her own offspring are worth more to her inclusive fitness than the offspring of a sister.

Figure 10.6. Relationships between the power curves of a mother (solid line) and her successive daughters (broken curves) in a population with the reproductive characteristics of the Cayo Santiago rhesus macaques (i.e. life expectancy $e_x = 14$ years for a female reaching sexual maturity, age at first reproduction of 4 years, mean age at the birth of the first daughter of 5 years, mean interval between successive daughters of 2 years). In (a), the heavy lines indicating each individual's power trajectories are plotted in real time; the power curves themselves are notional (but based on those of the gelada). The narrow lines above give the combined power of the mother and the daughter of the same line type if they form a coalition against the older daughter(s), assuming that power is additive. The mother's share of the gains from ranks 1 and 2 is plotted in (b) for the four options: mother acting alone on her own intrinsic power (solid line), mother acting in coalition either with her oldest daughter (long dashed line) or with her second-born daughter (short dashed line) or with her youngest daughter (dotted line). The mother's share of the gain from ranks 1 and 2 is assumed to be proportional to the ratio of her own power to that of the daughter with whom she forms a coalition. The mother will maximise her share of the gain from ranks 1 and 2 by opportunistically forming a coalition with the youngest daughter at any given moment

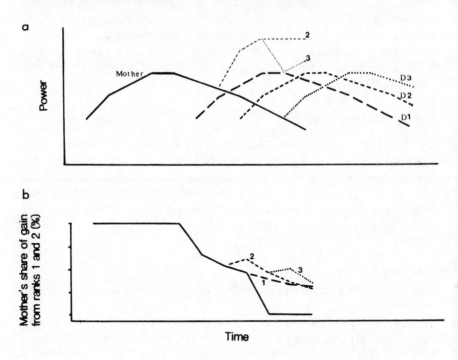

I am not sure that the evidence at present available is sufficient to allow us to distinguish conclusively between these three possible explanations. None the less, there is considerable circumstantial evidence to support the third hypothesis. Chapais and Schulman (1980) themselves noted both that younger daughters often provoke fights with their elder sisters (thereby precipitating attacks on them by the mother) and that these attacks build up in frequency as the younger sister approaches puberty (the time at which rank reversals normally occur). Datta (1983a) noted similar behaviour in another of the Cayo Santiago rhesus groups, and Walters (1980) observed much the same in baboons. In addition, when the inverse rule does fail in populations that normally operate it, the reason invariably lies either in the female having only a single daughter (and hence lacking an ally she could use to control that daughter's rise in rank: see Missakian 1972, Silk *et al.* 1981c) or in the mother's death resulting in the younger daughter having no one to support her against her older sister (e.g. Datta 1981, Furuishi 1983). Harcourt and Stewart (1987) point out that an inverse rule can emerge as a natural outgrowth of a tendency for the mother to support the more vulnerable of her daughters (invariably the younger): where such support is given consistently, it may encourage a younger sister to challenge an older sibling with the result that, providing the size asymmetry between them is not too great, it is able to dominate it.

There is, none the less, presumably a limit to the mother's ability to pursue this strategy. Once her own powers begin to decline, she will reach a point where she is unable to prevent her youngest daughter outranking her. Alternatively, there will come a point at which no coalition formed with the youngest daughter will be sufficiently powerful to outrank the combined power of the two older daughters. She would then either have to switch allegiance back to an older daughter or accept a loss of rank as inevitable. Presumably, any attempt to switch allegiance would be fiercely resisted both by the youngest daughter and by the older daughter's ally. In either case, we would expect far more rank reversals as the mother ages than we actually find in practice. There are two quite separate issues here, each requiring a different explanation, namely (1) why should the youngest daughter continue to allow the mother to dominate her? and (2) how can the two of them continue to dominate the older daughter(s)?

I can see only one reason why the youngest daughter should be prepared to allow the mother to continue to outrank her, and this is that the mother is able to hold her to ransom by threatening to withdraw her support if the daughter tries to outrank her. Such 'alliance fickleness' (Nishida 1983) has been noted in both wild and captive chimpanzees, albeit in the context of alliances between males (see below, p. 245). In both cases, the lower ranking ally was able to gain privileged access to restricted resources (oestrous females) by switching his support to the intrinsically more powerful rival whenever his coalition partner tried to prevent him doing so (de Waal 1982, Nishida 1983). Whereas such a strategy works well in a triangular context, it is a much less effective strategy where four individuals are involved. In such a case, the dominant alliance partner can always call its ally's bluff and form a new coalition with one of the other two individuals. An ageing matriarch is likely to be in a weak position in such a situation because the youngest daughter might be able to form an even more powerful coalition with one of her older sisters and still be dominant. Current explanations of coalition behaviour and rank determination in these species are clearly unable to account for these apsects of the situation, and further investigation is called for.

It is easier to see how a mother and her youngest daughter might continue to be able to outrank two older daughters despite the mother's declining intrinsic power. There are at least two possibilities. One is that in large families the older daughter's preoccupations shift from forming coalitions with her sisters to forming coalitions with her own daughters in order to maximise their chances of producing offspring since a daughter will always be more valuable than a sister's offspring. Figure 6.2 indicates that the oldest daughter's oldest daughter will, on average, reach maturity at about the same time that the mother's third-born daughter does. If females are limited to forming a coalition with just one female at a time, this would have the effect of dividing the opposition at a time when the youngest daughter has no such conflicts of interest because her own daughters are still immature. The other possibility is that the power of a coalition is not additive; rather, two animals acting together can exert more force than each of them brings to the alliance on her own. This extra multiplicative component may allow the mother and her youngest daughter to drive a wedge between the two older daughters to prevent them from actively allying against them.

So far, we have considered only species that operate the inverse rule. The same considerations might, however, also explain why the inverse rule should be rare in more slowly reproducing species like those of the Amboseli baboons and the gelada where family sizes tend to be small (see Tables 6.1 and 6.2). Figure 10.7 shows that, as the interbirth interval approaches the 2 years typical of the gelada, the likelihood of the mother having a younger daughter with whom to form a viable coalition against an older daughter is very much lower. By the time the second daughter is old enough to constitute an effective ally, the mother's power is so far into decline that the two together are unable to match the power of the older daughter. The long age gap between successive daughters (3.5 years on average in the Sankaber gelada population: see Dunbar 1984a) introduces a marked asymmetry in power between the two daughters that does not occur in a more rapidly reproducing population like the Cayo Santiago rhesus where successive daughters are much closer in age. In such circumstances, a mother should prefer to form a coalition with her older daughter since the younger is less likely to reach an age at which she can be an effective ally against an older sister during the mother's lifetime. One obvious consequence of this is that the mother will be unable to prevent her oldest daughter outranking her once their power trajectories cross over. Both these predictions are confirmed by observations for gelada (Dunbar 1984a). Chikazawa et al. (1979) provide further evidence to support this suggestion. In a small captive group of rhesus macaques, all nine of the first-born daughters outranked their mothers when they reached maturity. This group

Figure 10.7. (a) Relationship between the power curves of a mother (solid line) and her successive daughters (broken lines) in a population with the reproductive characteristics of the Simen gelada (i.e. life expectancy $e_x - 14$ years, age at first reproduction of 4 years, mean age at the birth of the first daughter of 5.75 years, mean interval between successive daughters of 3.5 years). (b) The mother's share of the gains from ranks 1 and 2 for the three possible options: mother acting on her own intrinsic power (solid line) and mother acting in a coalition with either of her two daughters (broken lines). Conventions as for Figure 10.6. In this case, the mother gains marginally from forming a coalition with her older daughter

a

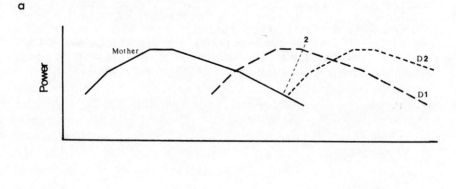

b

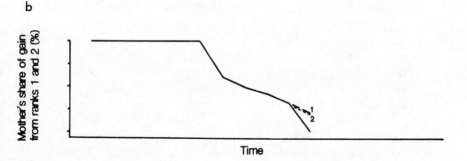

had been together for only 8 years, so that there had not been sufficient time for large matrilines to build up.

Taken together, these observations suggest that the same set of rules can generate radically different predictions in different demographic contexts. In other words, qualitative shifts in the preferred social strategy can owe their origins to ecologically driven differences in lifehistory processes. Once again. this should caution us against searching for universal rules that always find expression in the same form in all species (or even all populations of the same species).

Note

1. De Waal and Luttrell (1986) have recently developed a more complex version of this hypothesis. They argue for the pre-eminence of similarity of interests as the primary factor promoting alliance formation. Similarity can arise through membership of the same genetic family or social 'class' and similarity in age and/or dominance rank.

2. Data given by Ehardt and Bernstein (1986) for four episodes of inter-matriline rank reversal in a captive group of rhesus macaques suggest that rank reversals occurred either (1) following the maturation of several juvenile females at a time when all matrilines contained only a single adult female (1976), or (2) in years when a large number of juveniles matured (1982 and twice in 1983). The mean number of juvenile females maturing in years when a reversal occurred was 4.0 ($n=3$) compared with only 2.3 ($n=6$) in years when no reversals occurred (Mann–Whitney test, $z=1.896$, $p=0.058$ 2-tailed). If the two episodes in 1983 count as separate trials, then $z=2.207$, $n_1=4$, $n_2=6$, $p=0.038$ 1-tailed. This suggests that reversals occur when the existing asymmetries in power are altered by the maturation of immature females.

Chapter 11
Mechanics of Exploitation

Relationships develop as a result of animals interacting with each other. It is, therefore, in the detailed behavioural content of those interactions that we need to look in order to find out how animals prevail on other individuals to behave in the ways they want them to. The key to this lies in communication and in the processes of negotiation. This highlights, in many ways, the essentially political nature of primate societies and draws attention to the fact that primates are particularly intelligent, even by mammalian standards. A number of authors have argued that the super-intelligence of primates evolved not for ecological problem-solving (as has conventionally been assumed) but to enable the animals to develop (and negotiate their way through) complex networks of social relationships (Jolly 1966b, Humphrey 1976, Kummer 1982, de Waal 1982, Cheney and Seyfarth 1985a, Harcourt in press).

In this chapter, I first explore in some detail the behavioural mechanisms that are used to mediate social relationships. In doing so, I shall stress the essentially dynamic nature of social relationships: they exist to serve specific purposes in the animals' reproductive strategies, and when they have ceased to serve those ends they may be modified or even abandoned. Finally, I shall examine a number of cases in which one individual seeks to exploit another individual for its own personal gain.

Processes of Negotiation

An animal's social strategies are essentially investments made to solve problems of reproductive significance on a long-term basis. Temporary coalitions can solve problems of immediate urgency, but the permanent nature of primate societies suggests that the animals need a more stable basis from which to pursue their reproductive strategies. For most primates, social relationships are more than passing interactions: they need to be negotiated and, once established, serviced and maintained through time.

Negotiating relationships

A relationship will rarely be of exactly equal value to both the individuals involved, particularly in the initial stages. This asymmetry of a relationship's utility will usually mean that one member of the pair will be responsible for initiating most interactions because it has a greater interest in promoting the relationship. Smuts (1985), for example, found that special friendships between male and female baboons are invariably initiated by the male: he opens the negotiations by inviting grooming with the female. This is also true of gelada, where males seeking to take over a reproductive unit concentrate on soliciting interactions with individual females. They do this by peering closely at the female while lipsmacking, contact calling and giving what Smuts (1985) has termed the 'come-hither' face (a signal that combines elements of both submissiveness and invitation).

The problem faced by prospective social partners (or allies) stems from the risk that each runs of being exploited if the other reneges on repaying a service rendered at some cost in terms of the actor's own reproductive output. Initial interactions are thus often ambivalent and contain both positive and negative motivational overtones. Opening gambits in a developing relationship therefore invariably involve expressions of both submission and appeasement (as in lipsmacking and the flattening of the ears against the skull in baboons) because any approach by an animal that is not a regular social partner is potentially threatening until the animal's intentions are declared.

Kummer (1975) has suggested that the process of bonding passes through a series of stages at which the interactants reach what are essentially points of agreement. In order of increasing intensity these are: fighting, presenting, mounting and grooming. In order to reach the highest stage of bonding (grooming), an interacting pair has to pass through each of the preceding stages. While it is possible for a well-meshed pair to miss out one of the lower stages, animals never go straight to grooming or groom before the first present takes place. Each stage is the outcome of a process of mutual assessment during which the animals assess the value of the prospective relationship. Kummer's stages correspond, in effect, to levels of mutual acceptance and trust that provide an opportunity where withdrawal is still possible before too much has been committed.

Bonding depends on two key processes. One is the assessment of a prospective ally's probable intentions and power. In fact, assessment of an opponent's power is an important component in the decision processes dictating an animal's behaviour at many different levels (Parker 1974a, Maynard Smith 1982). In the context of coalition formation, animals need to be able to determine just how powerful a prospective ally is in order to know whether it is worth investing in an alliance with that particular individual. Cheney (1977) has presented evidence suggesting that juvenile baboons can identify which individuals will make the most powerful allies. The second process is what de Waal (1982) has termed *reconciliation*. In any relationship, an individual constantly teeters on a fine divide between exploiting and being exploited by its partner. This inherent instability creates a tension that, if unchecked, will eventually result in the dissolution of the relationship as the costs sustained by the more exploited partner exceed its gains by too great a margin. Because reconciliation forms so important a part of social interaction, it often involves highly ritualised forms of behaviour. In baboons, gelada, macaques and chimpanzees, males commonly engage in formalised 'greeting' ceremonies in which one animal presents its rear to the other, who will then either mount the presenter or reach through to touch its genitals (see Smuts 1985, Dunbar 1984a, de Waal 1982, 1984a, respectively). In Barbary macaques, males collect nearby infants which they present to each other and interact over in order to reduce tension following a fight (Deag and Crook 1971).

De Waal (in press) has pointed out that an animal who fails to go to an ally's aid in a critical situation places its relationship with that individual at risk. Such situations can arise either (1) when the animal has relationships with both protagonists in a fight and is forced to support one against the other or (2) when its ally's opponent is so powerful that any attempt at intervening on the ally's behalf would only result in the aid-giver itself incurring serious costs by being attacked. Situations of the first kind are likely to be common in fights within families, but are probably less serious in their consequences than situations of the second kind. As we have already noted (p. 219), there is unequivocal evidence demonstrating that primates evaluate the risk they run if they go to another individual's aid and will refuse to do so when the potential costs are too high. Reneging on a coalition in this way is likely to invite retaliation on a future occasion when the roles are reversed.

De Waal suggests that behavioural mechanisms have evolved that allow the renegade to repair the damage to the relationship in order to preserve its own interests in the coalition and he refers to such mechanisms as *consolation*. Where reconciliation restores the *status quo* in a relationship between two protagonists in a fight, consolation restores the *status quo* in a relationship where one member has been attacked by a third party. In chimpanzees, consolation may involve the renegade approaching its ally and placing an arm around its shoulder, with grooming as a possible sequel (de Waal, in press). In the gelada, we have noted contact calls being used under identical circumstances: in this case, the call apparently serves to attract the ally's attention, with the result that it approaches its partner to be groomed. Such behaviour may allow the renegade to avoid moving towards its ally in a way that might be misinterpreted by the opponent as a direct intervention on the ally's behalf. This is clearly an area that has considerable potential for research.

One important finding that emerges from Kummer's (1975) experiments is that some combinations of animals find it much harder to develop stable relationships than others do. In the gelada, for example, Kummer found that male–female dyads progressed most quickly through the bonding stages and typically achieved higher levels of bonding than female–female dyads, which in turn did better than male–male dyads. Rather similar results were obtained for hamadryas baboons by Stammbach (1979). These differences seem to correspond quite closely to the degree of competition between the component members of these dyads. In the presence of females, for example, males are rivals, whereas a male and a female share a common interest in forming a relationship. Note that, in the absence of females to compete over, males may also share a common interest in forming coalitions with each other, and their interactions will in consequence be more conciliatory: this flexibility allows gelada males, for example, to form very cohesive closely knit all-male groups (see Dunbar and Dunbar 1975).

This sex difference does not apply in all species, however. De Waal and van Roosmalen (1979) and de Waal (1984a) found that males had more highly developed mechanisms of reconciliation than females in both chimpanzees and rhesus macaques, respectively. The reason for this is fairly obvious in the chimpanzee: female chimpanzees are generally more asocial than males and tend to range alone, whereas males form coalitions aimed at giving them access to reproductive females (Wrangham and Smuts 1980). De Waal (1984b) has shown that the bases of coalition formation differ significantly between the two sexes in the chimpanzee. Coalitions among males are directly concerned with status competition, with access to reproductive females as the primary goal: their coalitions are flexible, often temporary and depend on the current rank of the individuals concerned. Female coalitions, on the other hand, seem to be more oriented to long-term reproductive goals, as we might expect: their coalitions are primarily concerned with protecting both the individuals and the relationships with which the animal has a particularly close association (principally kin and close 'friends'). This can hardly explain the absence of conciliatory behaviour among female rhesus, however, since rhesus females are highly social animals for whom coalition formation is a key social strategy. A likely explanation for this puzzling result is that de Waal's (1984a) study group was an experimentally convened group of juveniles who would have been largely unrelated to each other. Since the coalitionary behaviour of rhesus females is directed almost exclusively towards female kin (see Sade 1972), it is perhaps to be expected that little such behaviour would be shown when none of the animals was closely related to any of the others.

Similar considerations allow us to explain differences between such species as the gelada and the patas. Kummer (1974) found that male patas monkeys were significantly less tolerant of each other and maintained much larger inter-individual distances than male gelada did. Indeed, two male patas could not be housed together in the same compound with females for any length of time because the more dominant one would eventually kill the other. In contrast, male gelada fairly quickly arrived at a stage of reconciliation following an initial fight. Kummer noted that whereas gelada males commonly use sexual behaviour patterns like presenting and mounting, patas males never do so. Only when there are no females present can patas males coexist (as, for example, in all-male groups: see Gartlan and Gartlan 1973, Chism and Rowell 1986).

The difference between these two species can be understood in terms of the context in which males acquire their reproductive units. Gelada reproductive units live in large herds where there are many other males waiting for the opportunity to acquire units. Newly ensconced harem-holders are therefore at serious risk of attack by these males just at the time when their relationships with the unit's females are still very unstable. Analysis of the costs and benefits of whether or not the new harem-holder should allow his predecessor to stay on in the unit after his defeat suggests that the new male's primary gain comes from the support received from his predecessor during this critical early period in his tenure (see Dunbar 1984a). The process of reconciliation is highly ritualised in the gelada, with the two males formally establishing an effective working relationship once the new male's dominance has finally been established. In contrast, patas reproductive units are widely dispersed so that the risks of a new male being himself taken over are very much lower. He can therefore perhaps afford to dispense with the coalitionary services of his predecessor and force him to leave the group. Whether the differences in grouping patterns between these two species are cause or consequence of the contrasts in the males' tolerance of each other remains unclear, however. One plausible argument would be that gelada reproductive units need to form large herds in order to be able to exploit the high quality but predator-risky plateau habitats (Dunbar 1986) and this is only possible if harem-holders can develop sufficient tolerance of each other's proximity to form a herd. This would in turn provide the basis for the harem-holders' willingness to allow other males to join their units as followers as well as allowing their predecessors to stay on, given that the massed presence of potential competitors in such large herds provides a pressing reason to favour such an extension of this behavioural tolerance.

I do not necessarily wish to suggest that this is the correct explanation in this particular case, but I do wish to suggest that we need both to build and to test such historical hypotheses in order to be able to explain both how and why a species' social system has evolved to the form it has. I shall return to this in Chapter 13. In the present context, the important point to appreciate is that, in seeking to understand the coalitionary behaviour of animals, it is essential to know what their long-term strategic objectives are. Once we understand the reasons why animals behave as they do, we can then begin to appreciate why it is that they behave differently towards different social partners not only in different species, but also sometimes in different populations of the same species.

Negotiating group decisions

Negotiation can also play an important role in many other aspects of primate survival strategies. Kummer (1968) and Stolba (1979) have described the ways in which hamadryas baboon bands arrive at a decision about the direction in which a band should forage during the day. The process involves individual males indicating their preferred directions by moving out to the band's periphery and sitting there facing away from the rest of the group. Other males may then add their votes to a given direction by joining the male indicating that direction. During this stage, there may be considerable shifting of votes by males until, finally, a clear majority builds up in favour of one direction. At this point, the entire band departs in this direction.

A crucial feature of the decision-making process in this case is a behaviour which Kummer (1968) termed *notifying*. This involves one male approaching another with a rapid, rather nervous gait, stopping immediately in front of it to peer closely into its face, then turning suddenly on the spot to present its hindquarters, following which it moves rapidly away with a characteristic swinging gait. The seated male will often reach through between the presenter's legs to touch its genitals, but more importantly it always watches the departing male closely as it walks away. In a detailed study of the function of notifying, Stolba (1979) found that this behaviour occurred mainly between related males and that it occurred with particular frequency whenever decisions about which direction the band should travel in were being taken. In effect, notifying serves to draw the recipient's attention to the notifier's personal preference for a particular direction. This may happen repeatedly as the males seek to notify different individuals. Notifying can thus be interpreted as an attempt to recruit votes for the notifier's preferred direction. Notifiers tend to concentrate on related males of their own clan because these stay together during the day's foraging, so that a commonly agreed route among clan members in particular is essential if they are to sway the general vote in the band as a whole.

This process of decision-making is significant not so much because it concerns relationships between individual animals but because it is a particularly clear example of how groups of animals make a decision that affects all of them equally in circumstances where they cannot proceed unless a collective decision is agreed. Persistent refusal to accept the consensus decision would presumably result in a weakening of the bonds that hold the social group together, and eventually the group would disintegrate. In the hamadryas's habitat, failure to arrive at mutually acceptable decisions would have very significant consequences for the animals' survival. So critical is this consideration that, in cases where the harem-holders cannot reach a decision, one of the very old femaleless males in the band will sometimes make a unilateral decision and head off in a particular direction (not always one of those being suggested by the other males). Because these old males are in a very real sense repositories of historical knowledge about the distribution of resources under a wide variety of climatic conditions over many years, their decisions are invariably given considerable weight by the other band members. The significance of such decisions in the lives of the hamadryas is reflected in the fact that hamadryas make use of many more greeting behaviours than do the other species of baboons (Peláez 1982).

Other examples of the way in which decisions about the timing and direction of group progressions are taken (e.g. gelada: Dunbar 1983e; baboons: Norton 1986) indicate that this process is far from being unique to the hamadryas, though it has been developed to an unusual degree of complexity in their case. Collective voting has also been described in other contexts: one of these, for example, is the decision made by female gelada as to whether to desert their incumbent harem-holder in favour of a new male. In this case, voting is effectively carried out by the females accepting the prospective new male's invitations to groom him. The first female's decision to do so invariably precipitates active attempts by some females to prevent others from following suit (see Dunbar and Dunbar 1975, Dunbar 1984a). Similar attempts by females to prevent other females of their group from mating with intruder males have been described for blue monkeys (Cords *et al.* 1986; see also Rowell and Chism 1986a).

Dynamics of Social Relationships

Social relationships are not static features of an animal's life. Their value to an individual inevitably changes with time as its circumstances alter in response to advancing age and the occurrence of important demographic events like deaths and maturations. Servicing a relationship in order to maintain it through time can, therefore, be just as important as building it up in the first place.

The network of relationships that constitutes a group is often very sensitive to changes in group size and composition. In the gelada, for example, the extent to which the members of a reproductive unit form a cohesive integrated set depends very much on its size (see Figure 6.4). That the unit becomes increasingly fragmented in social terms as its size increases was confirmed by a comparison of the diversity of interactions within individual units before and after the loss or gain of a member (Dunbar 1984a). The cause lies in the fact that most females have a single preferred ally on whom they tend to concentrate their social time. The male, on the other hand, is concerned to interact with all his females, since only by doing so can he ensure their loyalty to him. So long as the number of females in the unit is fairly small, the male can interact with each one, thus providing a link between all members of the unit. But once group size exceeds five adults (four females and the male), he no longer has the time to groom with all of them to the extent necessary to guarantee each one's loyalty. Rather than divide his time equally among the females (so grooming each for less than the minimum criterion for loyalty), males generally prefer to concentrate their time on the few females they can groom to this criterion. This leaves most of the unit's females free to concentrate on the relationships that really matter to them, with the inevitable result that there is a significant contraction in the median size of grooming cliques that individuals are members of once group size passes the threshold of five adults (Figure 11.1). Because they are fragmented, units that are larger than this critical size are more likely to be taken over and to undergo fission (Dunbar 1984a). Studies of several species of macaques have also indicated that group fission is most likely to occur when one section of a group becomes socially isolated from the rest as a result of the death of a pivotal individual (for examples, see Chepko-Sade and Sade 1979, Grewal 1980, Sugiyama and Ohsawa 1982b, Yamagiwa 1985).

Figure 11.1. Median size of grooming cliques in gelada reproductive units, plotted against the number of adults in the unit. Vertical bars represent interquartile ranges. Sample sizes are 3, 1, 2, 14, 18, 7, 7, 3 and 8 cliques drawn from a total of 25 units

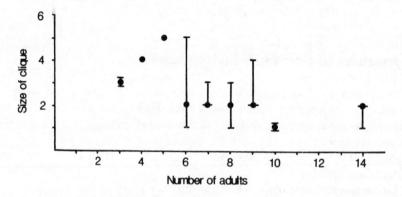

Source: Dunbar (1984a, Figure 36).

A number of experimental studies have also demonstrated significant shifts in the pattern of relationships as group composition has changed. Vaitl (1978), for example, found that when the preferred grooming partners of female squirrel monkeys were removed, they adjusted their pattern of grooming to interact with individuals whom they had previously ignored. These new relationships were later inhibited again when the preferred partner was returned to the group. Kummer (1975) used experimentally convened groups to explore the way in which preferred relationships inhibit interactions with other individuals among gelada. He found that not only did animals withdraw from interactions with other individuals once they had established a preferred grooming relationship, but they would also actively intervene in attempts by their partner to interact with other individuals (see also Dunbar and Dunbar 1975). Similar 'jealousy' reactions have been noted in titi monkeys: males intensified their relationships with their mates as a rival male's distance from the pair was experimentally reduced (Cubicciotti and Mason 1978). In a more detailed series of experiments with hamadryas baboons, Stammbach and Kummer (1982) were able to show that the relationships between pairs of females can change dramatically when other individuals are present. In isolated dyads, each individual tends to groom with a characteristic frequency irrespective of the partner's identity. But as soon as other individuals are added to the group, each individual's pattern of grooming changes to take account of the identity of the partner: grooming between preferred partners was significantly higher than it had been in isolated dyads. The presence of potential competitors apparently prompted females to intensify their relationships with their preferred partners in order to prevent them being appropriated by a rival. In fact, more than half a century ago, Maslow (1936) demonstrated that dominance relationships derived from pair-wise tests of individuals often failed to predict their ranks in a hierarchy when all the animals were placed together in a group. In the group context, he noted, new behaviours emerge that cannot be predicted from simple pairings because they appear to be a function of group size. One of these, of course, is the formation of coalitions that enable low-ranking individuals to defeat individuals who, in a simple dyadic context, had been able to dominate them.

One of the points highlighted by these findings is the fact that individuals will often attempt to defend relationships when these are under threat from a third party. In certain circumstances, however, the reverse can occur: a coalition partner can threaten to withdraw its support if an ally has been lax in maintaining its side of a coalition. In a captive group of chimpanzees studied by de Waal (1982), for example, the second-ranking male was able to become dominant by forming a coalition with the third-ranking male; in return for this support, the latter was permitted occasional access to oestrous females. From time to time, however, the male attempted to renege on this arrangement by denying his ally access to a female. Whenever he did so, the ally would support the rival on the next occasion when a fight broke out between them. The male rarely needed more than one such reminder of the precariousness of his position to restore the withdrawn 'privileges' to his ally. Similar behaviour has been described in wild chimpanzees by Nishida (1983).

Relationships are something that animals have to work at. Smuts (1985) found that male and female baboons who had a 'special relationship' behaved quite differently towards each other than they did towards other members of their group. The male, in particular, was willing to spend significantly more time grooming the female, whereas normally males were content to allow females to do most of the grooming. Among female gelada, grooming bouts were longer with a preferred partner than when grooming other femal (Dunbar 1983c). The gelada harem male was also groomed for longer b primary female partner than by the other females of his unit (Dunbar

There were analogous differences in the effort the male had to invest in order to elicit interactions with his females: several behavioural indices indicated that the male had to work significantly harder to get non-partner females to interact with him than he did in order to interact with his main partner female. Whereas responsibility for maintaining their relationship was divided equally between the male and his partner female, it is very clear that most of this burden fell on to the male in relationships with his other females (Table 11.1).

Table 11.1. Contrasts in the effort male gelada had to expend in order to interact with their main female social partner and with the other females in their units

	Partner female	Other females	Significance[a]
Mean number of grooming bouts per interaction	3.0	1.0	$p < 0.001$
Male grooms more than female (% of dyads)	33.3	71.4	?[b]
Mean number of presents by male per grooming bout by female	0.55	0.76	$p < 0.05$
Percentage of presents by male in which female refused to groom (mean)	14.6	25.0	$p < 0.01$

Source: Dunbar (1983d).
[a] Mann–Whitney or non-parametric t tests based on data from 7 units, each with 3–7 reproductive females.
[b] Sample size too small to test for significance.

One final important aspect is the influence of personality on an individual's ability to form and maintain relationships over time. By 'personality', I mean consistent styles of behaviour that characterise particular individuals and mark them out as different to other individuals of the same age and sex. Hitherto, the very notion of personality in animals has been anathema to behavioural scientists, but there is growing evidence that individual primates do differ considerably in characteristic ways (baboons: Buirski et al. 1973; chimpanzees: Buirski et al. 1978; macaques: Chamove et al. 1972, Stevenson-Hinde and Zunz 1978, Stevenson-Hinde et al. 1980, Nash and Chamove 1981). These studies have used modified versions of standard human personality rating questionnaires to show that not only do animals differ as individuals on various dimensions, but also that human observers can agree on their ratings of each individual. Moreover, these ratings have been shown to have considerable stability over time. Stevenson-Hinde et al. (1980) made repeated assessments of immature rhesus macaques and found correlations between ratings taken in successive years of around 70% for the personality dimensions *Confidence* and *Sociability*, though values on a third dimension,

Excitability, were rather lower (20–40%). Ratings of adults over a four-year period revealed even greater consistency (Table 11.2).

So far, few attempts have been made to evaluate the functional implications of personality differences, though they must inevitably be profound. None the less, there are clear hints in the literature that personality (and in particular those aspects of personality commonly referred to as 'confidence') can have important consequences for an individual's ability to manage social relationships in an effective way. Riss and Goodall (1977), for example, found that qualitative differences in personality played an important role in the manner and ease with which individual males achieved high rank in a community of wild chimpanzees. Smuts (1985) provides two particularly telling examples from her observations of baboons: in each of two cases where a pair of males joined her troop at the same time, one of the males operated with significantly more social finesse than the other, responding more sensitively to the females' initial nervousness. In both cases, the socially more adroit male was able to establish a firm relationship with one or more of the females and went on to lead a successful social and reproductive career within the group; the socially less sophisticated males, in contrast, failed to establish such a relationship (primarily because of a tendency to frighten off the females by chasing after them whenever they nervously withdrew from an attempt at interaction) and, as a result, failed to breed. Richards (1974) also noted that, among captive rhesus macaques, an individual's dominance rank reflected certain basic properties of the animal's complete personality. In a more detailed experimental study of personality in stumptailed macaques, Nash and Chamove (1981) found that when an animal's position in the dominance hierarchy was manipulated experimentally (by altering the composition of the group), its friendly social behaviour towards other individuals remained consistent irrespective of its rank, even though other behaviour patterns (e.g. vigilance) were significantly influenced by its rank at any given moment. Finally, several studies have noted that it is only possible to understand the nature of the relationships in a captive group if the differences in temperament and personality between the animals are taken into

Table 11.2. Consistency in the personality characteristics of adult rhesus macaques: mean Spearman correlation coefficients for scores on three personality dimensions sampled over a four-year period

	Personality dimension			Number of monkeys
	Confidence	Excitability	Sociability	
Adult males (> 8 years)	0.65	0.48	0.72	5
Adult females (> 6 years)	0.90	0.80	0.59	11

All correlation coefficients are significant at $p = 0.05$ level.
Source: Stevenson-Hinde (1983).

account (rhesus macaques: Michael and Saaymen 1967, Herbert 1968; greater galago: Roberts 1971).

While there are, of course, general rules that apply to gross classes of animals, in the final analysis it is individuals that pursue reproductive strategies. It is therefore only with recourse to individual differences that we shall in the end be able to offer detailed explanations for the behaviour of animals. As the level of analysis becomes more coarse-grained, the kinds of prediction we can make become increasingly limited and superficial. The study of personality therefore acquires considerable significance from both a psychological and functional perspective, and it will undoubtedly become an area of increasing interest during the years ahead.

Role of Communication

Interactions are mediated by behaviour. Without communicating, an animal cannot inform another of its intentions nor persuade that individual to behave in ways that best suit its own reproductive strategies. Successful manipulation of other individuals obviously depends on a system of communication that is both sophisticated and functional. Primates are essentially visual and auditory animals, with only the prosimians having a well developed sense of smell.

Although communication is a large enough topic to warrant a book to itself, I shall confine my remarks here to just two examples of the role communication plays in allowing primates to develop and service their relationships. It is worth stressing at the outset, however, that communication is often multi-channel (Goldfoot 1982). Not only is information transmitted simultaneously through several sensory channels (e.g. animals rarely give a threat vocalisation without an accompanying facial expression), but the sender closely monitors the response of the addressee to ensure that the message is getting through: if it is not, the sender may increase the intensity or switch to another channel (see Keverne 1982). In order to illustrate the way communication functions in social relationships, I shall concentrate on the roles that contact calls and grooming play in the bonding process. Before doing so, however, I want to say something about visual monitoring since it is a process that is very important in pacing and coordinating the behaviour of members of the same group.

Visual monitoring

An animal's ability to operate effectively within a social group depends as much on its ability to manipulate the behaviour of fellow group members as on its ability to anticipate what they are about to do. Thus, monitoring of the on-going stream of activity within a group plays an important role in allowing an animal both to anticipate imminent events and to learn from observing other individuals interacting. Cheney (1977), for example, found that juvenile baboons seemed to know which adult females were the most valuable allies (as judged by the frequency with which they tried to groom with them) even though they had never actually been involved in agonistic interactions with them. Presumably, the only source of information available was that gained by watching these individuals interacting with other group members.

It has been suggested that visual monitoring (repeated glancing up at other individuals) provides a sensitive index of an animal's nervousness and hence may be a more reliable guide to dominance rank than other behavioural indices (Chance 1967). Some studies have found a tendency for the rates with which individuals glance at other group members to be correlated with their dominance ranks, with most glances being directed 'up the hierarchy' to the more dominant animals (Keverne *et al.* 1978, Torres de Assumpçao and Deag 1979), but such consistency is by no means true of all populations (e.g. Altmann 1980, Torres de Assumpçao and Deag 1979). Even within a single population, individual groups may differ markedly in the extent to which glance rates correlate with dominance ranks (Dunbar 1983d).

These conflicting results are due to the fact that visual monitoring reflects two quite different motivational states in an animal: it glances at other animals both because it is tense in their presence and because it wants to keep track of the movements of its friends. Thus, Keverne *et al.* (1978) found that the structure of monitoring in talapoin groups depended on the group's composition. In single-sex groups visual monitoring was very closely correlated with dominance rank, but in mixed-sex groups this relationship was overlain by a tendency for sexual interests to dominate the monitoring behaviour of high-ranking males. In gelada groups, females tend to maintain a close watch on both their harem male and their own preferred female social partners (Dunbar 1983d). This division of attention apparently reflects the females' conflicting concerns both to avoid straying too far from the male and to maintain close contact with their main alliance partner.

An analysis of visual monitoring patterns can tell us a great deal about an animal's immediate concerns, though the data need careful interpretation. Which individuals an animal regards as most salient at any given time will depend on (1) the state of its relationships with its preferred partners (e.g. whether it is a new relationship in the process of development or an old one under threat), (2) the general level of tension in the group (e.g. following a series of rank changes or the replacement of a harem male), and (3) the animal's own current social machinations (e.g. its attempts to challenge a higher ranking animal's status). The one advantage of visual monitoring with species that live in more open habitats is that it is relatively easy data to record. In more forested habitats, both animals and observers experience much greater difficulty in monitoring the movements and behaviour of other animals, and in these cases auditory monitoring of nearby animals will often be more important.

Contact calling

Primates in general gain considerable information about what is happening around them from auditory cues. Recent field studies have demonstrated that animals can use vocalisations to exchange information about the location and nature of predators (Struhsaker 1967b, Seyfarth *et al.* 1980) and food sources (Dittus 1984) as well as on the identity of the caller (macaques: Hansen 1975; marmosets: Snowdon and Cleveland 1980; vervets: Cheney and Seyfarth 1980; mangabeys: Waser 1977b; chimpanzees: Marler and Hobbett 1975) and its immediate social context (Gouzoules *et al.* 1984). The information contained in calls can thus allow coalition partners to determine where an ally is, what its current situation is and whether it requires any assistance. The more sophisticated use of field experiments in recent years has also allowed us to explore in more detail the cognitive aspects of an animal's social relationships by using playback of vocalisations from hidden speakers (e.g. Cheney and Seyfarth 1980, 1985a). This approach is proving to be especially promising as a means of exploring primates' cognitive abilities in the social sphere, though so far we have done no more than scratch the surface. In the present context, I shall limit myself to a very brief account of the way in which primates can use a particular class of calls (contact calls) both to monitor what is happening in their group and to regulate their interactions. Many of the other significant findings that have emerged from this area during the past five years or so have, of course, been referred to in appropriate contexts elsewhere in the book.

In many species, contact calls serve to coordinate the movement of a group of individuals; but they can also be used to draw attention to the behaviour of third parties and perhaps to comment on events as they occur. In vervets, for example, structurally specific versions of the contact call can inform another individual that is is being approached by a higher or lower ranking animal, that the caller has left the safety of the trees and moved into an open area or that another group of vervets has been sighted (Cheney and Seyfarth 1982). Gelada also have a very complex series of contact calls, and these are used in conversational exchanges between preferred social partners both while engaged in grooming and while the two animals are apart feeding, as well as in a variety of other contexts which have social connotations (Table 11.3). Bibeen *et al.* (1986) also found that vocal exchanges in squirrel monkeys were more common between individuals that had an affiliative relationship than between those that did not. These last two sets of observations suggest that contact calling (and similar vocalisations) may be used as a form of vocal grooming in the maintenance and servicing of relationships.

It is already clear that we can learn a great deal about the ways in which primates see their world and their relationships with each other, from both observational studies and from carefully controlled field experiments using playback of particular individuals' calls (Cheney *et al.* 1986b). There are, inevitably, limitations to the kinds of information that we can glean from such

Table 11.3. Contexts in which female gelada give contact calls

Context	per cent
Subject feeding (no apparent stimulus)	31.0
Reply to contact call by coalition partner	19.7[a]
Reply to contact call by a non-partner female	11.7[a]
Reply to contact call by another animal	6.9
Subject looks at another animal	5.3
Adult animal moves near subject	2.4
Subject moves	10.7
Subject interacting	11.6
Miscellaneous	1.0

Source: 403 contact calls sampled from a single reproductive unit of three adult females.
[a] Subjects replied to their coalition/grooming partners (when not actually interacting with them) significantly more often than to other adult females of the unit ($\chi^2 = 46.270$, df=1, $p < 0.001$).

data. Indeed, so far they have been able to tell us only about the static cognitive aspects of an animal's behaviour. To learn about the dynamic strategic aspects of primate social life we need to examine the structure of relationships in the light of the uses to which they are put. Clearly, one of the methodological challenges of the future will be to find ways of exploiting field experiments using vocalisations to provide us with a new window on to the strategic nature of primate social biology.

Social grooming

All primates groom to some extent, but social grooming acquires particular significance as species become more social. Many, like baboons and macaques, devote up to 20% of their total time budget to grooming with other members of their group. The mechanics of grooming (searching through the fur to remove parasites and dirt) tend to imply a function in terms of hygiene. This suggestion has been reinforced by the finding that, when animals groom each other, they tend to groom those parts of the body like the head and back than an animal cannot easily groom for itself (Sparks 1967, Hutchins and Barash 1976). Such an explanation, however, begs the question as to why social groomers like baboons and macaques spend so much time grooming parts of the body that less social species like gorillas and galagos find it less essential to groom. Indeed, contrary to what would be expected on the hygiene hypothesis, bigger species do not necessarily spend more time in social grooming than smaller species (Figure 11.2).

In fact, several features of grooming concur in suggesting a function that has more to do with social bonding than hygiene as such. In the first place, analysis of the data given in Figure 11.2 shows that there is no correlation between body size and time spent grooming ($r_s = 0.086$, $n=16$, $p > 0.50$

Figure 11.2. Mean percentage of time spent in social interaction (mostly or entirely grooming) for 16 genera of primates, plotted against body size

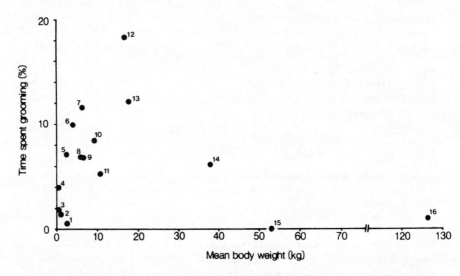

Sources: *1*, Cebus *(mean of 2 species: Terborgh 1983); 2*, Saimiri *(one species: Terborgh 1983); 3*, Callicebus *(one species: Terborgh 1983); 4*, Saguinus *(mean of two species: Terborgh 1983); 5*, Lemur *(mean of two species: Sussman 1977); 6*, Cercopithecus *(one species: Dunbar and Dunbar 1974b); 7*, Macaca *(mean of two species: Teas* et al. *1980, van Noordwijk 1985); 8*, Ateles *(mean of two species: Richard 1970, Klein and Klein 1977); 9*, Alouatta *(one species: Richard 1970); 10*, Colobus *(mean of three species: Clutton-Brock 1972, Dunbar and Dunbar 1974c, McKey and Waterman 1982); 11*, Hylobates *(one species: Chivers 1974); 12*, Theropithecus *(one species: Iwamoto and Dunbar 1983); 13*, Papio *(mean of four species: data collated from 10 populations: Dunbar and Sharman 1984); 14*, Pan, *(one species: Wrangham 1977); 15*, Pongo *(one species: Rodman 1977); 16*, Gorilla *(one species: Fossey and Harcourt 1977)*

2-tailed) but that there is a significant correlation between time spent grooming and group size ($r_s = 0.655$, $n=16$, $p<0.001$ 2-tailed). (These analyses are based on mean values for individual genera in order to avoid the problems of statistical bias created by a tendency for species of the same genus to be of similar size and to have similar social and ecological characteristics: see Harvey and Mace 1982.) This suggests that social complexity may have more to do with the amount of time spent grooming, particularly in those species that groom a great deal. Secondly, the fact that grooming often occurs as a sequel to agonistic interactions (notably the herding of females by males) and copulations is inexplicable on purely hygienic grounds, but fits well with the idea that it plays a role in servicing relationships (particularly in reducing tension: see Goosen 1981). Thirdly, the fact that animals are willing to invest heavily in grooming when relationships are being established or are under threat by rivals, combined with the fact that this willingness to groom declines with time in stable relationships (hamadryas baboons: Kummer 1968; gelada: Dunbar 1983d, 1984a; marmosets: Evans and Poole 1984), implies vested interests that have more to do with longer term benefits. Fourthly, several studies have reported a correlation between the amount of time a pair of animals spend grooming and their willingness to support each other in agonistic interactions (baboons: Strum 1983, Smuts 1985; gelada: Dunbar 1980a; vervets: Seyfarth and Cheney 1984). This suggests that the functional consequences of grooming are often other than mere hygiene. Finally, the fact that social time is commonly conserved in the face of demands made on the time budget by ecological factors suggests that the animals themselves regard grooming as something more than a mere favour that can be withdrawn when extra feeding time is needed (see Dunbar and Sharman 1984, Lee 1984, Dunbar and Dunbar in press).

This is not, of course, to suggest that hygiene was not the reason why grooming evolved *originally* in the primates. Nor is it to say that hygiene might not be a sufficient explanation in cases where grooming occupies only a tiny fraction of the animals' time budget (e.g. many New World ceboids and callitrichids). But the fact that hygiene may explain the occurrence of grooming in these species does not necessarily mean that this is the appropriate explanation in all cases.

Part of the problem with grooming has been the difficulty of relating benefits to costs when these are on different dimensions of an organism's biology. The common currency from an evolutionary point of view is, of course, numbers of genes contributed to future generations, but as yet we have no satisfactory way of translating parasites removed or coalitionary supports given into genes. There has consequently been a tendency to interpret the behaviour at face value and to assume that grooming must be an altruistic act in which the groomer gives up otherwise valuable time to remove parasites from the groomee (e.g. Dawkins 1976a). In some cases, this has resulted in some considerable ingenuity being expended to find plausible costs that would make grooming a genuinely altruistic act (e.g. Kurland 1977).

One alternative suggestion has been that animals exchange grooming benefits for coalitionary aid at a later date in a form of reciprocal altruism: this assumption is implicit in the analysis by Seyfarth and Cheney (1984), for example. But this is no more help. For one thing, grooming is often reciprocated by return grooming in those species where grooming partners commonly support each other in fights (e.g. gelada: Dunbar 1983c). Where this is not the case, the imbalance in costs would be so great that the groomer would gain disproportionately from the coalitionary support of the recipient of its grooming, while the recipient would give away far more in terms of risk of injury than it received in return from grooming. Moreover, if time spent grooming really is the only serious cost for the groomer, then it is one that is borne equally by the groomee, for the groomee is constrained to wasting time

that could usefully be spent in other activities. The most plausible interpretation of grooming, at least in the case of the more social species, is that it
functions in the building and reinforcing of the trust and familiarity between
animals that makes it possible for them to form reliable coalitions.

Having claimed that grooming functions in the context of bond formation,
however, we still need to explain in proximate terms how this is achieved.
What is it about grooming, if it is neither a commitment of valuable time nor a
hygienic favour, that allows it to function in this context to reinforce social
bonds? There are several possibilities, none of which are mutually exclusive (see
Goosen 1981). One is that being groomed is physiologically relaxing: in fact,
monkeys do often go to sleep while being groomed. This must considerably
reduce any tension or anxiety that an animal experiences in the company of a
threatening individual. Hence, individuals that groom together regularly may
feel more relaxed in each other's company and may therefore be more
motivated to go to each other's aid in order to preserve this relationship. A
second possibility suggested by Goosen (1981) is that such close interaction
exposes an animal to risk of attack at a time when its defences are most
vulnerable (e.g. because it has fallen asleep while being groomed). Thus
grooming essentially acts as a mutual demonstration under conditions of
minimal risk that the animals are both trustworthy and would be willing to
engage in forms of reciprocal altruism that incur more serious costs (i.e.
coalitionary aids). A third possibility that Goosen (1981) does not consider is
that coalition formation depends on an ability to predict an ally's future
behaviour, and grooming may permit animals to interact with sufficient regularity that they gain the necessary familiarity.

So far, most of the attempts to examine the proximate factors promoting
grooming have tended to focus on the hygienic aspects and on the environmental determinants of grooming (e.g. Troisi and Schino 1986). Aside from
the fact that they do not tackle the proximate *functional* causation of
grooming, most of these studies are uninterpretable without additional information on the way in which environmental variables affect the animals'
motivational priorities.

Exploiting Others

When two animals form an alliance, they necessarily do so by mutual consent.
It is virtually impossible for one animal to force another to form a reciprocated alliance, since it is too easy for an animal to renege on its commitments:
such alliances have to be built on trust. None the less, there is an important
category of relationships in which one individual makes use of another
without that individual's compliance and, perhaps, even against its expressed
wishes. Obviously, exploitation is only possible if the differential in power is
such as to leave the exploited individual no option but to comply.

The best documented cases of exploitation concern the use that the adult males of some species make of infants. These use infants in agonistic contexts in order to reduce the level of tension and/or prevent the confrontation escalating into a serious fight. This kind of exploitation was first documented in Barbary macaques by Deag and Crook (1971), who termed it *agonistic buffering*. Here, low-ranking males collect small infants and present them to higher ranking males in order to reduce tension following a fight. Since dominant males were presumed to sire most of the offspring in the group, it was assumed that the subordinate male brought the dominant's offspring to him. This can be interpreted as implying that the dominant will risk inflicting serious injury on its own offspring if it attempts to escalate the fight (Popp and DeVore 1979).

Subsequently, Packer (1980) and Busse and Hamilton (1981) independently showed that, in two different species of baboon, males normally use their own infants against other males, notably those attempting to immigrate into the troop. Packer (1980) found that a male's tendency to use infants in agonistic contexts correlated with both the length of his residency in the troop and with the extent to which he looked after the infant in other contexts. He suggested that, in terms of proximate causation, caretaking probably made it possible for the male to use an infant in a fight at a later time without incurring the mother's aggression because the infant (and its mother) and the male essentially trade benefits in different dimensions. Busse and Hamilton (1981) argued that, in functional terms, the male uses his own infant as an implied threat that he would be prepared to fight with particular ferocity to prevent his own infant getting injured. It is, they suggest, a mechanism for preventing infanticide by immigrant males, many of whom are young powerful males that can successfully immigrate into a troop precisely because they can easily dominate the resident males.

A further possibility, suggested by Dunbar (1984c), is that males use infants as an implicit threat that escalation by the opponent will result in the infant's mother intervening on its behalf, this in turn being likely to precipitate a concerted attack by all the other females in the group. In gelada, for example, young follower males commonly use infants in this way when threatened by their harem-holders; indeed, establishing a relationship with a female with a young infant at an early stage is a prerequisite for survival as a follower (Dunbar 1984a). The strategy apparently works because harem-holders are not powerful enough to withstand a concerted attack by all their females at once (Dunbar and Dunbar 1975). Rather similar conclusions were reached independently by Smuts (1985) and Collins (1986) from studies of infant-use in *Papio* baboons. Collins found that the infant's distress in an escalated fight was likely to precipitate an attack on an unfamiliar male by the mother. In all three of these studies, the males that used infants invariably used the infants of the females with whom they had particularly close grooming relationships. The infants themselves might or might not have been the user's offspring, but this was probably less important than the relationship between the user and the mother. Collins (1986) noted that males sometimes established a close relationship with a female soon after she had given birth, apparently with the specific intention of being able to make use of her infant.

This last explanation need not contradict those offered by Popp and DeVore (1979) and Busse and Hamilton (1981). In principle, it is possible for males to be relying on different threats when using infants in different contexts, while in some cases more than one explanation may apply simultaneously. This is suggested by the fact that the contextual variables relevant to infant-use by different gelada males (i.e. harem-holders, young followers and old followers) are very different (Dunbar 1984c). These differences include the age of the infant used and the manner in which it becomes involved (whether it initiates its own involvement or is collected by the male). Smuts (1985) also concluded that no single explanation was adequate to account for all cases of infant-use in olive baboons.

Infants are undoubtedly effective in terminating fights. Strum (1984) found that fights did not escalate further in 52% of the cases where an infant became involved. In the gelada, no fight in which an infant was used escalated beyond threats given at a distance; indeed, in every case, the aggressor walked away and resumed his previous activity (e.g. feeding) as soon as an infant became involved (though he might well return to the attack the moment the infant moved away: see Dunbar 1984c). In view of the difference in sexual dimorphism between gelada and *Papio* baboons (see Table 10.1), it is no doubt significant that the implicit threat of intervention by the mother was significantly more successful in the gelada than it was in baboons. Being relatively smaller, female baboons are that much less of a threat to the males.

Strum (1983) also found that baboons made direct use of adult females in these contexts as well as using their infants. In fact, using a female seemed to be even more effective in preventing the escalation of a fight: once a female became involved, 65% of the fights terminated. Unfortunately, no study has so far provided data on success rates in fights when neither infants nor females were involved. Without such data, we have no baseline against which to judge the effectiveness of these individuals in terminating fights: half of all fights might terminate short of physical injury anyway, irrespective of whether these individuals become involved. To test these hypotheses conclusively we would need data on the distribution of intensity levels in fights where infants are involved and those where they are not, as well as data on their outcomes.

Other forms of exploitation that are more subtle than the use of hostages in fights have been described by several observers. Kummer (1967), for example, found that hamadryas females will use their harem male as an unwitting source of support in squabbles with other members of the unit. During such encounters, a female may try to manoeuvre herself into a position close to the male from where to threaten her opponent. By presenting to the male (and occasionally glancing back at him) while threatening her opponent, the female is often able to incite the male to come to her aid. In addition, the opponent herself is likely to invite the male's attack because, every time she threatens the female, she unavoidably threatens the male sitting directly behind her.

Byrne and Whiten (1984) give a number of examples in which chacma baboons used deceptive tactics to exploit other individuals to their personal advantage. One form involved displacing a more powerful individual from a desired food source by screaming as though under attack by an opponent: the risk of intervention by a more dominant ally was often sufficient to persuade the opponent to abandon the food source. In other instances, an animal was able to defuse an opponent's aggression either by staring at the horizon as though a predator had been sighted or by soliciting the opponent's aid against an innocent bystander. Other examples of deception have been described in hamadryas and gelada baboons (Kummer 1982), macaques (Anderson and Mason 1978), chimpanzees (de Waal 1982) and vervet monkeys (Cheney and Seyfarth 1985b). Tactical deceptions of this kind probably occur widely among the primates, though they have not commonly been reported until recently (mainly, one suspects, because such behaviour was assumed not to occur).

Menzel (1978) and Woodruff and Premack (1979) have undertaken elaborate series of experiments on deception in captive chimpanzees in which they have been able to show that chimpanzees can selectively withhold information when it is to their advantage to do so. In some cases, they may even deliberately misinform other individuals when they know that the correct information would result in their being deprived of a hidden food reward. Although deceptions of this kind do occur in the wild, they are likely to be fairly rare in practice because excessive cheating will inevitably undermine the essential trust on which social life depends (Byrne and Whiten 1984). Krebs and Dawkins (1984) have also pointed out that a low level of cheating is likely to be evolutionarily stable, but that too high a level will invite retaliation.

Infanticide as a Reproductive Strategy

A particularly controversial example of exploitation involves infanticide by males who have recently taken over a reproductive group. Such behaviour has been observed (or surmised) in a wide range of species, including langurs (Sugiyama 1965, Mohnot 1971, Hrdy 1974, Wolf and Fleagle 1977), colobus (Marsh 1979b, Struhsaker and Leland 1985), guenons (Struhsaker 1977) and howler monkeys (Rudran 1979, Sekulic 1983), and similar behaviour has been noted in captivity in a number of other taxa (for recent reviews, see Hrdy 1979, Angst and Thommen 1977).

Hrdy (1977, 1979) has suggested that infanticide is a functional male reproductive strategy in that females who lose their infants prior to weaning tend to return to oestrus up to a year sooner then they would normally do (see Newton 1986). An infanticidal male thus gains (a) directly by reducing the delay before he can mate with these females and (b) indirectly by depressing the reproductive output of his predecessor. This claim has been disputed on two main grounds. One is that infanticide is abormal behaviour caused by environmental conditions such as overcrowding or human disturbance (e.g. Curtin and Dolhinow 1978). The other is that all the evidence for infanticide is either circumstantial or presumptive and that the phenomenon itself is in fact so rare in nature as to be unimportant (e.g. Boggess 1979, Schubert 1982). On the other hand, while recognising its adaptive nature, Rudran (1979) and Ripley (1980) have suggested that infanticide is a natural mechanism for reducing population density when this rises too high and/or for reducing the level of inbreeding within a population. Infanticide is known to be a mechanism used to stabilise populations below their habitat's carrying capacity in Australian Aboriginals (Birdsell 1979). Since this last explanation can only with difficulty avoid group selection when it is applied to non-human primates, I shall not discuss it further.

In a review of the field, Hausfater and Vogel (1982) concluded that although the quality of the data was not always what we might wish, there was none the less more than sufficient evidence to dismiss any suggestions that infanticide was not a real phenomenon. Moreover, they point out that many of the arguments that insist that infanticide is not an adaptive strategy are based on the false assumption that a behaviour pattern must occur in all members of a population if it is to be biologically adaptive. In fact, models of the evolution of infanticidal behaviour show that it is possible for a balanced polymorphism of infanticidal and non-infanticidal males to evolve to a stable equilibrium even when the frequency of infanticidal males is as low as 3% (Hausfater et al. 1982b).

Newton (1986) has attempted a direct test between the two competing hypotheses that infanticide is either (a) a reproductive strategy or (b) a behavioural aberration resulting from abnormally high population densities. He compared the population densities and proportion of one-male groups in populations of langurs where infanticide has been observed with those where it has not. He found that 'infanticidal' and 'non-infanticidal' populations did not differ in density, but did differ significantly in the frequency of one-male groups (Figure 11.3). 'Infanticidal' populations averaged 85% one-male groups compared with only 10% for 'non-infanticidal' populations: moreover, there was a clear dichotomy with no overlap in the two distributions, the break-point being in the region of 50–60% one-male groups. This is clear evidence against the social aberration hypothesis and offers some support for the reproductive strategy hypothesis since competition between males can be expected to increase as proportionately more of them are excluded from the female groups as the frequency of one-male groups increases. Struhsaker and

Figure 11.3. The percentage of groups that contain a single mature male, plotted against population density for 15 populations of hanuman langurs. Populations are differentiated into those in which infanticide has been observed and those where it has not been observed. Infanticide is not more likely to occur in high density populations than in low density ones, but is clearly related to the relative frequency of one male groups in the population

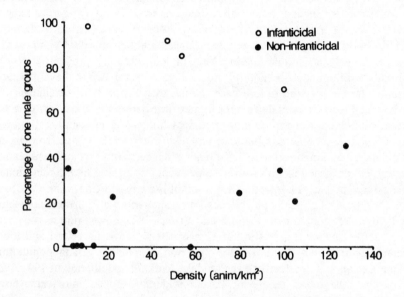

Source: redrawn from Newton (1986) with the permission of the publisher.

Leland (1985) likewise concluded from an analysis of the circumstances surrounding infanticide in a red colobus group that reproductive strategies were the only plausible interpretation.

Further evidence to support the reproductive strategy hypothesis derives from observations on the proximate factors involved. First, infanticide invariably occurs soon after a new male has taken over a group or, in multimale groups, following a major change in the male dominance hierarchy. Secondly infanticidal males kill only young infants that were born prior to or just after the takeover; they do not kill juveniles, nor do they kill infants that they themselves are likely to have sired when these are subsequently born. Thirdly, infanticide results in the termination of the mother's post-partum amenorrhoea and a resumption of oestrous cycling at the earliest possible opportunity. Most studies report a shortening of the interbirth interval by about 12 months following infanticide. In this respect, it is significant that males do not kill immature animals that are too old to have any effect on the mother's reproductive state.

It is clearly significant that, in the two species where infanticide has not been reported following the takeover of one-male groups, the females seem to come back into reproductive condition prematurely anyway, thereby obviating any need for the new male to behave infanticidally (gelada: Dunbar 1984a, Mori and Dunbar 1985; patas: Rowell 1978). It is not immediately clear why these two species should contrast so strikingly with the other species in which infanticide does occur. They do differ significantly from most langur populations in mean interbirth interval and mean male tenure as harem-holder, both being very short in patas (Rowell and Chism 1986b) and rather long in the gelada (Dunbar 1984a). Modelling of lifetime reproductive expectations for infanticidal and non-infanticidal males has shown that even though infanticidal males stand to lose part of their own fitness by their successor's infanticidal tendencies, they do have higher fitnesses on the average than non-infanticidal males at certain tenure lengths while non-infanticidal males have the advantage at others (Chapman and Hausfater 1979). Data from three populations of langurs where infanticide has been reported all had mean tenure lengths that fell within those time 'windows' where infanticidal males had an advantage. The length and spacing of these infanticidal windows does, however, depend critically on the mean interbirth interval. Further modelling by P. Winkler (quoted by Hausfater and Vogel 1982) showed that as the birth interval became shorter so the duration of the windows narrowed and their frequency increased. Data for the only non-infanticidal gelada population which we can match directly to the models (Sankaber population in 1971–72) show that the mean tenure of harem-holders following takeovers (56.4 months: Dunbar 1984a) falls unequivocally within one of the intervals where non-infanticidal males would have the advantage. The other two gelada populations (Sankaber in 1974–75 and Gich in 1973) had interbirth intervals (31.3 and 33.0 months, respectively) that were far beyond the range of those modelled so far, so we cannot use them to test the models directly until further modelling has been done. However, extrapolating from the graphs given by Hausfater and Vogel (1982) does suggest that they will probably fall within a non-infanticidal interval also. The tenure lengths of around 9 months given by Rowell and Chism (1986b) for a patas population with a 12-month interbirth interval also seems to fall within one of the periods where non-infanticidal males have the advantage, judging by Winkler's results from modelling a population with a 15-month interbirth interval. More detailed modelling of the actual costs and benefits that are incurred by infanticidal and non-infanticidal male gelada suggests that, given the lifehistory characteristics of the population, the non-infanticidal strategy does yield a net advantage (Dunbar 1984a).

So far, we have considered only the male's point of view. Even though infanticide may be to a male's advantage, it is still a serious cost to the females involved because they lose a large component of their reproductive output. This being so, we would expect females to evolve counter-strategies aimed at protecting the genetic investment represented by their offspring. Several studies have documented such strategies. Langur females, for example, attempt cooperative defence of infants when they are attacked by a male (Hrdy 1977), while female red colobus may transfer into a neighbouring group if they have an infant at risk (Marsh 1979b). Struhsaker and Leland (1985) noted that in multimale red colobus groups females attempted to form protective relationships with other males in the group when the dominant male began to behave infanticidally. Gelada and patas females seem to forestall the problem by coming back into oestrus prematurely following a takeover (Rowell 1978, Mori and Dunbar 1985). However, even in the case of the gelada, it seems that the costs to the female of doing so are not always negligible: whereas pregnant females abort and females with older infants return to oestrus prematurely, lactating females with infants less than six months old make no concessions whatsoever (Mori and Dunbar 1985). We can cost out the gains and losses that females incur from premature cycling, and doing so suggests that lactating females would suffer a net loss of fitness by returning to oestrus at a time when the current infant is not nutritionally independent (Dunbar 1984a).

★

Almost all the examples discussed in this chapter have been from Old World cercopithecine monkeys and apes. This is largely a reflection of the better observation conditions under which these species typically live. In such cases, the more subtle aspects of animals' interactions are more likely to attract an observer's attention simply because he or she is able to watch the whole build-up to an interaction over an extended period of time. Until we have comparable data from other arboreal species, we cannot, of course, assert that all primates behave in this way, but it seems implausible to suggest that they will prove to be significantly different.

Chapter 12
Socio-ecological Systems

The key lessons that have emerged from the analyses of the individual components of primate socio-ecological systems are: (1) that birth and death rates are determined by an interaction between environmental, demographic and social factors; (2) that birth and death rates filter through the demographic system to determine the social milieu in which individuals develop, so influencing the social skills they learn as well as constraining their choices of social partners; (3) that an individual seeks to maximise its lifetime reproductive output by forming relationships with those individuals that will be of most value in furthering these ends; (4) that its choice of social partners is limited not only by demographic constraints on the availability of partners, but also by those partners' own preferences for certain kinds of relationships; (5) that ecological and anti-predator strategies largely determine the way in which females are distributed and that this, in turn, determines what the males can do; and, finally (6) that primate social systems can be viewed as variations around a common theme (essentially, relationships between females).

In this chapter, I shall try to tie some of these findings closer together by integrating them into the framework of an interacting socio-ecological system. One important reason for doing this is to show how modelling can help us to achieve a better understanding of both the constraints that operate on individual animals and of the functional consequences (and hence evolutionary causes) of their behaviour.

Models as Descriptive Tools

In trying to model any system, we are endeavouring to put together the elements of that system in a coherent fashion such that the model will function in just the way the real system does. We do this not so much to be able to predict the future behaviour of the system (though this is an important test of the model's validity) but in order to elucidate the workings of the system. Our ability to predict the actual behaviour of a system is simply our measure of how well we have understood the functional relationships that make up the system.

Two factors have limited the number of attempts to build formal models of this kind. These have been (1) the fact that most field workers have concentrated on in-depth studies of a single group and therefore lack the breadth of data necessary to put such models together, and (2) a general reluctance to do so even when the data have been available, with a corresponding preference for the very much weaker kinds of conceptual (or verbal) models. The problem with these kinds of model is that they can only be tested at a qualitative level: this inevitably means a much less precise test that can all too easily mask errors and imprecise thinking. The precision of quantitative models, in contrast, forces us to think very carefully about the relationships between variables. Given that large quantities of comparative data are now available, it is important that serious efforts are made to develop formal quantitative models of both social and ecological systems.

The most detailed attempt so far to build such a quantitative model of a species' socio-ecological system is that developed for the gelada by Dunbar (1984a). Here, comparative data from three different study sites were used to generate functional equations relating key environmental variables to life-history parameters and, through these, to aspects of demographic structure. Finally, a link was established between demographic structure and male strategies of harem acquisition. The flow-chart for this network of cause–effect relationships is given in Figure 12.1. Each causal arrow in this flow chart corresponds to a functional equation relating cause and effect. In most cases, simple linear equations were found to be adequate to describe the observed data, thereby simplifying many of the computations.

It can be seen that even this attempt to model a species' socio-ecological system is far from complete. We know, for example, that group size is an important determinant of at least some aspects of behavioural ecology such as day journey length and time budgets (Chapter 3), and that these in turn have important implications for the animals' social strategies (Chapters 6 and 7). Despite these shortcomings, this particular model makes it possible to explore the relationships between environmental variables and male reproductive strategies in considerable detail. One important finding, for example, was that variations over time in key environmental variables can filter through a system to have a significant impact on the choices males make about how to acquire reproductive units. Moreover, due to the long maturation periods typical of primates, there is a marked lag such that the impact of changes in environmental variables on the behaviour of adults may not be detected until several years later. In this particular case, rainfall had a negative impact on birth rates due to the high altitudes at which the gelada live: low birth rates (and high death rates) occur in areas where ambient temperatures are low and rainfall is high. Because of the way the demographic processes work, low birth

Figure 12.1. Flow diagram for the relationships between the main variables in the socio-ecological system of the gelada, based on the model developed by Dunbar (1984a). Solid arrows indicate causal relationships for which functional equations were obtained from field data; broken arrows indicate relationships which are known to be causally effective, but for which functional equations were not determined. Positive and negative influences are indicated by the plus and minus signs adjacent to the arrowheads

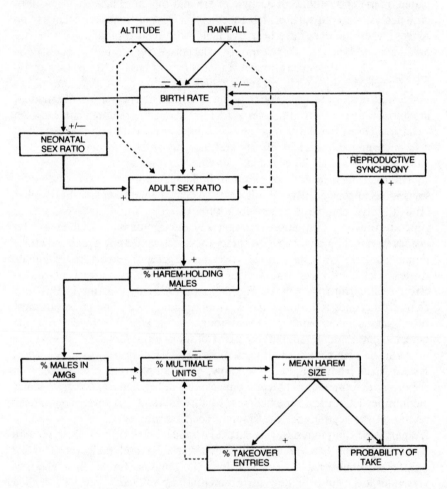

Source: reprinted from Dunbar (1984a) with the permission of the publisher.

rates will give rise to adult sex ratios some 5–6 years later that tend towards equality (the time lag being that required by the animals born under the influence of a given year's climate to mature into adults). This results in heightened competition among males for access to reproductive units, which in turn generates high rates of harem takeover (Table 12.1) and proportionately more males opting to acquire their harems by takeover rather than by

Table 12.1. Predicted probability of a reproductive unit being taken over during a year in habitats at different altitudes with different rainfall regimes, based on a model of the gelada socio-ecological system developed by Dunbar (1984a)

Annual rainfall (mm)	Altitude (m)					
	2000	2500	3000	3500	4000	4500
1000	0.124	0.129	0.137	0.151	0.173	0.209
1100	0.138	0.142	0.149	0.161	0.180	0.212
1200	0.151	0.155	0.161	0.171	0.188	0.216
1300	0.164	0.167	0.172	0.181	0.196	0.219
1400	0.177	0.180	0.184	0.191	0.203	0.223
1500	0.190	0.192	0.196	0.201	0.211	0.227
1600	0.203	0.205	0.207	0.212	0.219	0.230

the less aggressive strategy of follower entry.

We have to bear in mind, of course, that macro-models of this kind provide only the demographic context within which individual animals make their personal strategic decisions. They can tell us what the overall distribution of certain kinds of behaviour should be, but not necessarily what an individual animal will do. This overall distribution acts as a constraint on an individual's freedom of choice in its decision-making. In a gelada population living at an altitude of 3000 m under a rainfall regime of 1400 mm per annum, for example, only about 18.4% of the reproductive units will be successfully taken over (Table 12.1). Once these have been taken over, any other males that may be intending to start their reproductive careers will have to do so by some other strategy (e.g. by becoming a follower or migrating to another population with a more favourable demographic structure). Further analysis of these models suggests that in only about one-third of years will there be sufficient units to allow all the males that are looking for a unit to take one over.

For a male caught in the bind of not having a unit to take over, the consequences can be quite severe. Because males have to be fairly large to be able to withstand the fighting involved in a takeover battle, they have to delay entry until about two years after the optimum time at which a follower-strategist can begin his reproductive career (Table 8.3). Should he fail to succeed in taking over a unit at this point, he is already too old to have a long enough reproductive career as a follower to compensate for the lower instantaneous reproductive rate typical of followers. Consequently, we can expect the level of competition among the males for the few units that are available to rise, forcing those males with the least to gain and the most to lose by becoming followers to fight more vigorously for a unit.

Testing such models to establish their validity is not easy since this normally requires long-term data on particular populations. Tests are, none the less, essential otherwise we have little idea of whether or not the models are valid reflections of reality. There are two possible solutions to this problem. One is that the model should at least be self-consistent (i.e. it should predict correctly the data on which it is based). For models with many different components, each of which is based on a different set of data, this is a more severe test than might at first sight seem to be the case. While we can expect a single regression equation to predict quite well the data on which it is based, we cannot be so confident when the equation is repeatedly being transformed by further equations in an interacting system. An alternative and more powerful procedure is to search for consequences of the model that can be tested. This is possible because, in an integrated system, changes in any one component sub-system must disturb the balance of the whole system and thus have detectable implications for what happens in other sub-systems. So long as we can quantify the functional relationships between the various elements, we can generate testable predictions at various removes from the original component that is being varied.

One such implication of the gelada model, for example, is that the future demographic state of the population will always be unpredictable due to the randomising effects of several factors (notably rainfall patterns, neonatal sex-ratio biases and harem fission rates). Because of this, a male faced with the problem of deciding whether to enter a reproductive unit as a follower at the earliest possible opportunity or to wait for two or more years until he is old enough to effect a successful takeover is essentially engaged in trying to solve a problem without any information to aid his decision. Knowing this, we can predict the expected distribution of strategy choices at various ages if males make their decisions in the way that maximises their reproductive outputs. This is a fairly straightforward problem in Decision Theory (for the details, see Dunbar 1984a). The observed distributions turn out to be almost exactly those predicted by the theoretical analysis (Table 12.2). The closeness of the quantitative fit here not only implies that the decision problem was conceived in the right way, but also suggests that the assumptions on which it was based (which, in turn, were derived directly from the model of the socio-ecological system) were in fact true. This therefore provides valuable evidence that the model correctly describes the real biological system.

In this particular case, modelling has allowed us to see two important points that we could never otherwise have appreciated. One is the way in which environmental factors can influence aspects of the socio-ecological system as diverse as demographic structure and the behaviour of individual animals. The other is the fact that males have to make their decisions concerning their reproductive strategies in the absence of any knowledge about the future demographic structure of the population. Not only was this unexpected, but the magnitude of the male's problem was also surprising. Consequently, it is all the more remarkable that the males, as a whole, do apparently manage to solve their problem in a way that approximates an

Table 12.2. Number of gelada males attempting each of the two strategies of harem acquisition at different ages, together with the frequencies predicted by a decision model (in parentheses)

Age at which entry is attempted (years)	Number of adult males opting for:	
	Follower entry	Takeover
6	14	0
	(13.6)	(0.0)
8	1	10
	(< 0.1)	(11.6)
10	1	4
	(1.4)	(3.4)
Total	16	18

Expected values were calculated by determining the optimal age and strategy distribution given the lifetime reproductive output predicted for males pursuing each option (see Figure 8.7) and the likely expectations of a successful entry by takeover (see Dunbar 1984a for details). Comparison of observed and expected distributions: $\chi^2 = 0.595$, df $= 3$, $p = 0.90$. (There are 3 degrees of freedom in this test because, of the six cells, one is lost for the grand total, one for the cell that is *a priori* zero and one for pooling the cell with an expected value less than one.)

optimal solution. One important reason why we could not have foreseen these points without the aid of computer modelling is that thinking through the consequences of complex causal interactions is extremely difficult when more than two dimensions are involved. Purely conceptual models (or explanations) are invariably limited to one-cause/one-effect relationships. Not only do we then miss much of the intrinsic complexity of the natural world, but we also inevitably tend to oversimplify what we do see of it.

Models as Analytical Tools

In addition to using models to understand how a socio-ecological system works, we can also use them to elucidate the evolutionary forces that have created the particular systems as we see them now. If we have constructed a model that successfully predicts current behaviour, we can use it to explore the adaptive significance of the socio-ecological system's components by altering key variables and determining the consequences for the reproductive outputs of individual animals. If a change in any parameter always results in a loss of fitness, then we can be reasonably confident that we have identified the constellation of factors that has driven the evolution of the system into its present form. If a change in a parameter yields an increase in fitness, on the other hand, we need to look carefully at both the assumptions of the model and the biology of the species in order to be sure that we have not omitted factors from the model that are of crucial importance in constraining the animals' behaviour. Only after we are satisfied that no such constraints have been omitted can we conclude that the animals could be doing better. Although it is not impossible for animals to behave in a sub-optimal way, it is important when we do find such cases that we go on to ask why it is that they should continue to behave in this way, since evolutionary considerations should normally force them towards the optimal solution. I emphasise again here the role of evolutionary theory in providing us not so much with an explanation of behaviour as with a tool with which to explore natural systems by forcing us to keep asking questions rather than being satisfied with simple answers.

To do this, we can make use of a technique known as *sensitivity analysis* which is normally used to assess the robusticity of models in order to deter-mine how valid they are likely to be under a range of conditions. Sensitivity analysis has been used to assess the stability of the solutions generated by the demographic models of the reproductive strategies of male gelada illustrated in Figure 8.7 (see Dunbar 1984a). Evaluation of an average male's expected lifetime reproductive output for each of the two possible strategies suggested that, under moderately stable demographic conditions, a male ought to do equally well by either strategy. Each of the nine variables in the model (see Table 12.3) was then systematically varied one at a time in order to determine its effect on lifetime reproductive success. Because the two strategies occur with equal frequency (see Dunbar 1984a), we would expect them to yield identical lifetime reproductive outputs if they were in evolutionary equil-ibrium. We can therefore use the deviation from the equilibration of outputs as a measure of how close the animals are to the evolutionarily optimal

Table 12.3. Variables used in the model of gelada male reproductive strategies

Variable	Code[a]	Definition
Harem size at start	H/size	Number of reproductive females acquired at the start of male's breeding career
Female age distribution	F/age	Age distribution of females in harem
Fission rate	D/rate	Size-specific probability of a harem undergoing fission each year
Female survivorship	F/surv	Age-specific survivorship of adult females
Birth rate	B/rate	Age-specific birth rate for females
Reproductive suppression	R/supp	Size-specific reduction in fertility of females due to reproductive suppression of lower ranking females
Takeover rate	T/rate	Size-specific probability of a harem being taken over during a year
Male age	M/age	Male's age at start of reproductive career
Male survivorship	M/surv	Age-specific survivorship for males

[a] Codes used for variables in Figure 12.2.

Figure 12.2. Sensitivity analyses of the gelada male reproductive strategies model. Each of the nine variables in the model (listed in Table 12.3) was varied in turn to determine its effect on the lifetime reproductive outputs of males pursuing a takeover and a follower strategy of harem acquisition. In some cases, the variable was simply omitted, while in other cases the equation for the effects of that variable within the model was made steeper (indicated by a 'plus' sign) or less steep ('minus' sign) in its effects. The graphs give the relative deviation from the equality of outputs for the two strategies that would be expected if they were in evolutionary equilibrium. The open bar gives the results obtained from the basic model whose parameters and functional equations were obtained from observed data. The observed constellation of parameter values gives a very close fit to the expected equilibrium values. Only two sensitivity analyses produced better results, and both suggest that there might have been a tendency for observers to underestimate the true ages of animals in the field (as is known to be the case when observers use physical size with reference to captive animals as a basis for ageing animals in the field)

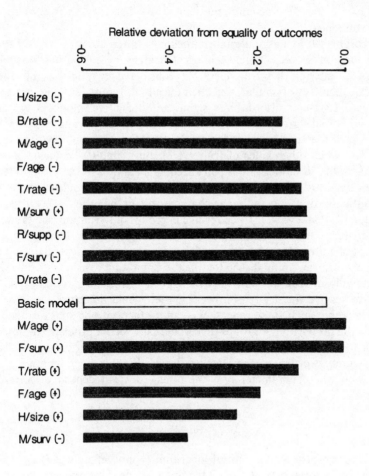

solution. Figure 12.2 shows that the observed constellation of values on those variables yields an outcome that is not significantly different from the expected optimum (open column). Moreover, only two of the sensitivity analyses produced outcomes that were better, and both these cases imply that the animals' ages might have been underestimated in the field. Since it is known that observers tend to underestimate the biological ages of animals whose births have not been witnessed (Altman *et al.* 1977, Eisenberg *et al.* 1981), these particular deviations only serve to strengthen the claim that the two strategies are in evolutionary balance. (Note that this would not have been true had there been a tendency to *overestimate* animals' ages in the field.) Given the demographic and lifehistory parameters to which they are subject, it seems that the gelada males could not do any better for themselves by altering any of those variables that are within their power to control (e.g. the size of harem they first acquire and the age at which they begin their reproductive careers).

Terborgh (1983) has undertaken the only other attempt to assess the adaptiveness of a species' socio-ecological strategy by evaluating the consequences of altering key variables. He considered two species of South American monkeys of similar size but radically different foraging strategies whose social systems also differed significantly. Squirrel monkeys typically live in multimale groups of 30–40 individuals in large ranges (250+ ha) within which they range widely in search of fruiting figs, notably during the dry season when other kinds of fruit are in short supply. Tamarins, on the other hand, live in monogamous pairs accompanied by 2–4 offspring in territories of about 30 ha. Terborgh asked how each species would manage if it lived in the size of group and range characteristic of the other.

If squirrel monkeys were forced to live in groups of five animals in tamarin-sized territories, there would, he concluded, be no problem so long as fruit remained plentiful. But the figs on which squirrel monkeys depend during the dry season are randomly distributed in the habitat at densities of only 1–2 trees actually in fruit per km^2 and he argued that many groups would be unable to survive the dry season because they would not have any fig trees available within such small ranges. We can, in fact, go one step further and estimate the proportion of groups that would face such a problem. If fig trees are distributed at random, then the proportion of groups that would have various numbers of fig trees in their territories, $P(x)$, is given by the Poisson distribution whose parameter m is the observed mean density of trees per territory:

$$P(x) = (m^x e^{-m})/x! \qquad (12.1)$$

where e is the base of the natural logarithm (approximately 2.718) and $x!$ signifies the factorial of x. With 1–2 fruiting trees per km^2, the average 30-ha territory would contain $m = 0.3–0.6$ trees. At these densities, between 55%

and 74% of groups would have no fruiting fig trees at all and would either starve or be forced to abandon their territories during the dry season. Only 4–12% of groups would have two or more fig trees in fruit. Clearly, squirrel monkeys could not pursue a tamarin-like ranging and grouping strategy with their existing dietary preference. If they were to do so, they would have to live at very low densities (about $0.7–2.0/km^2$ compared with the actual density in Terborgh's study area of $14/km^2$). Such low densities would inevitably mean that it would be difficult for individuals to find mates, with a resulting rise in the level of inbreeding and a greater susceptibility to local extinctions following random environmental 'catastrophes' of a kind that would not normally bother a squirrel monkey population.

Conversely, if tamarins were to live in squirrel-monkey-sized groups in large ranges, they would encounter severe time-budgeting problems. Under normal conditions, each tamarin group has to visit 8–10 trees a day to meet its nutritional requirements and, in doing this, the animals spend about 22% of their time travelling. If time spent travelling is proportional to the number of trees visited, then, in order to visit the 40–50 trees required by a group of 35, they would have to spend an impossible $(50/10) \times 22\% = 110\%$ of their time travelling. Even if there are some savings of scale, it is clear that the animals could not pursue such a strategy given their dietary niche because they would be left with insufficient time in which to feed.

Waser (1976) considered an analogous problem in the grey cheeked mangabey, an African monkey that also feeds predominantly on figs. He wanted to know how likely it was that a group would have enough fig trees within its ranging area to support it throughout the year. To solve the problem, we need to have some criterion that the monkeys might be trying to achieve. Waser suggested that this was that, since a fig tree is in fruit for about 10 days each year, there should be at most one occasion during the average mangabey's 10-year lifespan when no fig tree comes into fruit. In other words, $p(0)^{10} = 1/(10 \times 365) = 0.00027$, where $p(0)$ is the probability that no fig trees will come into fruit on any given day during one year. Given this, $p(0)$ itself must equal 0.44. From equation 12.1, we know that $p(0) = e^{-m}$, where m is the mean number of trees that come into fruit per day. Hence, $m=0.82$. Now, for a Poisson distribution such as this is assumed to be, $m=np$, where n is the total number of sampled events (here, trees) and p is the probability that any one of them will be in the state of interest (coming into fruit on any given day). Since each tree is assumed to come into fruit only once each year, $p= 1/365 = 0.0027$. So, the total number of trees required to obtain a mean number of fruiting trees of $m=0.82$ will be $n = m/p = 0.82/0.0027 = 303.7$. Waser (1976) estimated the density of fig trees as being 2.76/ha, which suggests that a group would need a total range of $(303.7/2.76) = 110.0$ ha. In fact, it had a range of 410 ha. Working backwards from this figure, we can see that if all the estimated 1131 fig trees in the group's territory fruited once a year, then an average of $m=3.10$ trees start to fruit each day, with a corresponding probability that no trees would set fruit on a given day of $p(0)=0.045$. The probability that no trees would be bearing fruit at all would obviously be infinitesimally small (it would in fact take 80 billion years to occur). That these mangabeys should use so large a ranging area when a very much smaller one would be more than adequate is therefore surprising. Two possible explanations that merit detailed investigation would be: (1) that not all the species of figs are used (or can be used) as food (Waser's estimate of fig density included all species), and (2) that other more sparsely distributed resources oblige the animals to range over a wider area. With regard to the second of these, it is perhaps significant that both red colobus and black-and-white colobus living in the same forest (Kibale, Uganda) also have range areas that are substantially larger than those recorded for some other populations (Marsh 1981a, Dunbar in press). Oates (1978) has suggested the availability of trace elements as the reason.

We can use Waser's approach to ask whether Terborgh's (1983) squirrel monkeys have ranges that are of sufficient size to meet their nutritional requirements. We can short-circuit some of the calculations by noting that Terborgh (1983) estimated there to be 1–2 fig trees per km^2 in fruit at any given time. This implies an average of $m = 2.5$–5.0 figs in fruit in the monkeys' 250-ha ranging area. Interpolating these values into equation 12.1, we find that the probability that a group would have no fruiting figs at all on any given day is $p(0)=0.082$ and $p(0)=0.007$, respectively. This would imply that figs would be unavailable on between 2.5 and 30.0 days each year. If the estimate of the density of fruiting trees is on the generous side (and Terborgh thought it was), then the squirrel monkeys can expect to encounter significant periods of food shortage. We can invert the question and ask how large an area they would need to meet Waser's criterion of no more than one period of 10 consecutive days without any tree coming into fruit during a 10-year lifespan. With a gross density of 14.58 fig trees/km^2 (estimated from Terborgh 1983, Figure 5.1), the group would need $303.7/14.58 = 20.8$ km^2, almost an order of magnitude larger than their actual range of 2.5 km^2.

There are three possible explanations for this shortfall: (1) that the animals make extensive use of resources other than fig fruits, (2) that the South American fig species bear fruit for significantly longer than the 10 days estimated by Waser for his Ugandan trees (Terborgh's data in fact suggest that trees must be in fruit for 25–50 days, otherwise his estimate of the density of fruiting trees is too high), or (3) that the animals accept a high risk of finding no figs and rely on stored fat and inferior foods to tide them over. In fact, most periods of shortage are likely to be quite short. With events that are Poisson distributed, the time to the first occurrence of the event is exponentially distributed with a parameter α equal to the probability of that event occurring (Meyer 1970). If there are 30 days without a fig tree in fruit each year, the probability of a fig being in fruit on any given day is $335/365 = 0.918$. The mean number of consecutive 'figless' days is therefore $E(x) =$

$1/\alpha = 1.1$ days. The probability of there being more than n consecutive days without figs is given by $p(x > n) = e^{-\alpha n}$: the likelihood that there will be more than two figless days in a row is quite small (0.160). Given that a fig tree in fruit is a superabundant food patch, the animals can probably make up any deficits incurred from such a short period relatively easily.

The Problem of Monogamy

So far, I have avoided saying too much about the evolution of monogamy, having preferred, instead, to concentrate on more general issues bearing on the evolution of different forms of polygamy. Among mammals as a whole monogamy is rare, being found in only about 5% of all species (Kleiman 1977). In contrast, monogamy is relatively common among primates where it is the predominant mating system for about 15% of all species (Rutberg 1983). Given the constraints on direct male parental care, this is surprising and suggests that mitigating factors must be militating against polygamous males in these cases.

Among birds, monogamy is the predominant mating system and occurs in about 90% of all species. Male birds are generally assumed to be constrained into being monogamous by the females' reluctance to mate with males who cannot offer them their undivided help in rearing the offspring. The classic Orians–Verner model for polygamy (Orians 1969) allows polygamy to occur only when resources are so patchily distributed that a female will do better by joining an already mated male on his territory (even if this means having to rear her chicks on her own) than by mating monogamously with a male on a poor-quality territory. Such an explanation is unlikely to be relevant in the case of primates for two main reasons. First, it assumes that females actively choose whom to mate with by selecting among males waiting on vacant territories: among primates, females do not normally have an opportunity to select among potential mates since animals pair for life and, in any case, the male is just as likely to join a female on her territory as the other way around. Secondly, it assumes that male help in *feeding* the young is critical for successful rearing: direct paternal investment of this kind is obviously not possible for male primates.

Two main alternative explanations have been offered as to why monogamy might occur in primates. One is that monogamy evolves when females are so widely scattered that a male cannot defend an area large enough to contain the range of more than a single female (Goss–Custard *et al.* 1972, Emlen and Oring 1977, Rutberg 1983). The other is that monogamy occurs whenever the female requires the male's help (either in carrying the young or defending a feeding territory) in order to rear her offspring successfully (Kleiman 1977, Clutton-Brock and Harvey 1977, Wittenberger and Tilson 1980). Van Schaik and van Hooff (1983) have argued that neither of these applies universally to primates. Instead, they suggest that the first applies to the large-bodied Old World monkeys and apes, while the second applies to the small-bodied New World monkeys, the distinction between them being based on the observation that direct male parental behaviour is common in the latter but rare in the former. I shall follow their example and consider the two cases separately.

Monogamy as a response to female dispersion

The suggestion that mammals are constrained to being monogamous by the dispersion of the females has been specifically advanced to account for the evolution of monogamy in the gibbons (Ellefson 1968, Mitani 1984). Van Schaik and van Hooff (1983) develop the argument further by suggesting that monogamy occurs whenever predation risk is low enough to allow the females to range independently in large ranges, such that the male gains by becoming monogamous and defending an exclusive food supply for her and the off-spring rather than by trying to be polygamous. They argue that when female ranges are substantially larger or smaller than this critical size, males opt for polygamy (as, for example, the males of many small semi-solitary prosimians do).

Unfortunately, this argument seems to be founded on an unsubstantiated assumption. In fact, there seems to be no difference at all in the sizes of range occupied by monogamous and polygamous females when differences in body size are taken into account (Table 12.4). For the purposes of these analyses, I have assumed that a normally monogamous female requires about two-thirds of the pair territory for herself and her offspring: this gives us a figure that is more directly comparable in ecological terms with the ranging area used by breeding females of polygamously mating semi-solitary species. When range size is adjusted for metabolic body weight (p. 34), we find that there is no significant difference between the two samples (mean range sizes of 8.3 and 9.9 ha/$kg^{0.75}$ for semi-solitary and monogamous species, respectively: Mann–Whitney test, $z = -0.603$, $p = 0.545$ 2-tailed).

Despite this, it might none the less be the case that, because many of the monogamous species are large-bodied, the ranges required by the females may be absolutely so large that a male cannot defend an area that is significantly larger. We can use the Mitani–Rodman formula for the defendability of an area (equation 3.1, p. 52) to ask how big an area males of conventionally monogamous species could in principle defend and hence how many females they could maintain exclusive access to. The maximum area that a male can defend, A_{max}, can be obtained by inverting the Mitani–Rodman formula and setting the defendability index $D = 1$ (the minimum required to permit terri-toriality) to give:

$$A_{max} = 0.25 \pi d^2 \tag{12.2}$$

Assuming again that a female and her offspring require two-thirds of the area used by a monogamous pair and that the male himself requires one-third, the maximum number of females that a male could expect to have within his maximally defendable territory, n_{max}, will be:

$$n_{max} = (A_{max} - 0.33A)/0.67A$$

The resulting values of A_{max} and n_{max} for all monoganous species for which the relevant data are available are given in Table 12.5. With few exceptions, the males could easily defend territories that gave them access to four or five females. This implies that the origins of monogamy have to be sought in factors other than the dispersion of the females, for there is nothing to prevent males defending larger areas should it be to their advantage to do so. The only plausible explanation is that the males must be contributing indirectly to the costs of rearing offspring in order to maximise the female's reproductive rate.

One way in which males might do this is to defend an exclusive food supply for the female's use. However, this seems an implausible explanation for monogamy for at least three reasons. First, in many of these monogamous species, the female is as active in advertising and defending the territory as the male is (e.g. gibbons: Ellefson 1968, Chivers 1974, Tenaza 1975, Mitani 1984). Secondly, if defence of an exclusive food source is crucial, then Table 12.5 suggests that there is no reason why a male cannot defend a larger area so as to provide several females with equally exclusive access to resources at the same time: in effect, this is exactly what the males of most semi-solitary species do. In any case, even if the male was obliged to defend a feeding territory for a single female, there is no reason why this should oblige him to maintain so strong a pair bond with her given that such close meshing of behaviour is expensive in terms of time and energy. Such behaviour would only be necessary during the breeding season (i.e. for a few weeks once every year or so), and a male that searched so small a territory randomly would encounter his female with sufficient regularity to have considerable advance warning of her impending return to oestrus.[1] In other words, if the function of territoriality is to defend a food supply, then we must seek an explanation for pair bonding in some other aspect of the animals' biology. The third and perhaps most serious consideration is that no one has in fact ever demonstrated that the females of any of these species actually need a territory to provide them with any resource. This assumption seems to have been imported from the ornithological literature without any further consideration. Yet, as Ralls (1977) has pointed out, the biological constraints operating on birds and mammals are very different, so that we should not simply assume that what applies in one case will necessarily apply in the other. In order to entertain this argument, we need to be able to show that females who lack males suffer serious costs from feeding competition and, as a direct result, produce fewer surviving offspring.

Table 12.4. Body weights and range areas for polygamously and monogamously mated females of primate species in which females habitually range independently of each other

Species	Body weight (kg)	Metabolic weight[a] (kg)	Ranging area[b] (ha)	Female's range[c] (ha)	Range/ BMR-wt (ha/kg)	Source
A. *Polygamous species*						
Mouse lemur	0.6	0.12	—	0.2	1.7	Martin 1973
Sportive lemur	0.64	0.72	—	0.2	0.3	Hladik and Charles-Dominique 1974
Dwarf galago	0.06	0.12	—	1.0	8.3	Charles-Dominique 1977
Zanzibar galago	0.14	0.23	—	1.4	6.1	C. Harcourt and Nash 1986
Lesser galago	0.30	0.41	—	3.0	7.3	Bearder and Doyle 1974
Greater galago	1.00	0.75	—	7.0	9.3	Bearder and Doyle 1974
Golden potto	0.20	0.30	—	10.0	33.3	Charles-Dominique 1974
Potto	1.10	1.10	—	10.0	9.1	Charles-Dominique 1974
Slender loris	0.18	0.28	—	1.0	3.6	Petter and Hladik 1970
Orang utan	37.0	15.0	—	70.0	4.7	Table 13.4

Species	Body weight (kg)	Metabolic weight[a] (kg)	Ranging area[b] (ha)	Female's range[c] (ha)	Range/ BMR-wt (ha/kg)	Source
B. Monogamous species						
Indri	12.5	6.6	23.0	15.4	2.3	Pollock 1977
Mongoz lemur	1.8	1.6	1.0	0.7	0.4	Clutton-Brock and Harvey 1979
Common marmoset	0.45	0.5	1.0	0.7	1.4	Stevenson 1978
Pygmy marmoset	0.15	0.2	0.3	0.2	1.0	Castro and Soini 1977
Emperor tamarin	0.5	0.6	30.0	20.0	33.3	Terborgh 1983
Saddlebacked tamarin	0.3	0.4	30.0	20.0	50.0	Terborgh 1983
Cottontop tamarin	0.6	0.7	17.5	11.7	16.7	Neyman 1977, Dawson 1979
Night monkey	1.0	1.0	3.1	2.1	2.1	Wright 1978
Paleheaded saki	1.1	1.1	7.0	4.7	4.3	Buchanan 1979
Monk saki	1.1	1.1	50.0	33.5	30.5	Izawa 1976
Dusky titi	1.5	1.4	4.0	2.7	1.9	Mason 1979
Widow titi	1.1	1.1	20.0	13.4	12.2	Kinzey 1977
De Brazza monkey	4.0	2.8	6.6	4.4	1.6	Gautier-Hion and Gautier 1978
Mentawi langur	6.5	4.1	19.0	12.7	3.1	Tilson and Tenaza 1976
Siamang	10.7	5.9	23.0	15.4	2.6	Chivers 1974
Whitehanded gibbon	5.5	3.6	39.0	26.1	7.3	Ellefson 1968
Agile gibbon	5.9	3.8	22.0	14.7	3.9	Gittins 1980
Kloss's gibbon	5.8	3.7	7.0	4.7	4.3	Tilson 1981

[a] Metabolic body weight = (body weight)$^{0.75}$; data on body weights mostly from Clutton-Brock and Harvey (1977, 1979).

[b] Group ranging area.

[c] For monogamous species, this is the female's share of the group's range, assuming that she requires 67% of the group's range to support her and her offspring.

Table 12.5. Maximum defendable area and maximum number of females that a male can maintain control over for various species of monogamous primates (see text for details)

Species	A	d	D	A_{max}	n_{max}	Source
Saddlebacked tamarin	0.30	1.220	3.194	1.17	5.4	Terborgh 1983
Emperor tamarin	0.30	1.420	3.718	1.58	7.4	Terborgh 1983
Widow titi	0.20	0.684	1.370	0.36	2.2	Kinzey 1977
Dusky titi	0.005	0.640	8.800	0.03	9.1	Mason 1965
Dusky titi	0.09	0.670	1.979	0.35	5.4	Wright 1986
Night monkey	0.09	0.708	2.091	0.39	6.0	Wright 1986
De Brazza monkey	0.07	0.530	1.775	0.22	4.2	Gautier-Hion and Gautier 1978
Whitehanded gibbon	0.39	1.600	3.222	2.01	7.2	Ellefson 1968
Whitehanded gibbon	0.57	1.490	2.053	1.74	4.1	Raemakers 1979
Whitehanded gibbon	0.40	1.300	1.822	1.33	4.5	van Shaik et al. 1983a
Siamang	0.23	0.900	3.073	0.64	3.7	Chivers 1974
Siamang	0.47	0.738	1.233	0.43	0.9	Raemakers 1979
Siamang	0.25	0.600	1.063	0.28	1.2	van Shaik et al. 1983a

Note: symbols follow those defined by Mitani and Rodman (1979); A_{max} is the maximum area that a male could defend given the length of day journey, d; n_{max} is the maximum number of females that a territorial male could control access to with a territory of size A_{max} and a female range that is 67% of the group range, D.

If the exclusive access to resources is not the cause of monogamy, what alternative hypotheses relating to indirect parental investment can we suggest? Clearly, monogamous groups are too small to function as a serious anti-predator deterrence, and it is doubtful (given the high predation rates incurred by vervets: see Table 4.5) that a 5-kg male gibbon could offer a serious defence against most predators. But it might be able to provide the female with a predator monitoring service that allows her to feed without risk of being taken unawares. There are a number of arguments in favour of this suggestion.

First, we need to distinguish clearly between habitats where predators can get close to prey using cover and those where cover is lacking such that the distance at which approaching predators can be detected is very great. In the latter case, animals have ample time in which to take appropriate evasive action once the predator has been detected, whereas species in poor visibility conditions are more likely to be surprised by a predator and so have to rely on active defence. The one feature that characterises many of the large-bodied monogamous species (and in particular the gibbons) is that they occupy arboreal niches in the forest canopy where (1) they are more or less secure from attack by cursorial and arboreal predators of any size, and (2) their position in the terminal branches of emergent trees (Gittins 1983) gives them an unimpeded view of their surroundings so that approaching avian predators cannot easily surprise them. Some evidence to support this derives from field experiments by van Schaik *et al.* (1983a) which showed that gibbons were able to detect the approaching observers at significantly greater distances than did two species of more terrestrial macaques when these were in groups of comparable size (median detection distances in groups of 1–4 animals: 32.5 m versus 19.5 m and 15.0 m, respectively; Mann–Whitney tests, $p < 0.01$ in each case).

In the second place, where detection alone is the primary consideration, large groups are rarely an advantage in such habitats. Both theoretical analyses and empirical studies suggest that the improvement in the rate of detection diminishes rapidly as group size increases (see Pulliam 1973), while at the same time larger groups incur an increasing cost in terms of generalised competition between group members. Hence, where vigilance alone is of interest, the optimal group size will often be the minimum, especially if one member can afford to take on a disproportionate share of the vigilance (as males will usually be able to do because of their lower costs of reproduction).

Thirdly, a female who finds it advantageous to forage alone for ecological reasons but needs some help with predator detection may benefit by forming a group with a male rather than another female because males (unlike females) have no other commitments to distract them from maintaining vigilance. Not only do males need to spend significantly less time feeding than females because they do not have to bear the heavy energetic costs of gestation and lactation (e.g. indri: Pollock 1977; siamang: Chivers 1974), but they will also in any case be anxious to monitor the environment in order to detect rivals that might pose a threat to their hegemony over the female. The fact that, in a monogamous relationship, males can guarantee the paternity of their off-spring means that they will have an additional vested interest in providing such a service for the female.

Precisely such an argument has been advanced to explain the evolution of monogamy in small antelopes that live in predator-dense habitats where there is excellent long-distance visibility (e.g. klipspringer: Dunbar and Dunbar 1980, Dunbar 1985). Although we must beware of transposing arguments uncritically from one taxonomic group to another, there are a number of similarities in the underlying biology of gibbons and klipspringer that make the analogy plausible.

The problem with any hypothesis involving predation risk is that it is often difficult to test. What we can do, however, is to ask how serious the risk of predation has to be for it to be worth a male's while forgoing the option he has of being polygamous. Assuming that each female produces just one infant a year, a monogamous male will gain one infant. If he defends the 4–7 females that he could in theory do (Table 12.5), he would gain proportionately as many offspring. But a polygamous male would be forced to defend his territory more vigorously because polygamy would automatically increase the number of males who did not have mates. This would tend to shorten the average tenure for a breeding male so that the initial gain is partially offset by a reduced reproductive lifespan. In the limiting case, these losses and gains would cancel out so that polygamous males would, on average, gain just as many offspring as monogamous males since it is the females that ultimately impose a limit on the number of offspring that can be produced overall. Polygamous males would gain over monogamous males only if polygamy caused some males to die before they had a chance to acquire a territory: this would then allow fewer polygamous males to share out the same number of offspring produced by the females. Even so, the mean reproductive output of polygamous males is unlikely to be more than double that of monogamous males. A monogamous male moving into a population of polygamists would thus have to gain only a moderate advantage in terms of the survival of its mate and/or its offspring to make monogamy worth its while.

There is some indirect evidence to suggest that these kinds of advantages might exist. All genuinely solitary primates, for example, are either very large (e.g. orang utans) or nocturnal (e.g. many prosimians). The fact that there are no small to medium-sized diurnal species that are solitary can probably be taken to imply that predation risk during the day is significant. Moreover, there is some evidence from other taxonomic groups to suggest that lone individuals are more susceptible to predation than animals in groups (e.g. hunting dogs: Frame and Frame 1977; antelope: Gosling 1974; buffalo: Sinclair 1977). What we need, therefore, is data on the survival rates of both females and their offspring that have lost their mates for comparison with those of pair-bonded animals. Alternatively, a comparison of the time budgets of paired and unpaired animals may show that females who have lost their mates incur severe costs in terms of an increase in the amount of time devoted to vigilance and/or a decrease in the amount of time spent feeding. The latter variables may be particularly important to determine because females who have lost their mates may be able to offset, at least in part, the increased risks from predation by taking on the additional costs of vigilance themselves. In the short term, this would mask the increased risk of predation; only in the long term would the female incur a cost in terms of her genetic fitness because she would be diverting energy away from reproduction.

If the male confers *any* advantage on the female by associating closely with her, then we can expect the female to behave in such a way as to make it more difficult or more costly for the male to desert her. She can do this in several ways: (1) by synchronising her reproductive cycles with those of neighbouring females so that the male cannot mate with more than one female during each breeding season (see Knowlton 1979); (2) driving other females away so that the male cannot provide a vigilance service simultaneously for more than one female; and (3) showing a positive interest in rival males so as to increase her own mate's assiduousness in guarding her. There is some evidence to suggest that the reproductive cycles of female gibbons are synchronised, with seasonal and annual variations in fruit production being the proximate causal factor (Chivers and Raemakers 1980). Evidence on the second and third possibilities is less equivocal. In almost all species of gibbons, females behave territorially only towards other females (Brockelman and Srikosamatara 1984, Mitani 1984). Moreover, playback experiments carried out by Mitani (1984) suggest that females are very much less concerned about intrusions by other males on to their territories than their mates are, so that even though they may not actively encourage rival males, they none the less do not support their mate in driving them out.

Monogamy as male parental care

While the same argument may also apply to the monogamous callitrichids and cebids, there is a further unique consideration that arises in their case. Leutenegger (1973) has shown that, among primates, the allometric relationship between fetal weight and maternal weight scales in such a way that the smallest species carry the largest fetal weight at term. Whereas fetal weight represents around 5–7% of maternal weight in Old World monkeys, it constitutes 20–25% in the callitrichids. In addition, below about 600 g maternal weight, the scaling of the dimensions of the cervix (and hence the limit on head size in the newborn) is such that the female cannot pass a neonate of maximum weight. At this point, she effectively has the choice between giving birth to a single infant of less than maximum weight (and hence wasting a significant proportion of her reproductive capacity) or dividing the full fetal weight into two or more smaller infants. Solitary prosimians that use nests as refuges are able to divide the fetal weight into as many as three offspring at little extra cost to the mother, but non-nesting species encounter a problem over carrying and feeding more than one offspring at a time. Squirrel monkeys, which at a body weight of 0.6 kg fall on the critical borderline, opt for a single infant, but in doing so incur a high cost in terms of birth difficulties: in captivity, approximately 50% of infants die from the consequences of birth complications (Leutenegger 1980). The callitrichids habitually twin, and it is generally supposed that they can afford to do so only because the male assists the female by caring for the offspring (Kleiman 1977, Leutenegger 1980).

We can use Altmann's (1980) model of maternal time budgets to examine this problem in more detail by asking whether a female callitrichid could in fact look after twin infants without help from the male. The horizontal lines in Figure 12.3 plot the basic time budget of a female saddlebacked tamarin based on data given by Terborgh (1983), namely 32% of time spent feeding, 43% spent resting, 20% moving and 5% in other activities. To calculate the additional feeding time requirement as lactation progresses, we can interpolate into equation 9.1 (p. 189) to determine A (the constant that converts energy requirement into time spent feeding) for a normal (non-lactating) female with a body weight of 0.3 kg (Terborgh 1983). It turns out to be $A = 78.942$. We can then use this in conjunction with equation 9.1 to calculate an approximate feeding time requirement for a fetal mass for twin infants that totals 25% of the female's weight (Leutenegger 1973). An infant baboon approximately doubles its birth weight over the course of the period of lactation (Altmann and Alberts 1987) and I have assumed that a tamarin infant does the same, but over a shorter time course since infant callitrichids cease to be dependent on the mother's milk at about two months of age (Ingram 1977, Soini 1982). The additional feeding time requirements for a female nursing either one or two infants are given by the broken lines in Figure 12.3. Feeding twin infants clearly uses up a very considerable proportion of the female's resting time, even if the additional feeding time does not oblige her to spend more time moving between feeding sites.

Estimating the additional costs of carrying infants is more problematic. Taylor et al. (1980) have demonstrated experimentally that the total energy consumed by a weight-carrying animal is directly proportional to the ratio of the load being carried relative to the unloaded body weight. Hence, a female carrying a newborn infant that weighs 12.5% of her own body weight will consume an additional 12.5% energy. Plotting these additional costs on the graph (dotted lines) demonstrates that while a female tamarin can comfortably feed two infants despite the heavy costs of lactation, she would be hard pressed if she had to bear the additional costs of transporting them as well. She would not be able to give up all her resting time (gelada seem to require a minimum of 5% resting time: Iwamoto and Dunbar 1983, Dunbar and Dunbar in press) and so large an increase in food requirement would almost

Figure 12.3. Maternal time budget for the saddlebacked tamarin. The normal time budget (indicated by the horizontal lines) is based on Terborgh (1983). The expected increases in the mother's feeding time due to the energetic costs of lactation for one or two infants as predicted by Altmann's model are indicated as dashed lines; the additional energetic costs incurred by having to carry the growing infants as well as feeding them are indicated by the pecked lines. See text for further details. Although only approximate, the analysis suggests that a female tamarin has insufficient free resting time available for conversion into feeding time if she has to carry and feed two infants: if she has to do both, she is limited to a single infant

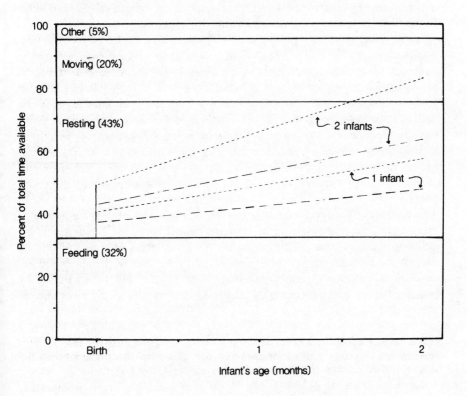

certainly oblige her to travel further, thereby increasing the moving time component. Whatever time she had to devote to resting and extra moving would then be the shortfall by which she could not achieve her energy intake target. In other words, if she has to transport the infant, then her time budget will only allow her to rear a single infant. Terborgh's (1983) data suggest that larger species of tamarins spend even more time feeding than saddleback tamarins do, so that the time-budgeting problems will be compounded as maternal size increases. A 0.5-kg emperor tamarin which normally spends about 50% of its time feeding would come perilously close to devoting all its time to feeding by the end of lactation. This suggests that another reason why females might produce single infants rather than twins once their body weight exeeds 0.6 kg is that the energy costs of feeding twin infants become prohibitive after this point.

These analyses, though crude, do suggest that a female can only rear two infants if she has help from the male. Does it, in fact, pay the male to help the female rather than accepting a lower birth rate from each female in order to be able to mate with more of them? It is easiest to begin with the simpler question of whether it would be worth the male's while to mate with a second female. Having established the general picture in this case, we can then consider the case where the male can mate with more than two females.

Table 12.6 gives the payoff matrix (in terms of numbers of offspring produced each year) for the male and the female under two alternative strategies for each sex: monogamy versus bigamy for the male and litter sizes of one versus two infants for the female. If the male is monogamous and the female twins, the female can produce two litters each year and the payoff to both of them is four offspring. If the female produces only a single infant, she can still produce two litters a year if the male helps out, so that they both gain two offspring. However, if the male deserts her and mates with a second female, it would be impossible for her to produce a second litter since she normally has to conceive her second litter at most one month after the previous litter has been born (the reproductive cycle length is 140 days' gestation plus 60 days' lactation = 6.5 months). Since infants are normally carried for a further two months after weaning (Hoage 1977), she would be unlikely to achieve more than a single litter each year. (The various figures suggest a minimum interbirth interval of 10.5 months.) With a litter size of one, she would now produce just one infant a year, but the male would produce two by being bigamous. If she tried to twin, she would be unable to

Table 12.6. Payoff matrix for male and female marmosets for the alternative strategies of monogamy or polygamy for the male and single or twin births for the female; in each cell, the payoff to the male is given below, that to the female above

Male strategy	Female strategy	
	singleton	twinning
monogamy	2 / 2	4 / 4
bigamy	1 / 2	0 / 0

rear either infant on her own since she would overtax her time and energy budgets through having to transport them: neither she nor the male would contribute any offspring to the next generation.

Given these payoffs, it is not hard to see why the male should opt for monogamy: providing the female twins, both of them do very much better. Even if the female produces only single infants, the fact that she can produce two litters a year with the male's help means that he does no worse. Only if the male could hold a large enough territory to encompass the ranges of more than four females could he do better by being polygamous. Table 12.5 suggests that males would only just be able to manage this, given the limits on territoriality. Since they would then suffer some reduction in reproductive lifespan as a result of the higher levels of competition, it is clear that the net balance is likely to come down in favour of monogamy providing the female is willing to twin. Note that the distribution of payoffs in Table 12.6 suggests that monogamy evolved before twinning: females who tried to twin before males became monogamous would have been heavily selected against.

Given that the female gains disproportionately from monogamy, we can expect her to pursue strategies that would make it even less advantageous for the male to be polygamous. Such strategies might include the synchronisation of reproductive cycles so as to make it difficult for males to mate with more than one female at a time (Knowlton 1979) or an insistence on a prolonged courtship. There is evidence to support both suggestions: marmosets have two quite restricted breeding seasons each year (Soini 1982), and pair bonding (and hence successful breeding) does depend on a considerable investment of time in contact and interaction (Evans and Poole 1984). Experimental studies of the slightly larger titi monkeys suggest that, in these species at least, the attachment between the male and female of a monogamous pair is much stronger than the bond between either of them and their offspring (Mendoza and Mason 1986).

Whereas an explanation based on the need for parental care works well for the callitrichids, the situation is less clear in the case of those New World cebids that are monogamous while producing only a single offspring at a time (e.g. titi and night monkeys). Wright (1986) has attempted to evaluate the factors predisposing these species to monogamy by comparing key aspects of their behavioural and ecological strategies with those of the squirrel monkey, a polygamous species of similar body size. These analyses demonstrated that the monogamous species (1) fed in smaller trees with significantly shorter inter-tree intervals in smaller ranges with shorter day journey lengths, and (2) invested heavily in male parental care (the male is largely responsible for the transport and care of the infant). Although the first set of observations meets the Orians–Verner criterion that monogamy will occur when food sources are small, abundant and evenly distributed, the direction of causality remains unclear. Are these species able to exploit small resource patches because other extrinsic factors oblige them to be monogamous, or are they monogamous because competition from other species obliges them to concentrate on small resource patches? Without knowing the causal order, it is difficult to evaluate the significance of these results. That male titi and night monkeys take as prominent a role in parental care as male callitrichids suggests that factors intrinsic to the animals' reproductive biology may make it worth their while helping the female out. Wright (1986) suggests that predation risk is the key factor, but it seems more likely that the energetic costs of lactation may be of greater importance. A more detailed comparison of maternal time budgets in monogamous and polygamous species seems to be called for, as well as a more detailed evaluation of the implications of different causal relationships between the ecological and social behaviour of the animals. Until these data are available, the small non-twinning monogamous species remain somewhat anomalous and it is difficult to provide a convincing explanation for the evolution of monogamy in their case.

Some unresolved problems

These analyses consider only a single birth season. In principle, there is nothing to prevent a pair changing mates between successive litters. However, if females did change mates between litters when producing two litters a year, the new mate would find himself raising another male's litter. Only if the female reduced her birth rate to a maximum of 1.4 litters a year (and this would assume immediate conception at the end of the period of infant transport) would the second male be able to sire the next litter without being compelled to invest in the previous male's offspring. Clearly, this would not be to the advantage of either the male or the female, both of whom would lose a significant proportion of their reproductive output.

None the less, there have been reports of considerable mobility among callitrichids. Garber et al. (1984) and Goldizen and Terborgh (1986) report frequent intergroup transfers in moustached and saddlebacked tamarins, respectively. Soini (1982) also noted frequent changes of territory in a population of pygmy marmosets, though in this case mated pairs always remained together. The significance of the transfers in the two tamarin studies remains unclear. The population studied by Garber et al. (1984) had only been released into its habitat four years prior to their field study, and it is possible that relationships among the animals might not yet have stabilised. In the wild population studied by Goldizen and Terborgh (1986), there was little evidence to suggest that established pair bonds were commonly broken up other than by the death of one of the partners. No female transferred to another group when she was the only female in her group. Only two females transferred within the study community, and both of these were from groups that contained two adult females: one of these females was known to have been born into the group. A failure to breed successfully is known to be an important factor precipitating mate desertion in other monogamously mating species (e.g. kittiwake gulls: Coulson and Thomas 1983; see review by D.R.

Rasmussen 1981b), so that more detailed data on individual lifehistories may be required before we can interpret the significance of these observations satisfactorily.

One other feature of interest has emerged from these studies. Both Garber et al. (1984) and Terborgh and Goldizen (1985) have reported a high proportion of polyandrously mated females in their populations. Evidence provided by Terborgh and Goldizen (1985) reveals that both males may be sexually active in such groups. From the female's point of view, the number of males she has on her territory is not important, so long as she has help in rearing her offspring. More serious questions arise for the males, however, since it is not immediately clear what they gain from sharing a female. To assess the significance of such behaviour, we need information on at least three points. First, we need to determine just how common polyandry really is: if it occurs only rarely or is a transient state, then it may not require an explanation. Terborgh and Goldizen (1985) indicate that tamarin groups contained a single female and two or more males for 62% of the time, while strict monogamy occurred for only about 17% of the time. Secondly, given that polyandry does seem to be common, this implies an adult sex ratio that is heavily biased in favour of males. In seven censuses of their study population over a four-year period, there was an average of 1.8 adult males per adult female, with all seven censuses being male-biased. Since this is unlikely to be due to chance, we need to know the cause of the biased sex ratios. Polyandry may, for example, be a natural response to sex ratios that are biased by high female mortality, so that its prevalence may vary with local ecological conditions. Or it may be that neonatal sex ratios are themselves biased, perhaps in order to ensure that a female has several male helpers to assist with rearing her young infants. Immature animals are known to take on a significant amount of infant transport in callitrichids (Ingram 1977, McGrew and McLuckie 1986). In populations that are heavily male-biased, the level of competition for access to breeding females may be so great that males prefer to share a female in order to benefit from the increased security of tenure offered by a larger coalition (as occurs, for example, in lions: Bygott et al. 1979, Packer and Pusey 1982). Thirdly, we need to know just how stable polyandrous groups are and what role the two (or more) males play in both mating and infant-care. Evidence presented by Goldizen and Terborgh (1986) suggests that, although both males in polyandrous groups are sexually active, one of them is usually responsible for a higher proportion of the matings and may, in some cases, actively defend access to the female. In populations where the number of breeding females is severely limited, younger or lower ranking males may find it advantageous to join an incumbent male as a satellite with a view to taking the female over when the male dies. Such a strategy of mate acquisition has been documented in a wide variety of species, including frogs (Perrill et al. 1978), birds (Smith 1978) and antelope (Wirtz 1981).

Detailed studies of both wild and captive cottontop tamarins suggest that sons remain on the parental territory much later than daughters do (e.g. Neyman 1980, McGrew and McLuckie 1986). McGrew and McLuckie (1986) suggest that sons gain (1) by helping to rear younger siblings (47% of all carrying of dependent infants was done by sons in their colony: McGrew 1987), and (2) by inheriting the parental territory on the father's death. McGrew (1987) has suggested that polyandry in the marmosets may be similar to the kind of cooperative breeding found in birds like the Florida scrub jay (Wolfenden and Fitzpatrick 1984). It is not obvious why sons but not daughters should benefit in this way, but it is clear that some very interesting issues are being raised by these recent observations.

In many ways, the callitrichids do seem to be a special case. Their small body size appears to be a derived state, having been evolved from a body of more conventional proportions (Szalay and Delson 1979, Ford 1980; but, for a contradictory view, see Hershkowitz 1977). That twinning is a secondary adaptation in response to this newly evolved dwarfism is indicated by the fact that the uterus is simplex and of the kind that is normally associated with a single embryo (Ford 1980).

Comment: The Function of Territoriality

The results of these analyses call into question the conventional assumptions about the function of territoriality in primates. It is commonly assumed, following the ornithological literature, that primates defend territories in order to provide themselves with exclusive access to resources (usually food). But, as Ralls (1977) points out, the assumptions that make this a valid claim for birds are directly violated by mammals, notably the assumptions (1) that females choose to mate with particular males because of the quality of the territories which they defend, and (2) that the need for direct male help in the rearing process is the primary factor opposing the evolution of polygamy. When the reproductive or digestive strategies change, the ground rules of the evolutionary game are often changed too so that we cannot necessarily transpose arguments derived from one case directly to another. Rather, we have to go back up to a higher level of analysis and ask what it is that animals are trying to maximise so as to be able to see how the new constraints affect their decisions.

One alternative reason why primates might be territorial is the defence of mates, a resource that, at least for males, is perhaps more important than the quality of the habitat on which the females have to rear their young. We can adduce five sets of observations in support of the mate-defence hypothesis, three of these being negative evidence against the resource-defence hypothesis and two being evidence that directly supports the mate-defence hypothesis.

First, territoriality often disappears altogether when the pressure on a limited ecological resource reaches a critically high level as the population density rises (see p. 52), yet this ought to be just the point at which defence of resources becomes most beneficial. At this point, the animals usually stop defending the territory as a whole and instead switch to defending the group in which they live. This can be interpreted just as easily as a defence of the group's integrity as it can as a defence of whatever resource the group happens to be occupying at the time. Secondly, the available evidence tends to suggest that, as a general rule, intra-group competition for food is significantly greater than inter-group competition, even in non-territorial species (e.g. capuchin monkeys: Janson 1985). Moreover, given the importance of inter-specific competition for food (e.g. Strum and Western 1982), we might have expected inter-specific territorial behaviour to be more common than it actually is (see p. 108). Thirdly, it seems odd that folivores should bother to defend territories as vigorously as they do (e.g. howler monkeys: Chivers 1969; colobus: Oates 1977b, Marler 1969), given that leaves are generally a much less patchily distributed resource than fruiting trees. While it is possible to argue that leafage is by no means homogeneous in its nutritional value, it none the less remains the case that folivores have a much broader selection of food sources available to them at any given moment, and that they can generally get by on a much smaller number of species of food tree than frugivores (see p. 38). One consequence of this is that folivory permits animals to reduce the ranging area they need to sustain them to a size where active territorial defence is a viable proposition. In other words, diet may be an enabling factor not a cause of territoriality.

On the positive side in favour of mate defence, there is evidence that, in at least some of the monogamous species, both males and females actively defend their pair bond against intruders of the same sex, even in captivity (titi monkeys: Mason 1971; marmosets: Evans 1983; gibbons: Mitani 1984). Geissmann (1986) has shown experimentally that singing behaviour in siamang (normally regarded as a form of territorial advertisement or defence: see Brockelman and Srikosamatara 1984) increases dramatically when an individual is paired with a new mate. These results suggest that singing is essentially a statement about the paired status of the caller rather than about territorial ownership: in other words, it may be functioning as a form of mate-guarding. In the wild, unmated females of the moloch gibbon do not give the loud call (quoted by Brockelman and Srikosamatara 1984), suggesting again that it might function in advertising the female's paired status. Secondly, in almost all species, animals are mainly territorial only towards members of their own sex (gibbons: Ellefson 1968, Chivers 1974, Tenaza 1975, Mitani 1984). This is also true in those polygamous species where only one sex (usually the male) is territorial (e.g. colobus: Marler 1969; mangabeys: Waser 1976; vervets: Cheney 1981; guenons: Cords et al. 1986). Moreover, even in those species that are not normally territorial, males are often reluctant to tolerate strange males near their groups (baboons: Cheney and Seyfarth 1977, Packer 1979b; macaques: Hausfater 1972, Sugiyama 1976). Females, in contrast, are often readily accepted (baboons: Kummer 1968; macaques: Angst 1973) even though the addition of other females must presumably increase the costs borne by females already in the group. In fact, female transfer between groups is probably the rule rather than the exception in primates (Moore 1984). Given that the natural sequence of group formation in primates is for females to band together and males to join them afterwards (Chapter 7), it is odd that females should rely so heavily on males to do their territorial defence for them if it is so important to their reproductive success. The fact that it is males rather than females who generally take on the responsibility for territorial defence suggests that this behaviour serves male interests rather than female ones. In ungulates, territoriality is also mainly concerned with the processes of mate acquisition rather than with the defence of food resources (see Owen-Smith 1977).

Taken together, these observations cast doubt on the conventional assumption that territoriality evolved as a means of protecting access to resources. As yet, however, no tests have sought to differentiate conclusively between this and any competing hypotheses. It is therefore essential that detailed studies of the function of territoriality in primates focus more closely on testing between competing hypotheses. Note that simply demonstrating that group members benefit from priority of access to some resource as a result of territorial behaviour is not an adequate test of the resource-defence hypothesis (though it is a necessary condition for it to be true): there is no reason why animals should not gain subsidiary benefits in this respect when defending territories for reasons of mate defence. As I have stressed repeatedly elsewhere in the book, unequivocal tests come only from forcing the competing hypotheses into direct conflict such that they must yield incompatible predictions. Such tests may be difficult in this case because of the extent to which the resource-defence and mate-defence hypotheses are confounded by common gains. One of the challenges of the future must therefore be to identify conditions under which strong tests can be made.

A Note on the Use of Modelling

The kinds of model discussed in this chapter are neither particularly sophisticated nor difficult to work with. The gelada model in Figure 12.1, for example, omits many key elements from the full biological system, and many of those relationships that were included could have been modelled in better ways. None the less, models like these provide us with a valuable (and, in some cases, the only) opportunity to explore the workings of complex socio-ecological systems in an evolutionary context. We cannot easily alter reproductive or digestive strategies by conventional experimental means, but we can do so in model systems. Similarly, we can estimate the consequences of changes in key environmental variables like rainfall or temperature where experimental manipulations would require a massive logistic exercise as well as a great deal of time and money. To this extent, the heuristic value of models is almost limitless.

However, we should beware of being seduced by the power of modelling to the point where we either ignore the constraints of the animals' natural biology or fail to check that our model is a fair representation of reality. We can use well constructed models to enhance our understanding of a species' biology and, perhaps more importantly, to point to aspects of the animals' biology on which particular kinds of data would allow us to distinguish unequivocally between competing hypotheses. But models whose parameters have been estimated from too little original data or whose functional relationships are based on theoretical preconceptions will in the end serve only to confuse.

Models should, therefore, always be used to generate quantitative predictions that can be tested against real data from the field. Such checks strengthen the power of any model and greatly increase the plausibility of any deductions made from a model that cannot otherwise be tested directly. This is particularly important in the case of deductions about evolutionary processes.

Note

1. Averaging the values for whitehanded gibbons in Figure 12.5, a male who can search an area 100 m either side of his route of travel with a mean day journey of 1.463 km and a total range area of 0.453 km² would be able to search 64.6% of his territory each day. With a probability of locating his female of $\alpha = 0.646$, his mean return time (i.e. the interval between successive encounters with the female) when searching randomly is $1/\alpha = 1.55$ days.

Chapter 13
Evolution of Social Systems

A species' social system does not materialise out of nowhere fully formed. It evolves by modification from the social systems of its immediate ancestors. This being so, an important test of our understanding of primate social systems is whether we can use our theoretical principles to reconstruct phylogenies for the social evolution of particular taxonomic groups. Such phylogenies must be consistent with what is known both of the species' phylogenetic history (i.e. the sequence of speciation events) and with the palaeo-environments through which they passed. Moreover, to be really useful, such reconstructions have to be able to specify not just why changes took place but also how they were brought about. In other words, we need to be able to point to the relationships that changed and show both why these changes might have been necessary and how their occurrence would have caused the appropriate changes in the ancestral taxon's social system.

My primary concern in this final chapter is to show that such an enterprise can be more than just a speculative exercise. Rather, we can use these analyses both to explore the evolution of particular species' social systems and to generate specific tests that point the way to new questions for research, thus allowing us to develop a better understanding of the way in which socio-biological principles interact with ecological contexts to produce the kinds of behaviour we observe. Since my intention is mainly to illustrate how we might set about such an exercise rather than to provide a comprehensive account, I shall confine my remarks to two particularly well studied groups, the baboons and the great apes. In each case, I shall outline the group's evolutionary history, trying in doing so to interpret not only the sequence of speciation events but also the ecological contexts in which they occurred and the selection pressures to which they were subjected.

My analysis is founded on the assumption that species (or genera) of a particular taxonomic grouping can be viewed as an adaptive array (cf. C. Jolly 1970). In other words, I argue that we can gain a better understanding of such species' social systems if we view the species as being the end-product of a radiation from some ancestral condition. To do this, we need to determine the probable social system of the common ancestor. As a first approximation, I shall use the social system of the least specialised member of the taxonomic group. With this providing us with a baseline, we can then go on to ask what specific changes would have led to the social systems of the other living species in the grouping and how these could have been brought about by changes in the species' ecological strategies. Ideally, we would want to generate several different phylogenetic sequences so that we could test between them to find the one that fits the known set of facts best. Where two such phylogenies interpret the observed data equally well, a principle of parsimony would dictate that we choose the one that involves the smallest number of changes. For reasons of space, I shall not attempt such a detailed analysis here, but I would emphasise the value of doing so as a heuristic device.

Social Evolution in Baboons

Table 13.1 summarises the main features of baboon socio-ecology. These data are mean values for sets of populations, each population's own value in turn being the mean for a number of groups in most cases. Two points deserve comment. One is the remarkable lack of variation in both body size and (with the exception of the gelada) diet. All baboons are medium-sized primates with omnivorous diets with a marked preference for fruits (Table 13.2). The other is that there is, none the less, considerable variation in both group size and mating system. Clearly, these differences in behavioural features cannot be attributed to simple differences in sexual dimorphism, gross diet or population density. Note also the extremely high densities of the gelada compared with those typical of *Papio* baboons. As noted in Chapter 3, this reflects the gelada's ability to exploit a resource (grass) that is particularly abundant, but which cannot easily be exploited without resort to certain specialisations (Dunbar 1984b).

If we exclude the gelada as a special case, many aspects of baboon behavioural ecology show just the kinds of correlation that we would expect. Thus, mean range size and day journey length correlate well with mean group size (cf. Chapter 2). These correlations can be attributed largely to the fact that, dietetically, baboons are a homogeneous group. This is very clear from a comparison of the dietary profiles for populations of the various *Papio* species (Table 13.2). (No data are available for the *Mandrillus* species, so their status remains unclear.) A comparison of the correlations in diet structure between populations of the same species with those between populations of different species reveals that intra-specific variation in diet is not significantly different from inter-specific variation (mean $r_{s(within)}$ = 0.419±0.284, n=6; mean $r_{s(between)}$ = 0.397±0.449, n=8; t=0.100, df=12, n.s.). (Independence of samples was preserved in the within-species comparisons by sampling only

Table 13.1. Behavioural ecology of baboons

Species	Body weight (kg) male	female	Diet[a]	Group size (mean)	Range size (mean, km²)	Day journey (km)	Range density[b]	Sample[c]	Mating system
Olive baboon	25.8	13.9	O/F	41.2	9.6	3.14	4.3	11	multimale
Yellow baboon	23.5	12.4	O/F	67.5	40.2	4.99	1.7	4	multimale
Chacma baboon	26.6	14.9	O/F	48.0	15.1	5.68	3.2	8	multimale
Guinea baboon	–	13.0	O/F	184.0	29.0	7.9	6.3	1	multimale
Hamadryas baboon	18.9	10.8	O/F	68.5	21.5	9.4	3.2	3	one-male
Mandrill	28.2	11.3	F	c.25.0[d]	40–50	8.0	0.6–3.3	1	one-male
Drill	20.0	–	F	c.15.0[d]	–	–	–	1	one-male
Gelada	23.3	13.8	G	144.7[e]	1.9	1.26	76.2	3	one-male

Sources: baboons based on data collated by Dunbar and Sharman (unpublished); mandrill on Jouventin (1975), drill on Gartlan (1970), gelada on Iwamoto and Dunbar (1983).

[a] O, omnivore; F, frugivore; G, graminivore (eater of grasses).
[b] Density of animals within their ranging area (i.e. mean group size divided by mean range size to give n/km²). The field data are generally too unreliable to provide unbiased estimates of true population density in more than a few cases.
[c] Number of populations sampled.
[d] One-male groups apparently form large foraging bands of up to 150 animals.
[e] Mean size of constituent one-male units is 13 animals.

Table 13.2. Dietary profiles of baboon populations

Species	Population		Percentage of diet[a]					Source
		Leaves	Flowers	Fruits	Roots	Other	Animal	
Olive	Bole, Ethiopia	41.1	11.7	40.8	1.4	1.1	3.9	Dunbar and Dunbar 1974c
	Gilgil, Kenya	52.9	2.6	9.5	27.0	5.7	1.9	Harding 1976
	Gombe, Tanzania	13.9	2.2	48.6	6.6	14.3	13.1	J. Oliver, pers. comm.
	Shai Hills, Ghana	7.5	4.6	58.8	16.5	12.4	0.0	Depew 1983
Yellow	Amboseli, Kenya	15.2	4.7	27.1	32.5	20.1	0.6	Post 1978
	Ruaha, Tanzania	18.7	1.4	15.5	51.6	4.4	8.7	D.R. Rasmussen 1978
Chacma	Suikerbosrand, South Africa	8.0	7.3	43.3	39.0	0.0	2.6	C. Anderson, pers. comm.
	Cape Point, South Africa	25.0	12.0	42.0	16.0	0.5	2.5	Davidge 1978
	Drakensberg, South Africa	25.7	14.3	2.5	53.4	0.0	3.9	Whiten et al. in press
Hamadryas	Erer Gota, Ethiopia	28.0	21.8	44.5	2.0	4.0	0.0	Kummer 1968
Guinea	Mt. Assirik, Senegal	7.8	8.6	73.5	4.0	3.9	2.5	Sharman 1981
Mean		21.0	8.3	36.9	20.1	8.9	3.6	

[a] Based on time spent feeding on different categories of food.

successive pairs of populations in the list, i.e. *A* with *B*, *B* with *C*, *C* with *D*, etc.; for between-species comparisons, each population of a given species was paired only once with a population of each of the other species, this population being determined by a random number generator.) The only plausible explanation for this is, of course, that baboons are dietetically very flexible, and each population's diet is determined largely by what is available in its habitat. Since, from an ecological point of view, we can consider the baboons to be a single homogenous taxon, gross dietary differences can be excluded from the list of variables that might influence the evolution of major differences in social system. This does not, of course, mean that diet cannot influence behaviour at the local population level where the energetic implications of different diets can have important consequences for time budgets and lifehistory variables.

As far as the general behavioural characteristics of baboons are concerned, we can draw a clear distinction between those species whose basic social unit is a multimale group (the three species of common baboon) and those that appear to have several types of group, generally a one-male-group mating system nested within a multimale ecological unit. The accumulating evidence reviewed during the course of this book is beginning to suggest that, once differences in demographic structure are taken into account, the social behaviour of the three species of common baboons is rather similar. The flexibility and lack of specialisation in the social systems of these species suggest that we should consider a simple multimale group of this kind as representing the ancestral condition. Given this, our problem is to account for (1) the evolution of the standard *Papio* multimale group, and (2) the exceptions that have more complex societies that include both one-male and multimale groupings. I shall begin by trying to reconstruct the early social evolution of the papionines and then deal with the exceptions individually.

Common baboons

General considerations suggest that the ancestral Old World monkey stock from which the papionines derived was almost certainly forest-living. Since folivory is a derived strategy that requires morphological specialisation, it seems most likely that the ancestral papionines were arboreal frugivores, probably occupying a niche not unlike that of the forest guenons (Szalay and Delson 1979). This being so, small one-male groups would have been typical, but the inherent flexibility to produce multimale groups under the appropriate demographic conditions (see Rowell and Chism 1986a) would obviously have been present.

Towards the end of the Miocene (i.e. some time prior to the point at which the macaque and mangabey lineages diverged from the papionine stock), a move to a more terrestrial way of life took place. This seems to have involved a minimum two-step transition. The first step, which may well have been taken quite early on (Kay and Simons 1980), was a transition from an arboreal way of life to a terrestrial one within the confines of the same forest environment. The second step occurred only later with the invasion of more open habitats by the *Papio* lineage itself, probably as the African forests began to give way to more open woodland habitats during the Pliocene after the macaque and mangabey lineages had diverged (van Couvering 1980).

Table 13.3 suggests that the first step probably involved an increase in group size from the 10–20 typical of arboreal guenons (see, for example, Struhsaker 1968) to the 20–30 typical of most forest-living macaques and baboons, presumably as a direct result of the increased susceptibility to ground predators. This shift is mirrored in an identical shift in mean group size between the arboreal mangabeys (greycheeked and white-collared mangabeys) and their more terrestrial congeners (agile mangabey) (Table 13.3). The small reproductive groups of the drill and mandrill (arguably the product of the same speciation event: Grubb 1973) may represent a retention of this early stage, but too little is known of their natural history to warrant serious comment. On the whole, the easiest way for larger groups to form is for related females to remain together as their matrilines increase in size over time rather than undergoing fission followed by emigration. Figure 7.9 (p. 142) suggests that even such a small change in group size would inevitably have precipitated a shift from one-male to multimale groups. A move into more open country would have resulted in a continuing trend in this direction (Table 13.3) under pressure from the significantly higher predation risks that animals encounter in more open habitats where large trees in which to escape from predators are rare (Chapter 7). Only one primate species has larger groups than these, and this is the gelada, the only surviving member of a genus that, at about the same time, abandoned the safety of the woodlands altogether to invade the open grasslands where they adopted a totally terrestrial way of life (Jolly 1972, Dunbar 1984b). (The gelada's terrestrial adaptation is so complete that its foot is unable to grip branches or tree stems that are more than 75 mm in diameter: Jolly 1972. Even where trees are available, they very seldom climb into them: Dunbar 1977, 1984b.)

These moves, but especially the second, would have had two very important consequences for the ancestral baboons. First, the increase in group size could only be accomplished at the cost of dramatically increasing the length of the day journey (Figure 3.5), with a consequent increase in both the amount of time committed to travel and the energy costs (and hence additional feeding time) required to fuel this. This would have been exacerbated by the move into poorer quality habitats (Figure 3.4). Secondly, it must have increased significantly the direct reproductive costs incurred as a result of the higher levels of social stress in large groups. The costs of reproductive suppression, however, can be offset by the formation of coalitions, so that we can expect a major development of coalitionary strategies following the invasion of open country habitats. This would be achieved most easily by an

Table 13.3. Mean group sizes of baboons and their allies, listed in descending order of dependence on trees

Taxon	Habitat	Habitus	Group size[a]	n[b]	Sources
Mangabey	Forest	Arboreal	17.8	2	Clutton-Brock and Harvey 1977
Mangabey	Forest	Terrestrial	26.0	1	Clutton-Brock and Harvey 1977
Macaques	Forest	Terrestrial	24.3	13	Caldecott 1986b
Mandrill, drill	Forest	Terrestrial	20.0	2	Gartlan 1970, Jouventin 1975
Baboon	Forest	Terrestrial	31.1	2	Figure 7.6
Baboon	Woodland	Terrestrial	38.1	7	Figure 7.6
Baboon	Open savannah	Terrestrial	48.4	18	Figure 7.6
Gelada	Grassland	Terrestrial	144.7	3	Iwamoto and Dunbar 1983

[a] Mean of all sampled population means.
[b] Number of sampled populations or species.

intensification of the relationships between closely related females (i.e. kin groups).

This increase in group size seems to mark an important shift in relationships within groups, for the need to form kin-based coalitions inevitably imposes a significant degree of stability that would previously have been lacking. The importance of such coalitions would automatically make it more difficult for females to move between groups, since in attempting to do so they would not only put themselves into conflict with the resident females but would themselves lack closely related individuals on whom to call for support. Whereas predation risk only requires that individuals remain together on a temporary basis, coalition formation as a response to size-dependent stresses demands a more permanent form of social life. On the whole, there does seem to be a good correlation between medium to large group size, low female migration rates, long-term kin-based female coalitions and a terrestrial or semi-terrestrial way of life among primates in general.

The invasion of a more open habitat also seems to be associated with a marked increase in body size. Male macaques and forest-living mangabeys typically weigh between 7 and 15 kg and are only 2–3 times the weight of the average male guenon (see Clutton-Brock and Harvey 1977), but the baboons and their allies are substantially larger (Table 13.1). This trend reaches its peak in the extinct sister species of the gelada which occupied open lakeside grassland habitats in eastern Africa during the Pleistocene: the largest of these was about the same body size and sexual dimorphism as living orang utans (Jolly 1972). This trend was almost certainly promoted by the higher predation risk in more open habitats, but it had an important consequence. Due to the allometric relationship between body size and sexual dimorphism (see Clutton-Brock and Harvey 1977), the move to a larger body size would have generated relatively larger males (Table 10.1) who would then have become more valuable as coalition partners for females. This in turn would have created a trend towards the development of 'special friendships' between individual males and females that are less likely to occur in the smaller macaques.

With this as a starting point, we can now try to fit the exceptional species into the same scheme.

Hamadryas baboon

The hamadryas is a relatively recent offshoot of the *Papio* stock: the point of divergence between the hamadryas and olive baboons has been estimated at about 340 000 years ago on the basis of mutation rates for serum antigens (Shotake 1981). We can therefore expect the hamadryas to exhibit traits for which precursors can clearly be identified in the behaviour of conventional baboons. Thus, although the hamadryas's multi-level social system (p. 12) seems to be very different, we can point to corresponding features in conventional baboon society which function in exactly the same way. Bands, for instance, are clearly a direct homologue of the conventional *Papio* troop: they are of exactly the size we would expect given the relationship between rainfall and troop size for *Papio* baboons in general (Figure 7.8). Like all baboon troops, each band has its own ranging area (Sigg and Stolba 1981). The band also seems to serve the same anti-predator function as the baboon troop, since in areas where there are no predators at all (e.g. Saudi Arabia), the band structure disintegrates completely and one-male units forage alone (Kummer *et al.* 1985). In certain areas, these bands assemble into the large sleeping troops described by Kummer (1968). Kummer was able to point, fairly convincingly, to the shortage of suitable night-time refuges in a treeless habitat as the cause of these large assemblages, a claim that is given considerable weight by the subsequent findings that (1) each of the constituent bands occupies its own part of the sleeping cliff, and (2) each band has its own exclusive ranging area that fans out on its own side of the cliff (Sigg and Stolba 1981). Such assemblages of troops in areas where suitable sleeping sites are rare are also known to occur in other baboon species (Altmann and Altmann 1970).

The substructuring of hamadryas bands into distinct one-male units has usually been regarded as a major structural change from the *Papio* system. In fact, it is easy to derive hamadryas one-male units from the special friendships that have been so widely described in *Papio* troops. It requires only a two-step change, the first of which is a dispersion of the troop under conditions of poor food availability when predator density is low to moderate. This is known to occur in other baboon populations under similar conditions (Aldrich-Blake *et al.* 1971, Anderson 1981, D.R. Rasmussen 1983). (Note the marked contrast here with the Amboseli baboon population which, although living in an even more impoverished habitat than the hamadryas as judged both by their lower rainfall and lower population density, is none the less prevented from dispersing in the same way by the significantly higher predator density.) Second, once the females have been forced to disperse into groups of minimum size (the mean number of adult females per unit is only 1.8 in the hamadryas: Kummer 1968), individual males would be able to defend them against other males. This would be enhanced by a marginal increase in the male's possessiveness and tendency to herd consort(s) away from other males. Herding is a well established behaviour in all *Papio* species (e.g. Hausfater 1975, Cheney and Seyfarth 1977, Byrne *et al.* in press). It differs in hamadryas males only in being a more highly coordinated activity, possibly as a result of a specific genetic change (Nagel 1971, Müller 1980). The importance of herding in maintaining the integrity of hamadryas one-male units through time is emphasised by a series of translocation experiments in which female hamadryas were released into olive baboon troops and olive baboon females were released into hamadryas groups. The results showed that the female's responses to their males are learned: olive baboon females soon responded to the herding behaviour of the hamadryas males and rapidly learned to follow their new 'owners', while hamadryas females gave up attempting to follow males in olive baboon troops as soon as they realised that the males were not punishing them for failing to do so (Kummer 1970).

If this interpretation is correct, then hamadryas bands should become more cohesive and *Papio*-like in habitats where predation risk is higher and/or food density greater. It was certainly my impression that hamadryas bands were much more cohesive in the Simen Mountains where feeding conditions are significantly better than in conventional hamadryas habitat and where humans and dogs occur at much higher densities, but the prediction remains to be tested at a quantitative level. A second pediction arises from lifehistory considerations. Although the hamadryas has often been quoted as an example of a species in which females habitually migrate (e.g. Wrangham 1980, Moore 1984), this claim rests on a misunderstanding. The evidence provided by Sigg *et al.* (1982) shows that females transfer between units only *within* the same band, and only males have been found to emigrate from their natal bands. Moreover, there is considerable evidence to suggest that when mothers and daughters do get separated as a result of kidnapping by males, they often end up back together again at a later date (Sigg *et al.* 1982). This suggests that the small size of hamadryas one-male units might in fact be a reflection of the size of female kin groups in these relatively slowly reproducing populations. Although survival rates are high, birth rates are quite low, so that effective birth intervals (i.e the intervals between successive *surviving* daughters) are in fact similar to those in the Amboseli baboon population (about 30 months). It may be no accident, therefore, that mean harem size in the hamadryas (1.8 females) is almost the same as the mean female kin-group size in the Amboseli baboons (1.2 females: Table 6.2). If this is so, then we can expect larger harem sizes in habitats with higher reproductive rates, and this in turn should alter the whole structure of hamadryas society. There is some evidence to suggest that large hamadryas one-male units do behave very differently from the conventional two-female units (Kummer 1968, pp. 78–9), but, again, the prediction remains to be tested.

Guinea baboon

Superficially at least, the Guinea baboon seems to present more of a problem. Its fission–fusion social system based on very large groups that repeatedly disperse into small unstable parties (Dunbar and Nathan 1972, Byrne 1981, Sharman 1981) seems to be quite unlike the conventional *Papio* troop familiar from the literature. However, I suggest that Guinea baboon society is puzzling only if we assume, as the literature does, that *Papio* troops are more coherent and structured than they really are. Given the tendency for other *Papio* troops to form semi-independent subgroups while foraging (e.g. Aldrich-Blake *et al.* 1971, Anderson 1981), we could argue that the Guinea baboon's social system is simply that of an unusually large *Papio* troop whose very size makes it difficult for the constituent members to maintain any form of cohesion while foraging (cf. the Mikumi yellow baboons studied by D.R. Rasmussen 1983). This suggestion is reinforced by the observation that the most common form of subgroup among Guinea baboons consists of a single adult male with two or three females and their dependent young (Dunbar and Nathan 1972, Boese 1975). These subgroups do not differ in size or composition from either those into which conventional *Papio* troops fragment in impoverished habitats (see Aldrich–Blake *et al.* 1971) or the reproductive units of the hamadryas (see Kummer 1968).

If this interpretation is correct, then two questions need to be answered: (1) Why do Guinea baboons form such large troops? and (2) Why do these troops fragment in the way they do?

The answer to both questions seems to lie in the peculiar nature of the habitats occupied by all populations that have been studied to date: they all live in the dense woodland savannah that borders the upper reaches of the Gambia river system in eastern Senegal. This habitat has two important characteristics. First, it lacks an upper canopy of tall trees that can be used as safe sleeping sites; secondly, it has a dense understorey where visibility is very poor but which is rich enough to support much higher population densities than are typical of most other baboon habitats (see Table 13.1: Guinea baboon population densities are roughly double those of all other *Papio* species).

Because safe sleeping sites are confined to the isolated groves of tall trees that occur along the River Gambia and its tributaries (Dunbar and Nathan 1972, Byrne 1981), animals from a relatively large area are obliged to converge on the few refuges available to them at night, rather as hamadryas bands converge on sleeping cliffs when these are scarce. But whereas hamadryas bands are forced to disperse during the day by the patchiness of the food supply in their habitat, those of the Guinea baboon disperse only because the poor visibility in the dense woodlands makes it impossible for such large groups to coordinate their movements well enough to stay together. Complete dispersal, however, is prevented by the relatively high predation risk. Byrne (1981) has pointed out that the moderate to high densities of predators in these habitats, combined with the relative scarcity of ungulate prey, makes the baboons particularly prone to predation under visibility conditions that are so poor that detection distances are seldom better than 20 m (Bert *et al.* 1967, Dunbar and Nathan 1972).

The significance of predation risk to the animals is suggested by two sets of observations. First, a comparison of the size distribution of night- and day-resting parties censused by Dunbar and Nathan (1972) shows that the resting parties that form on or near the ground during the day are significantly larger than those formed at night in the sleeping trees (medians of 5.5 and 3.0 animals, respectively: nonparametric $t = 2.297$, df=98, $p < 0.05$). In other words, party sizes are larger in the habitat with the higher predation risk (cf. van Schaik and van Noordwijk 1985b). Sharman (1981) also found that mean party size was greater during the wet season (when predators were less easy to detect because the vegetation was significantly more dense) than during the dry season. Secondly, subgroups foraging through the bush during the day are sufficiently concerned to maintain coordination of travel that they keep vocal contact with each other by means of the males' loud calls (Byrne 1981). These are given repeatedly at intervals, so making it possible for isolated parties to regroup from time to time during the day. The use of loud calls to re-establish contact after part of a troop has become separated in poor visibility conditions also occurs in olive baboons (personal observations in three different areas; see also Aldrich–Blake *et al.* 1971).

If this interpretation is correct, then we can make a number of specific predictions about Guinea baboon society. One is that populations that live in more open savannah or more heavily forested habitats should live in groups that resemble conventional *Papio* troops more closely. A second is that the high population growth rates implied by both the high densities and moderate rainfall should give rise to large female kin groups (Figure 6.2). Bonding within these kin groups should be particularly intense, both because of the high levels of stress incurred from living in such large groups and from the need to ensure that individuals do not become separated during foraging in conditions of poor visibility. Neither of these predictions can be tested with the data available at present. However, the mean proportion of time spent in social interaction given by Sharman (1981) is 18.9%: this is more than any other population of *Papio* baboons (see data summarised by Dunbar and Sharman 1984) and is significantly greater than the mean for all other *Papio* populations (mean 10.8% of time in social interaction, $n=11$ populations; $z=2.011$, $p=0.022$). Only the gelada spend as much time as this in social interaction; they also live in unusually large groups (herds).

Gelada

The gelada stand somewhat apart from the other baboons both taxonomically and ecologically. The ancestral theropithecines made a dramatic change of niche by moving right out on to the open savannah grasslands in order to exploit a grazing niche that is quite unique among the primates (Jolly 1972, Dunbar 1984b). Any such move necessarily faces a major difficulty: open grass plains have very high predator densities combined with a chronic shortage of trees for use as refuges. The obvious solutions to this problem would be either to form very large groups or to grow to a very large physical size. Both options would be impossible for a frugivore because fruiting bushes and trees are too thinly distributed to sustain large numbers or large animals. A change of diet to a more abundant resource would have been necessary, and only two resources are at all common in such habitats, namely grass and the herbivorous animals that feed on the grass. Of these, only grass is sufficiently abundant and evenly distributed to allow day journey length and range size to be reduced to a level commensurate with the maintenance of large groups. This is the strategy adopted by the theropithecines as a group, all of whom are adapted to a grazing niche (Jolly 1972). The extant gelada have the largest group sizes and some of the highest biomass densities of any naturally occurring primate populations (see Table 13.1).

One consequence of forming such large groups is likely to be a significant increase in the costs to the females in terms of both ecological disruption and reproductive suppression. In living gelada, inter-individual distances decline significantly as herd size increases (Dunbar 1986), resulting in a compression of individual units and an increase in stress. One way to overcome this problem is for related females to form coalitions: these would buffer their members against stress and harassment by keeping the rest of the herd at a reasonable distance without driving them away altogether (Dunbar 1986). With a *Papio*-like social system as the starting point, these coalitions would arise quite naturally out of the matrilineal groupings which are known to exist in *Papio* populations with reproductive parameters as high as those of the gelada (see Table 6.2). Thus, the ancestral theropithecines' move out on to the open savannah would have resulted in the evolution of large highly structured groups which were probably quite coherent and stable.

The living gelada did not, however, remain for long in the ancestral habitat. At a fairly early stage in the genus's history, their ancestors invaded the high-altitude montane habitats on the Ethiopian plateau (Jolly 1972, Szalay and Delson 1979). This habitat has a unique structure. It consists of two radically different sectors: a very steep, rather impoverished, cliff face of very considerable height on which the animals are virtually immune to predation (other than by large raptorial birds) and a flat plateau top above that supports rich grasslands but lacks any form of refuge into which the animals can escape from predators (see Table 7.1). To exploit the rich plateau habitat, gelada have to form large herds as their only form of defence against predators. But once back on the safety of the cliffs, the risk of predation is too low to overcome the tendency for reproductive units to drift apart under pressure from social stress. Hence, the small female kinship cliques that made up the large groups of the ancestral theropithecines would tend to drift apart from each other on the cliffs, thereby making it possible for individual males to establish hegemony over them (Dunbar 1986).

An alternative possibility would be to argue that the gelada social system arose directly from the one-male groups of the ancestral forest cercopithecines whose small one-male groups coalesced into herds in response to predation pressure as they moved out into the open grasslands. An obvious analogy here would be the patas which also occupy an open grassland savannah habitat while living in one-male groups (Hall 1965a). There are two difficulties with this suggestion. One is that it presumes a direct move from an arboreal forest niche to life on the open plains, which seems implausible. One or two intermediate stages as suggested for the *Papio* lineage seem almost unavoidable. In any case, if such a move were possible, it would invalidate our account of papionine social evolution and so require a completely new explanation for the evolution of multimale groups in mangabeys, macaques and baboons. Taken as a whole, the hypothesis that gelada one-male groups evolved by condensing out of a large multimale grouping requires fewer overall steps and makes for a generally more coherent account. The second difficulty lies in the fact that patas groups are considerably larger than the one-male units of the gelada (means of 18.9, $n=9$, and 12.0, $n=48$, animals, respectively; mean number of females: 6.2 and 4.1, respectively; Hall 1965a, Dunbar 1984a). This perhaps suggests that patas also went through a multimale group stage as they moved out from the guenon-like niche in the forests to the open grass-lands, but have since evolved one-male groups because of the unique character of both their habitat (it is more arid than that of the gelada) and their ecological strategy. Some evidence in support of this is provided by Chism and Rowell (1986) who found group sizes as large as 74 animals in a Kenyan population of patas. The largest gelada one-male unit ever recorded is 28 individuals (Dunbar 1984a).

The gelada social system provides a particularly clear example of the way in which occupancy of a new niche (itself presumably a solution to the prob-lem of dwindling forest habitats during the late Miocene) creates a new problem (increased predation risk), which has to be solved by a new solution (larger groups) which in turn generates a further problem (stress and repro-ductive suppression) that has to be resolved with yet another solution (female coalitions). It is this give and take that marks out the intrinsic complexity of the social systems of many primates.

Social Evolution in the Great Apes

The socio-ecological characteristics of the various species of great apes are summarised in Table 13.4. For comparison, I also list four species of gibbons. Three points are particularly striking about these data. One is that, with the exception of the siamang and the gorilla, all the apes are frugivorous. The second is that all, without exception, live in small groups. Among these, however, two distinct grouping patterns can be distinguished: the relatively stable groups of the monogamous gibbons and the polygamous gorillas and the unstable groupings of the more solitary orang utans and chimpanzees. The third point is the marked dichotomy in physical size: there are small apes (gibbons and siamang weighing 5–10 kg) and large apes (chimpanzees, orang utan and gorillas weighing 30–170 kg) but no medium-sized apes filling the niche occupied by the baboons. This curious bimodal distribution has attracted surprisingly little attention, most people apparently having been content to take the large body size of the great apes for granted. Since body size has important implications for many reproductive and ecological charac-ters, our understanding of their current adaptations may be greatly enhanced

Table 13.4. Behavioural ecology of the apes

Species	Body weight (kg)[a] Male	female	Diet[b]	Group size mean	Females	Range size (mean, km²)	Day journey (km)	Habitus	Mating system
Whitehanded gibbon	6.1	5.5	F	3	1	0.50	1.5	Arboreal	Monogamous
Agile gibbon[c]	5.9	5.9	F	4	1	0.22	—	Arboreal	Monogamous
Kloss's gibbon	5.8	5.8	F	3	1	0.07	—	Arboreal	Monogamous
Siamang	10.7	10.7	L	3	1	0.36	0.7	Arboreal	Monogamous
Chimpanzee	43.0	33.2	F	2–12[d]	1	5–13	3.9	Terrestrial	Multimale
Bonobo	45.0	33.2	F	8–17[d]	3–6	22	2.4	Terrestrial	Multimale
Orang utan	57.0	37.0	F	1	1	0.7–5.2	0.5	Arboreal	Area polygamy
Gorilla	170.0	80.0	H	10	2.5	4–12[e]	0.4	Terrestrial	One-male

Sources: unless otherwise indicated, all data derive from Wrangham (1979, 1986).

[a] Weight data for African apes from Jungers and Sussman (1984), Asian apes from Clutton-Brock and Harvey (1977).
[b] F, frugivore; L, folivore; H, herbivore.
[c] From Brockelman and Srikosamatara (1984).
[d] Mean number of individuals travelling together.
[e] From Fossey and Harcourt (1977).

if we can determine why they came to be so large.

Whatever its origins, however, large body size would have had two important consequences for the evolving ape lineage: (1) it would have increased the size of the area required to support one individual, and (2) it would have lengthened the interbirth interval and the period of infant dependency (in great apes, an interbirth interval of about 4 years is typical: Tutin and McGinnis 1981, Harcourt *et al.* 1981a, Galdikas 1981). Wrangham (1979) has pointed out that these have profound consequences for ape social evolution since they both increase the likelihood of feeding competition and reduce the predictability of oestrous females.

Taking this as our starting point, then, we can ask how differences in dietary strategies might influence the behavioural ecology of the various ape species. I shall not say anything here about the gibbons, partly because we have already discussed the evolution of monogamy in some detail in the previous chapter. More importantly, however, it is essential to concentrate on a group of species that are sufficiently closely related to each other phylogenetically to be considered an adaptive array. The great apes (and in particular the African apes) provide such a grouping (see Jungers and Susman 1984). The inclusion of the gibbons would significantly reduce the coherence and homogeneity of the group and thus complicate our analysis. Among the great apes, the chimpanzees are generally considered to be the closest to the ancestral Miocene apes since they are anatomically less specialised and more primitive than the other species (Tanner 1981, Zihlman 1984). Moreover, the chimpanzee is also less specialised ecologically and the general features of its social system provide the kind of flexibility that allows us to derive the social systems of the other species fairly easily. I shall therefore begin by examining the chimpanzees in some detail.

Chimpanzees

Detailed studies of a number of populations living in East and West Africa show that chimpanzees live in fairly stable communities that consist of between 3–6 (Nishida 1968, Sugiyama and Koman 1979) and 12–18 (Sugiyama 1968, Suzuki 1979, Goodall 1983) adult males and a comparable number of adult females and their dependent young. Each community occupies its own ranging area, variously estimated to be between 5–8 km^2 (Sugiyama 1973, Sugiyama and Koman 1979) and 10–13 km^2 (Wrangham 1977, Nishida 1979), with exceptional communities in poorer quality habitats occupying ranges as large as 100 km^2 (Suzuki 1969, Baldwin *et al.* 1982). Although community size appears to be similar in the bonobos or pygmy chimpanzees, range sizes apparently tend towards the larger end of the distribution (22 km^2 for the one community for which data exist: see Kuroda 1979, Badrian and Badrian 1984b).

Despite these large group sizes, however, chimpanzees spend most of their time in very small parties whose composition changes constantly. Ghiglieri

(1984) found a median party size of just 2.6 in the Kibale (Uganda) population, whereas Suzuki (1969) gives a median of 11–12 individuals at Kasakati in Tanzania. The largest parties recorded are for the bonobos, with mean party sizes of 7.9 and 16.9 being recorded for two different populations (Badrian and Badrian 1984b, Kuroda 1979). Within populations, however, there is considerable variance in party size, with groups of up to 30 individuals congregating at particularly good food sources. In fact, most studies report that party size is dependent on seasonal changes in the size of the food patches, with larger parties being found during the fruiting season than at other times of year (see, for example, Wrangham 1977, Badrian and Badrian 1984a).

From a detailed analysis of chimpanzee ranging patterns, Wrangham and Smuts (1980) concluded that whereas the male members of a chimpanzee community defend a large common range the females have their own individual ranges within the males' communal range. At Gombe, the communal range is of the order of 10 km², but the individual ranges of the females are only about 2 km² in area. They argued that although females occasionally come together at good food sources, they are essentially solitary animals, and that males form what amount to coalitions to defend the ranges of as many females as they can effectively cover.

Male mating strategies.   The problem from a male's point of view is this: should he wander at random through his range searching for females in oestrus, or would he be better off attaching himself to one female (or a group of females if they form more stable groups)? The best approach to this problem is to consider a simple model and then vary its parameters so as to gain some idea of how a male's mating success will vary under different conditions. Consider a Gombe male who can move at random through a 10 km² range that contains the ranges of five females, each of whom occupies a non-overlapping 2 km² range. What we want to know is the average number of oestrous females that a male will locate (and therefore presumably fertilise) during the time it takes the average female to complete one reproductive cycle if either (1) he searches randomly, or (2) he attaches himself to a group of females for the whole time.

The expected number of oestrous females for a social male is obviously the number of females in the group he associates with, since each female will come into oestrus just once during the sample period. The expected number of oestrous females that a randomly searching male encounters will depend on the density and dispersion of the females and on the area that the male can search during the sample period. To determine his rate of success, we can use a modified version of the 'gas' model that Waser (1976) used to analyse encounter rates between groups of mangabeys. (It is called a 'gas' model because it derives from the equations used to determine collision rates between atoms in a gas cloud.) The number of oestrous females that a randomly searching male will encounter, $E(f)$, is:

$$E(f) = K\frac{2dv}{A}(p_n.ng) \tag{13.1}$$

where K is the length of the interbirth interval (in this case, $K = 4 \times 365$ days $= 1460$), d is the male's detection distance (the radius around him within which he can detect a female), v is his velocity per day (i.e. the length of his day journey), A is the total area of his range (in the same units as d and v), p_n is the probability of finding a group of females of size n (i.e. the density of female groups), and g is the proportion of females in a given group who will be in oestrus on any given day. Assuming that a female is in oestrus for just 10 days in each of three menstrual cycles before conceiving during a full 4-year reproductive cycle, $g = 30/(4 \times 365) = 0.0205$. These variables yield a daily rate of encounter with oestrous females, so to obtain the number he will encounter during the full reproductive cycle we need to scale the equation up by $(4 \times 365) = 1460$. Note that this equation will slightly overestimate the number of oestrous females that a male encounters because it ignores the fact that, once a female has been fertilised, any further encounters with that female are of no real interest to him. However, since the likelihood of encountering any one female is quite small, the length of time before he encounters her again will be quite long (around 50 days under the most favourable conditions), so that the error introduced by ignoring this factor will be small enough to ignore for present purposes. We can now consider the male's expected gains for various values of d, v and n. In the latter case, I assume that, when females form groups, their density remains the same: p_n will, however, vary inversely with n.

Table 13.5 gives the number of oestrous females that a randomly searching male can expect to encounter during a 4-year period for various male search resolutions (dv) and various female group sizes (n). Initially, I have considered the case where the male's detection distance is $d = 0.5$ km: in other words, he can search an area 1 km wide along his route of travel. Although it may seem rather large, we need to remember that auditory cues may allow males to detect females at very much greater distances than they can detect them visually. Moreover, the hilliness of the terrain and the use of trees may allow males to detect females at distances that are significantly greater than those at which they could be detected on a flat surface under identical vegetation conditions. A value of $d=0.5$ km may not, in fact, be too unrealistic given that Waser (1976) found that small mangabey groups could detect each other at distances of 500 m on a significant proportion of occasions. Table 13.5 also gives the results for two other conditions: (1) assuming that the females occupy smaller ranges (roughly equivalent to those of female orang utans observed by Horr 1972 and Rodman 1973, so that the density of females is effectively doubled), and (2) assuming that the male's search

Table 13.5. Number of oestrous females that a male can expect to locate in a 4-year period if (1) he searches randomly and (2) he attaches himself permanently to a group of females in relation to female group size, n, his day journey length, v, and the distance at which he can detect females, d, (see text for details)

Female density (x/km²)	Parameters of model[a]				$y=$	Number of oestrous females obtained by:				
	n	$P_n{}^{b}$	d (km)	A^c (km²)		Randomly searching male with $v = y$ km				Social male
						0.5	1.0	2.0	4.0	
0.5	1	0.5	0.5	10.0		0.8	1.5	3.0	5.0	1.0
0.5	2	0.25	0.5	10.0		0.8	1.5	3.0	5.0	2.0
0.5	5	0.10	0.5	10.0		0.8	1.5	3.0	5.0	5.0
0.5	1	0.5	0.25	10.0		0.4	0.8	1.5	3.0	1.0
1.0	1	1.0	0.5	5.0		3.0	5.0	5.0	5.0	1.0
1.0	1	1.0	0.25	5.0		1.5	3.0	5.0	5.0	1.0
0.4	1	0.2	0.25	5.0		0.6	1.2	2.4	2.5	1.0

[a] The model assumes that the probability of a female being in oestrus on any given day is (30/4 × 365) = 0.0205 if the female is in oestrus for 10 days in each of three menstrual cycles before conceiving on each 4-year reproductive cycle.
[b] Probability of locating a group of n females when searching at random (i.e. the density of female groups).
[c] Male's ranging area.

resolution is poorer and that he can only search an area 250 m either side of his travel path.

Two points may be noted about the distributions in Table 13.5. First, whether or not females live in groups does not affect the distribution of the number of oestrous females that a searching male can find. This is intuitively obvious on reflection, since although a male will take longer to locate a large group of females, this will be offset by the fact that the probability of locating an oestrous female (ng) will be proportionately greater. Secondly, both the density of females and the male's search resolution have a dramatic impact on the likelihood that the male will find an oestrous female. As an aside, we may also note that the length of the female's reproductive cycle itself is not all that important: it simply acts as a scaling factor that influences social and searching males equally.

Table 13.5 shows that whether a male will find it worth his while to join a female group depends critically on the size of the female groups. If females forage alone (as the Gombe females are supposed to do), then it will always pay the male to search randomly, unless his day journey length is very short or his search resolution very poor. As female group size increases, however, the advantage of a searching strategy declines, so that even with groups of two females it would pay a male to stay with them unless he is able to travel very long distances each day (greater than about 3 km). Were females to form groups of five, the male would have to be able to search the whole of his ranging area each day to break even: in other words, his day journey length would have to approach 10 km, a value that only the hamadryas baboons have been found to manage at all consistently. The female density (i.e. their range size) has the same effect: as density declines (or range size increases), so the advantages of a searching strategy are reduced.

Given the behavioural ecology of the Gombe chimpanzees, it would generally seem to pay a male to search randomly almost no matter how narrow his search resolution. With females ranging separately in 2 km² ranges, the break-even point for the male comes at a daily search area, $2dv$, of about 0.75 km². For males that normally travel about 4 km a day (Wrangham 1977), this implies that the width of their search path is only $0.75/4=0.19$ km: in other words, a male only has to have a detection distance of 100 m to do better by a random search strategy. An important test of the model would therefore be to determine the distances at which males can detect females in the wild. Note, however, that this parameter will be habitat-specific: the model will yield different predictions under different search resolutions. More precise knowledge of the male's search resolution in different habitats would, none the less, help us to develop a quantitatively more refined version of the model.

Tests of the model.   Despite its rather crude form, does the model nevertheless correctly predict the ranging strategies of males from communities other than the Gombe population? Data for bonobos in the Congo, for example, indicate that the animals do form larger groups (mean number of females per group is 5.9: Kuroda 1979) and travel less far each day (mean day journey of 2.4 km: Kano and Malavwa 1984). In addition, the population density is about half that of the Gombe chimpanzees (c. 2.0 versus 3.5 animals per km², respectively: Wrangham 1986; the bonobo range area is about double that of the Gombe chimpanzees: see Table 13.4). This would make the last line of Table 13.5 the appropriate basis for comparison. With a female group size of about six females, there is no question that a bonobo male in this population will benefit by staying with the females rather than searching randomly for female groups, as seems to be the case (Badrian and Badrian 1984b).

The model also makes one other prediction. Because chimpanzee parties are very unstable, with fission and fusion taking place throughout the day, any male who joins a party of females may well, later on, find himself accompanying a smaller group. Table 13.5 suggests that we should expect males to alter their searching strategy as female group size changes on a daily basis: they should stay with parties that contain more than the critical number of females and leave those that contain fewer. In addition, if dominant males are able to monopolise access to any oestrous females in a group, then low-ranking males will gain little by staying once group size falls below a critical limit; instead, they should leave to search for a new group. This will give rise to a positive correlation between the number of females in a party and the number of males. Table 7.4 confirms that there is a significant correlation between the number of females in a party and the mean number of males associated with it in the two populations for which such data are available. Further evidence to support the prediction is given in Figure 13.1. This shows, for all chimpanzee populations that provide such data, that the proportion of parties that contain both adult males and adult females ('bisexual parties' in most classifications) increases with the mean party size, with an asymptote at about 85%.

Female grouping strategies.   So far, we have simply assumed that females form foraging groups of a certain size. What determines the party sizes that females are found in and, more specifically, why is it that females often forage alone? One likely explanation lies in the length of the typical day journey. The data given in Table 13.4 suggest that all frugivorous apes (gibbons, chimpanzees and orangs) have *per capita* day journeys in the order of 0.5–1.0 km. This is exceptionally high: the five *Papio* species (which are also preferentially frugivorous) have a *per capita* day journey that averages only 0.09 km (Table 13.1). Why should the large apes travel so far? There are two reasons. One is their size: with a metabolic body weight (p. 34) that is approximately double that of the average baboon, they will inevitably require double the energy intake to maintain metabolic processes (p. 33) and hence

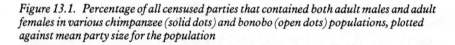

Figure 13.1. Percentage of all censused parties that contained both adult males and adult females in various chimpanzee (solid dots) and bonobo (open dots) populations, plotted against mean party size for the population

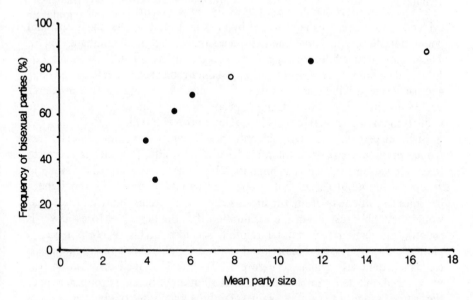

Sources (in order of increasing party size): Gombe, Tanzania (Goodall 1965); Budongo, Uganda (Sugiyama 1968); Mt. Assirik, Senegal (Tutin et al. 1983); Mahale, Tanzania (Nishida 1968); Lomako, Zaire (bonobo: Badrian and Badrian 1984b); Kasakati, Tanzania (Suzuki 1969); Wamba, Zaire (bonobo: Kuroda 1979).

will be obliged to search double the area each day to find the required quantities of food (p. 47). The other reason is the difference in diet: baboons are omnivores that will exploit other kinds of food besides fruits if they need to, whereas the apes (and particularly the chimpanzees) are almost entirely dependent on fruit (Wrangham 1986). In effect, they face the same problem as squirrel monkeys do in the dry season: because fruiting trees are irregularly distributed, it is often necessary to travel considerable distances from one to the next. Some indication of this is given by the fact that there is a correlation between the size of food patches and/or seasonal abundance of fruits and the size of chimpanzee parties in almost every population that has been studied (see Wrangham 1986).

Wrangham (1979, 1986) has argued that feeding competition forces the large parties to disperse when food patches become smaller, as in the non-fruiting season. As evidence of this, he has pointed to the fact that time spent feeding correlates negatively with party size.[1] This could be due either to direct competition for access to individual feeding sites or because the patchiness of the food supply forces animals to travel further when they are in groups. Wrangham (1977) has shown that feeding parties are smaller at smaller sites. However, that in itself is probably less important than the fact that the sites themselves are dispersed in space: if many small sites are close together, then there would be no need for large groups to break up. Some evidence to support this is provided by data on time budgets, party size and population density at Kibale and Gombe summarised by Wrangham (1986). These show that while time spent travelling remains constant at about 12–14% of day time, both density (an approximate measure of habitat quality) and party size at Kibale are half those at Gombe. This suggests that in poorer quality habitats party size declines proportionately in order to avoid the animals incurring excessive costs by having to travel too far.

Such dispersion, however, is only possible providing the animals can reduce predation risk to a tolerable level. Essentially, this means access to trees. Hence, in well wooded habitats, chimpanzees can afford to move in groups of minimum size, but group size should increase as the habitat becomes less wooded despite the increased costs of extended day journeys that would inevitably result. On the assumption that the degree of forest cover is roughly proportional to annual rainfall (see also McGrew *et al.* 1981), I plotted mean party size against local rainfall for all chimpanzee populations for which data are available (Figure 13.2). The U-shaped distribution was unexpected, but its resemblance to the distribution of baboon group sizes over the same range of rainfall values (right hand side of Figure 7.8) is striking. The relationship, however, is, on face value, weak and a skeptic would point out that a curvilinear regression set through the data would produce a nearly linear equation (slope $b=0.404$ with a negligibly small quadratic term). As a counter to this, we can point out that the equation is clearly being weighted by the fact that most of the data points lie to the right of the presumed trough of the U-shape and that the slope parameters of the linear equation are in any case not significantly different from zero ($F_{2,6}=0.078$). Such a result therefore ignores the steeply rising distribution on the right of the trough where party size is significantly correlated with rainfall (Spearman $r_s = 0.929$, $n=7$, $p < 0.01$ 2-tailed, compared to $r_s = 0.533$, $n=9$, $p > 0.05$ for the whole sample). One way to resolve this dilemma is to ask how well the observed distribution fits different kinds of possible curves. Because a parametric regression produces a linear relationship, the easiest way to do this analysis is non-parametrically using Spearman correlation analysis (for another example of this procedure, see Dunbar 1986).

The four theoretical options would seem to be (1) a positive linear correlation, (2) no correlation at all, (3) a U-shaped relationship and (4) a J-shaped relationship (with the long tail to the right). We can calculate expected ordinal distributions for group size plotted against rainfall for each of these by reassigning the group size data to give the closest fit to the desired theoretical relationship. The first is obviously exactly the same order as rainfall: the smallest group size is paired with the lowest rainfall value, and the largest

Figure 13.2. Mean party size for various chimpanzee (solid dots) and bonobo (open dots) populations, plotted against mean annual rainfall for the habitat. Regression line fitted by eye

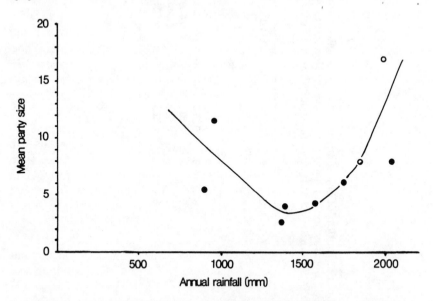

Sources (in order of increasing rainfall): Mt. Assirik, Senegal; Kasakati, Tanzania; Kibale, Uganda (Ghiglieri 1984); Gombe, Tanzania; Budongo, Uganda; Mahale, Tanzania; Lomako, Zaire; Wamba, Zaire; Okorobiko, Equatorial Guinea (Sabater Pi 1979). (Unless otherwise indicated, sources are as for Figure 13.1.)

group with the highest rainfall. The second is essentially a random distribution: this was determined by assigning a group size rank to each rainfall rank using a random number, the final ranking being based on the average of five rankings determined in this way. (Note that we could have analysed significance of fit in these two cases simply by determining the likelihood that the correlation between the observed party size and rainfall ranks differed from theoretical correlations of $r=1$ and $r=0$, respectively.) The third distribution was determined in exactly the same way as we did for the baboon and howler analyses in Figure 7.4, namely by assigning party sizes in descending order alternately to the right and left sides, working inwards from the outermost rainfall values. The fourth theoretical distribution was determined in the same way, except that the first three party sizes were assigned to the three largest rainfall values before alternating sides. We can now correlate the observed ranks of party size against each of the four theoretical distributions. The observed and expected rank orders and the resulting statistical analyses are given in Table 13.6. The best fit is clearly to a J-shaped distribution (i.e. a truncated-U), this being the statistically most significant correlation. The fit

Table 13.6. Analysis to determine the most likely shape of the distribution of chimpanzee party sizes given in Figure 13.2: Spearman correlations are used to compare the observed distribution with expected distributions obtained by re-assigning the observed party sizes among the various habitats in order to obtain specific relationships with the distribution of rainfall (see text for details)

Population[a]	1	2	3	4	5	6	7	8	9	r_s[b]	t_7	p(1-tailed)[c]
Rainfall (mm)	955	975	1360	1495	1570	1760	1830	2005	2110			
Mean party size:												
Observed: mean	5.3	11.5	2.6	5.6	4.4	6.1	7.9	9.9	16.9			
rank	3	8	1	4	2	5	6	7	9			
Expected: linear	1	2	3	4	5	6	7	8	9	0.533	1.667	0.070
random	7	1	8	6	4	5	3	9	2	−0.533	−1.667	0.930
U-shape	8	6	4	2	1	3	5	7	9	0.600	1.984	0.044
J-shape	6	4	2	1	3	5	7	8	9	0.683	2.475	0.021

[a] Souces as for Figure 13.1.
[b] Spearman correlation between observed and expected rank order of party size.
[c] One-tailed probability of goodness-of-fit of observed to expected.

to a random distribution is, by comparison, particularly poor. Thus, pending data from more study sites, some form of U-shaped distribution is the most likely.

Clearly, increased predation risk in less wooded habitats would explain the rising curve to the left, but predation risk hardly seems an appropriate explanation for the rising curve to the right. This leaves only one real possibility: habitat quality. We could suggest that this reflects greater food availability on the grounds that a number of studies have reported that group size increases with habitat quality (see p. 77). But a correlation does not imply a causal relationship. Indeed, as we have already noted (p. 116), food supply cannot cause animals to form groups, it can only facilitate the formation of groups when these are required for other purposes. Only when the food supply is very patchy in its distribution will animals be forced to clump and form groups. Can we justify this alternative interpretation, then?

Several studies have noted that climax primary forest is often less productive than secondary or colonising forest because it contains a lower diversity of plant species and a less dense understorey (see, for example, Oates 1977a). Thus, as forest tends towards climax, its quality in terms of the abundance and patchiness of food sources is likely to decline. The animals will then be forced to converge on those few food sources that are available. As a general rule, it is the case that primary forest occurs in areas of high rainfall. Wrangham (1986) has pointed out that the three populations with the highest rainfall (the two bonobo study sites and the West African lowland forest population at Mt. Okorobiko in Equatorial Guinea) feed much more heavily on herbage than do other chimpanzee populations in more woodland habitats in eastern Africa. (Indeed, bonobos exhibit a degree of dental adaptation to a more herbaceous diet: Kinzey 1984.) Herb beds often constitute substantial concentrated food sources that the chimpanzees rely on as a staple food when fruits are not available (Kuroda 1979, Badrian and Malenky 1984). Although Wrangham (1986) suggests that these might be rich food sources, this seems implausible for a species that is not digestively adapted to a folivorous diet for the reasons discussed in Chapter 3 (p. 38).

I would interpret these data as suggesting that female chimpanzees are forced into groups in areas where *either* predation is high *or* food sources are limited and highly clumped in their distribution.

When females form groups, males will tend to join them in order to be able to monitor their reproductive state better. Wrangham (1986) has offered a different interpretation, arguing that females prefer to associate with males because this reduces the risk that they run from attack by other males. To decide conclusively between these two hypotheses, data would be required on the frequencies with which males and females join and leave parties and on the exact relationship between male and female stay times in parties of different size. Such data are not yet available, but it is possible to carry out an indirect test with existing data.

If it is males that join parties of females in order to acquire access to oestrous females, we would expect, from Table 13.5, that males will be anxious to prevent females leaving groups that are over the critical size. One way males might persuade females not to leave is to groom with them. Hence, we would expect a correlation between mean group size and the proportion of the males' grooming bouts with other adults that are with females. Conversely, if it is females that join groups of males for protection, then we would expect that females will benefit by preventing males from leaving and so will work harder. There are, however, two ways of viewing the female's dilemma: (1) they might be most worried about losing the last male in the group (in which case we would expect a negative correlation between mean group size and the proportion of the females' grooming bouts with adults that are with males) or (2) they might be more concerned to maximise the number of males in the group (in which case we would expect a positive correlation). Table 13.7 presents data on mean group size and the proportion of the males' and females' grooming bouts that are with members of the opposite sex. There is no correlation between mean group size and the proportion of the females' bouts that are with males ($r_s = -0.107$, $n=7$, n.s.), whereas there is a significant positive correlation between mean group size and the proportion of the males' bouts that are with females ($r_s = 0.857$, $n=7$, $p < 0.02$ 1-tailed). It seems most likely that males join females as predicted by the mating access model.

Community structure.   One final question remains to be answered. Why should all chimpanzee populations exhibit clear evidence of a community structure, especially among the males? Wrangham (1979) has argued that this is because it pays males to form coalitions that can defend the ranging areas of a number of females. In this way, they can maximise priority of access to

Table 13.7. Mean group size and the proportion of an adult's grooming that is with members of the opposite sex for various chimpanzee populations

Population	Mean group size	Grooming bouts with opposite sex (%) males	females	Source
Kibale, Uganda[a]	2.6	41.4	37.7	Ghiglieri 1984
Budongo, Uganda[a]	4.4	15.4	83.2	Sugiyama 1968
Gombe, Tanzania[a]	5.6	23.7	75.1	Goodall 1968
Mahale, Tanzania[a]	6.1	45.9	79.7	Nishida 1979
Lomako, Zaire[b]	7.9	78.1	59.5	Badrian and Badrian 1984b
Yalosidi, Zaire[b]	8.6	100.0	70.8	Kano 1983
Wamba, Zaire[b]	16.9	82.9	61.7	Kuroda 1979

[a] Common chimpanzee populations.
[b] Bonobo populations.

females in oestrus. In the worst case in which females occupy individual non-overlapping ranges (e.g. Gombe), what is to prevent a dominant male from defending a large territory of his own which would give him exclusive access to the many females living within it? Chimpanzee day journeys are often quite substantial by ape standards (3.3–6.4 km in the Gombe population according to Wrangham 1977, 2.8–3.5 km according to Wrangham 1986), so that, even with the shortest of these, equation 12.2 suggests that an individual male ought to be able to defend an area of 6.2–9.6 km^2, with the larger community ranges being well within the capabilities of some individuals. With a mean female range at Gombe of 2.1 km^2, a territorial male ought to be able to monopolise 3–5 females.

The reason why chimpanzee males do not do so almost certainly lies in the sheer size of the female ranges. No species that actively defends a territory has a ranging area greater than 1 km^2 (see Mitani and Rodman 1979). If a dominant male can only defend an area of 1–2 km^2, then he would only ensure access to at most one female, in which case he would always be better off ranging more widely in order to gain access to more females. Under such conditions, a male's best strategy may be to share the costs of area defence with other males so as to be able to minimise the number of rivals he has to compete with (Wrangham 1979). Given this, it is no surprise to find that the males of a chimpanzee community are closely related to each other (Wrangham 1979) or that males within a community tend to form defensive coalitions preferentially with their brothers (Riss and Goodall 1977). In this way, a male will at least gain through kin selection even if he fails to mate with any of the females living within his range.

One obvious prediction here is that dominant males will be able to monopolise a larger proportion of the matings in populations that form larger, more compact groups than when females are more dispersed (Chapter 8). This prediction is confirmed by a comparison of data on mating rates in two of the Mahale Mountains (Tanzania) communities. Hasegawa and Hiraiwa-Hasegawa (1983) found a significantly higher proportion of 'restrictive' (i.e. monopolised) matings in the population with the larger mean group size (K-group: 27%; M-group: 6%; χ^2 test, $p < 0.005$). Moreover, the dominant male achieved a higher proportion of all copulations in the population with the larger groups (K-group: 32.5%; M-group: 10%; χ^2 test, $p < 0.001$).

We can pursue the logic of this a little further and argue that if female chimpanzees formed very cohesive groups (perhaps because they ventured right out on to the open savannah), we would expect dominant males to be able to monopolise them more easily. With the long interbirth intervals, males would be able to defend quite large groups. Even if females bred seasonally, the probability of a female being in oestrus on any given day of a 3-month breeding season during an average 4-year reproductive cycle is only $p = 0.042$ (assuming as before that she has an average of three menstrual cycles to conception). Interpolating these values into the graphs in Figure 7.9 suggests that a dominant male would be able to monopolise groups of 15–20 females. This clearly has significant implications for the evolution of hominid social systems.

Orang utan

We can view the orang utan as the end-point of a trend towards solitary foraging by female apes. With the risk of predation minimised by their large size and arboreal lifestyle, there is little to prevent female orangs foraging alone if the costs of group-living are at all significant. Wrangham (1979) has suggested that they are obliged to forage separately by severe costs incurred from foraging in groups. Again, we can attribute this either to direct competition for access to feeding sites or to indirect costs incurred by having to travel further when in groups. Mackinnon (1974) noted that, unlike chimpanzees, which eat only ripe fruit, orang utans will eat fruit at any stage of growth: indeed, their ability to discriminate bitter from neutral tastes seems to be poor (Chiarelli 1963). Thus, although the predominantly dipterocarp forests that they inhabit are relatively poor in terms of fruit production (see, for example, Caldecott 1986a,b), their ability to exploit a wider range of foods than are exploited by chimpanzees should allow them to reduce the day journey and range size required to support an animal proportionately. In fact, *per capita* day journey length is no greater than that for chimpanzees (Table 13.4). The more likely cost of grouping may thus lie in the costs of travel for an animal that is both as heavy and as arboreally adapted as the orang. For such an animal, movement through the trees is necessarily ponderous and slow since the animal is too heavy to jump gaps between trees or make use of terminal twigs to move from one tree to the next. Solitary foraging by females thus seems unavoidable.

Given that females forage alone, what should the males do? To be able to make use of equation 13.1, we need to know what size of territory the female has. Horr (1972) and Rodman (1973) give figures of about 0.7 km²; Mackinnon (1974), on the other hand, found that groups of females occupied communal ranges with densities in the order of 2.5 km² per female. All three studies agree that males occupy ranges of about 5 km², however. Interpolating these values into equation 13.1 gives us the last three lines in Table 13.5. With the female range sizes observed by Horr (1972) and Rodman (1973), a male orang utan gains by searching randomly for oestrous females no matter how short his day journey length or his detection distance. In Mackinnon's (1974) study area, the payoffs are more finely balanced: so long as his detection distance is $d=0.5$ km, a male that can travel only 0.5 km per day just gains by a searching strategy, but if detection distances are any less than this it would pay a male to join a female in a monogamous relationship.

If population densities in Mackinnon's (1974) study area are atypically low, we can plausibly intepret his males as finding themselves trapped in an impossible situation. Ideally, it would pay them to be social and join a female but they are prevented from doing so by their own massive size which makes it all but impossible for them to keep up with the female as she moves through the trees (large males are obligatorily terrestrial). Their large physical size has presumably evolved in response to intense competition among males in a context in which area-defence polygyny (Emlen and Oring 1977) is a profitable strategy. When male orangs do meet, the result is often a violent battle, especially if one of them is consorting with an oestrous female (Galdikas 1981).

Though such a strategy is effective in competition with males of comparable size, it does leave mature males exposed to the risk that immature subadult males can out-manoeuvre them in the trees and so gain access to females under the very nose of a territorial male. Subadult males do, in fact, seem to do this (Galdikas 1981, 1985, Schürmann and van Hooff 1986) and this may account for the high rate of forced 'rapes' by younger males noted by Mackinnon (1974). A rather similar age-based shift in ranging and mating patterns has been reported for ibex, with younger males ranging widely in search of females and older males remaining within more delimited areas (Dunbar and Dunbar 1981). To some extent, territorial males may be able to rely on the females' preferences for mating with a male whose 'fitness' has been demonstrated by his survival to become a territory-holder. The female's resistance to the advances of younger males often does attract the resident male's attention, and he may rush across to them to chase the offending subadult away (Galdikas 1981). The territorial male's frequent 'loud calls' may serve to advertise his whereabouts to the females within his territory, thereby attracting them to him (Horr 1972, Mackinnon 1974, Galdikas 1981, 1985).

Gorilla

The gorilla represents something of an anomaly among the great apes in so far as it is the only species that lives in permanent stable groups. However, we can view the gorilla as the logical end-point of a trend towards increasing body size initiated by the orang utan. Again, the sheer size of the gorilla, as the largest of all the primates, suggests a response to predation. But unlike the orang and the chimpanzee, the gorilla has pushed beyond the limits of arboreality to achieve a size at which it is no longer capable of exploiting an arboreal refuge. Quite why gorillas should be *so* large is puzzling, because most species that are large for their taxon are open-country species (e.g. patas among guenons, baboons among papionines, the gelada's extinct sister species among theropithecines, living zebra and horses among the equids and even our own species among the hominids); yet the gorilla's other anatomical characters clearly suggest that it has never passed through such a phase. An alternative possibility is that it opted to exploit the relatively abundant herb layer within tropical forests. Exploiting a terrestrial resource of this kind, even within the confines of the forest, necessarily increases the risk of predation.

Only the small secretive antelope can exploit a terrestrial niche while remaining solitary (e.g. Feer 1979). An animal that was probably already as large as a chimpanzee would have had little choice but to increase its physical size or increase its group size. Neither strategy would have been easy for a frugivore. The metabolic body weight of an 80-kg female gorilla is almost exactly double that of a 33-kg female chimpanzee, so that, with a *per capita* day journey of 0.5–1.0 km for the latter (Table 13.4), a frugivorous gorilla would need to travel around 1–2 km a day. Since a group of 10 gorillas would have to travel 10–20 km a day and solitary foraging would be far too risky (even animals as large as buffalo sustain significantly higher predation rates when on their own: Sinclair 1977), a terrestrial frugivorous niche is clearly out of the question for so large an animal. However, a herbaceous diet makes it possible to sustain much higher biomasses (p. 45) so that animals can reduce day journey length while increasing group size without incurring serious costs. The obvious comparison here is the gelada: these are able to maintain very large group sizes at very high population densities on very short day journeys because they feed exclusively on grass (Table 3.3; see also Table 13.1). The ratio of metabolic body weights for an 80-kg female gorilla and a 13.8-kg female gelada is 3.74:1, while their *per capita* day journeys scale at a ratio of 4.59:1 (Tables 13.4 and 13.1, respectively). Given that large physical size is itself a deterrence to predators, animals the size of the gorilla may only need to live in relatively small groups to gain from the predator-detection advantages of grouping (Pulliam 1973).

Wrangham (1979) has offered an alternative explanation as to why gorilla females should form groups, namely that they are seeking the protection of dominant males from sexual harassment by males in the community at large. Since dominant males could be regarded as a resource over which the females compete, we would predict an inverted-U-shaped relationship between reproductive rate and female group size (Figure 7.1). In contrast, a predator-defence hypothesis would predict a negative relationship between group size and reproductive rate (Figure 7.1). In fact, the relationship between the two variables is negative (Harcourt *et al.* 1981c), as predicted by the predator-defence hypothesis.

With females clustered into groups of 2–3 (Harcourt *et al.* 1981c), Table 13.5 suggests that a male would always benefit by staying with them rather than pursuing a random search strategy. Only if he can manage very long day journeys and/or search a wide area will he do better by random searching. Since the gorilla's bulk and morphology seem to preclude the kind of sustained rapid travel required to achieve a day journey of 3–4 km, small harem groups defended by a single breeding male seem inevitable. Thus, once again, the observed constellation of parameter values seems to suggest that the gorilla's social system is an optimal solution to the problems they face.

Conclusion

We have learned a great deal during the past decade about the ecology and demography of primates. This new information has given us a very much more detailed picture of how the main components in an animal's biological system interact, which, in turn, has provided us with a more sophisticated understanding of socio-ecological relationships than those developed during the 1960s and 1970s. One key finding to emerge from these new data is the immense flexibility of most primates' ecological and social strategies. During the 1960s, we tended to assume that all aspects of an animal's biology were determined by its species-specific biological strategy and ecological niche. As a direct result of this assumption, variation around the species-typical mean was regarded as no more than trivial biological error. We have since come to realise that this variance is actually a reflection of the animals' responsiveness to environmental and demographic conditions that are changing continuously throughout an individual's lifetime. Different populations of the same species may, in fact, have radically different social systems not because they are genetically different, but because a given fundamental sociobiological principle can generate radically different optimal solutions in different ecological contexts. The degree of reproductive synchrony among females and the rates of population growth turn out to be important considerations in this respect. This flexibility underlines the point that we must beware of assuming that an animal's behaviour is a simple expression of some fundamental principle. What it does is a consequence of that principle finding expression in a particular biological context: the biological state variables describing that context will play an important role in dictating the optimal solution that any such principle predicts.

The emphasis on individual strategies stemming from sociobiology has made field workers more aware of just how variable the social strategies of different individuals (or even the same individual at different times) can be. This in turn has directly prompted a shift of perspective away from an essentially behaviourist perception of animals as genetically programmed machines to a more cognitive view in which animals are conceived as evaluating strategies in a rational way. Inevitably, this has raised the thorny question of consciousness, an issue that has come to dominate studies of animal cognition in recent years (see Humphrey 1976, Griffin 1981, 1982). I do not wish to discuss the question of consciousness in primates in the present context, but it is important to recognise that the assumption of consciousness may be an indispensible tool in aiding our explorations of how animals behave.

Kummer (1982), in particular, has stressed the extent to which our understanding of primate social strategies is restricted by the behaviouristic assumption that animals are no more than machines responding automatically to stimuli presented to them by the outside world. None the less, ethologists continue to adhere to a neo-behaviourist position as an essential safeguard for scientific validity. Such a view may be fine for the relatively simple ecological strategies pursued by most animals, but it will not do when we come to consider the very much more sophisticated social behaviour of primates interacting in complex social systems. Here, we need to know not just what an animal does, but also what it is *trying* to do: what it actually ends up doing is a compromise between what it wants to do and what it can do (Dunbar 1984a; see also Seyfarth 1980, Kummer 1982).

Animals as advanced as primates undoubtedly do evaluate their options and choose those which are most advantageous. This is not to say, however, that they make their decisions on the basis of the way in which their options affect their genetic fitness, even though that is the ultimate consequence of what they decide to do. Rather, their choices are made in terms of more proximate cues: these may be cues such as how hungry they are, the extent to which their sexual needs are satisfied or even such nebulous considerations as how comfortable or stressed they feel. Their immediate concerns are to solve specific biological problems of survival and reproduction as these become of primary importance on a moment-by-moment basis in the light of the general state of the biological state-space.

What makes it possible for us, as observers, to intepret an organism's behaviour in terms of its contribution to genetic fitness is that its proximate goals have come to be entrained to genetic fitness over evolutionary time. In other words, evolutionary processes have selected those individuals whose proximate goals are most in tune with genetic fitness precisely because they have been the ones that have reproduced most successfully in the past. This need not imply that an animal's behaviour is genetically determined, only that its goals are programmed into it (directly or indirectly) and that it has a genetically inherited mechanism for evaluating the mismatch between this goal and its current state so as to be able to make a decision as to how to rectify the situation. Consciousness then emerges as a mechanism for making more effective decisions that are sufficiently flexible for fine tracking of environmental changes to be a realistic possibility for such a long-lived organism (see, for example, O'Keefe 1985). Grasping the nettle not only allows us to see a great deal more of the inherent complexity in the social behaviour of primates (and indeed other higher vertebrates) but also provides us with a powerful tool for the exploration of that social world.

Note

1. Wrangham (1986) offers two sets of data to substantiate this claim. One is that Wrangham and Smuts (1980) and Bygott (1974) found that females and males, respectively, spent significantly less time feeding when in groups than when alone (though this might, of course, be influenced by what other activities animals form groups in order to do — e.g. mate). The other is that he found a significant negative regression for time spent feeding plotted against group size, apparently based on

pooling data given by Wrangham (1977) for three different seasons. If each season's data are analysed separately, there is a significant negative correlation for one set of dry season data ($r_s = -0.587$, $n=12$, $p=0.022$ 1-tailed) but not for the following dry season ($r_s = -0.067$, $n=9$, $p=0.432$ 1-tailed), whereas the intervening wet season yields a near-significant *positive* correlation ($r_s = 0.323$, $n=17$, $p=0.897$ 1-tailed). Fisher's procedure for combining significance levels from different tests of the same hypothesis (see Sokal and Rolf 1969, p. 621) indicates that there is no overall underlying trend ($\chi^2 = 9.529$, df=6, $p > 0.10$). Although it seems likely that the Gombe chimpanzees may sometimes face significant disruption to their feeding time budgets by foraging in groups, this is by no means always the case (suggesting perhaps that it is only when there is competition for preferred foods that problems arise: see Wrangham 1977).

Appendix

Sources for Data in Figure 7.8, page 136

Sources are listed in order of increasing rainfall (left to right on Figure 7.8); n gives the number of groups sampled in each case (with a minimum of $n=2$ being required for inclusion in order to avoid the problems of sampling bias noted by Sharman and Dunbar 1982).

Kuiseb, Namibia, $n=3$ (Hamilton *et al.* 1976)
Honnet Reserve, South Africa, $n=6$ (Stoltz and Saayman 1970)
Amboseli, Kenya, 1963, $n=51$ (Altmann and Altmann 1970)
Okavango, Botswana, $n=5$ (Hamilton *et al.* 1976)
Awash Station, Ethiopia, $n=2$ hamadryas (Nagel 1973)
Masalani, Kenya, $n=2$ (Maxim and Buettner-Janusch 1963)
Cape Reserve, South Africa, 1958, $n=3$ (Hall 1963)
Cape Reserve, South Africa, 1975, $n=3$ (Davidge 1978)
Metahara, Ethiopia, $n=8$ (Aldrich-Blake *et al.* 1971)
Awash Falls, Ethiopia, $n=5$ (Nagel 1973)
Erer Gota, Ethiopia, 1961, $n=5$ hamadryas (Kummer 1968)
Erer Gota, Ethiopia, 1973, $n=3$ hamadryas (Sigg *et al.* 1982)
Gilgil, Kenya, $n=7$ (Harding 1976)
Suikerbosrand, South Africa, $n=4$ (Anderson 1981)
Mikumi, Tanzania, $n=5$ (D. Rasmussen 1979)
Nanyuki, Kenya, $n=7$ (Berger 1972)
Kimana, Kenya, $n=6$ (van Citters *et al.* 1967)
Amboseli, Kenya, 1959, $n=15$ (DeVore and Washburn 1963)
Kariba, Zimbabwe, $n=18$ (Hall 1963)
Manyara, Tanzania, $n=2$ (Altmann and Altmann 1970)
Shai Hills, Ghana, $n=5$ (Depew 1983)
Drakensberg, South Africa, 1958, $n=5$ (Hall 1963)
Nairobi Park, Kenya, $n=9$ (DeVore and Hall 1965)
Bole, $n=4$, and Mulu, $n=2$, Ethiopia (Dunbar and Dunbar 1974b)
Murchison Falls, Uganda, $n=7$ (Hall 1965b)
Drakensberg, South Africa, 1982, $n=19$ (Whiten *et al.*, in press)
Ishasha, Uganda, $n=2$ (Rowell 1966a)
Gombe, Tanzania, $n=6$ (Packer 1979a)
Budongo, Uganda, $n=4$ (Patterson 1976)
Ruaha, Tanzania, $n=4$ (J. Oliver, personal communication)

Scientific and Common Names

Primates

Lemuridae	*Lemur catta*	ringtailed lemur
	Lemur fulvus	brown lemur
	Lemur mongoz	mongoz lemur
	Lepilemur mustelinus	sportive lemur
Cheirogaleidae	*Cheirogaleus* spp.	dwarf lemurs
	Microcebus murinus	mouse lemur
Indriidae	*Indri indri*	indri
	Propithecus verrauxi	white sifaka
Daubentoniidae	*Daubentonia madagascarensis*	aye-aye
Lorisidae	*Loris tardigradus*	loris
	Nycticebus coucang	slow loris
	Arctocebus calabarensis	golden potto
	Perodictus potto	potto
	Galago senegalensis	lesser galago
	Galago crassicaudatus	greater galago
	Galago demidovii	dwarf galago
	Galago zanzibaricus	Zanzibar galago
Tarsiidae	*Tarsier* spp.	tarsiers
Cebidae	*Cebus appella*	brown capuchin
	Cebus capuchinus	white-fronted capuchin
	Saimiri sciureus	squirrel monkey
	Callithrix jacchus	common marmoset
	Cebuella pygmaeus	pygmy marmoset
	Saguinus fuscicollis	saddleback tamarin
	Saguinus imperator	emperor tamarin
	Saguinus mystax	moustached tamarin
	Saguinus oedipus	cottontop tamarin
Atelidae	*Ateles geoffroyi*	blackhanded spider monkey
	Ateles belzebuth	longhaired spider monkey
	Alouatta palliata	mantled howler monkey

	Alouatta seniculus	red howler monkey
	Callicebus moloch	dusky titi
	Callicebus torquatus	widow titi
	Pithecia pithecia	paleheaded saki
	Pithecia monachus	monk saki
	Cacajao spp.	uakari
	Aotus trivirgatus	night monkey
Cercopithecidae	*Macaca sylvanus*	Barbary macaque
	Macaca nemestrina	pigtail macaque
	Macaca nigra	Celebes black macaque
	Macaca maura	moor macaque
	Macaca fascicularis	longtailed macaque
	Macaca mulatta	rhesus macaque
	Macaca fuscata	Japanese macaque
	Macaca sinica	toque macaque
	Macaca radiata	bonnet macaque
	Cercocebus albigena	greycheeked mangabey
	Cercocebus torquatus	whitecollared mangabey
	Cercocebus atys	sooty mangabey
	Cercocebus galeritus	agile mangabey
	Papio anubis	olive baboon
	Papio cynocephalus	yellow baboon
	Papio ursinus	chacma baboon
	Papio papio	Guinea baboon
	Papio hamadryas	hamadryas baboon
	Mandrillus sphinx	mandrill
	Mandrillus leucophaeus	drill
	Theropithecus gelada	gelada
	Cercopithecus aethiops	vervet
	Cercopithecus mitis	blue monkey
	Cercopithecus neglectus	De Brazza's monkey
	Cercopithecus nictitans	spotnosed guenon
	Cercopithecus ascanius	redtail monkey
	Miopithecus talapoin	talapoin
	Erythrocebus patas	patas
	Colobus guereza	black-and-white colobus
	Colobus satanas	black colobus
	Colobus badius	red colobus
	Presbytis entellus	hanuman langur
	Presbytis senex	grey langur
	Presbytis pileatus	capped langur
	Presbytis potenziani	Mentawi langur

Hominidae	*Hylobates lar*	whitehanded gibbon
	Hylobates moloch	moloch gibbon
	Hylobates agilis	agile gibbon
	Hylobates pileatus	pileated gibbon
	Hylobates klossi	Kloss's gibbon
	Hylobates syndactylus	siamang
	Pongo pygmaeus	orang utan
	Pan troglodytes	common chimpanzee
	Pan paniscus	bonobo (pygmy chimpanzee)
	Gorilla gorilla	gorilla

References

Abbott, D.H. (1984). Behavioural and physiological suppression of fertility in subordinate marmoset monkeys. *Am. J. Primatol.* 6: 169–86

Abbott, D.H. and Hearn, J.P. (1978). Physical, hormonal and behavioural aspects of sexual development in the marmoset monkey, *Callithrix jacchus. J. Reprod. Fertil.* 53: 155-66

Abbott, D.H., Keverne, E.B., Moore, G.F. and Yodyinguad, U. (1986). Social suppression of reproduction in subordinate talapoin monkeys, *Miopithecus talapoin. In* J. Else and P. Lee (eds.) *Primate ontogeny,* pp. 329–41. Cambridge University Press, Cambridge

Abrams, J.T. (1968). Fundamental approach to the nutrition of the captive wild herbivore. *Symp. zool. Soc. Lond.* 21: 41–62

Aldrich-Blake, F.P.G. (1970). Problems of social structure in forest monkeys. In: J.H. Crook (ed.) *Social behaviour in birds and mammals,* pp. 79–101. Academic Press, London

Aldrich-Blake, P., Bunn, T., Dunbar, R. and Headley, M. (1971). Observations on baboons, *Papio anubis,* in an arid region in Ethiopia. *Folia primatol.* 15: 1–35

Alexander, R.D. (1974). The evolution of social behaviour. *Ann. Rev. Ecol. Syst.* 5: 325–83

Altmann, J. (1979). Age cohorts as paternal sibships. *Behav. Ecol. Sociobiol.* 6: 161–9

Altmann, J. (1980). *Baboon mothers and infants.* Harvard University Press, Cambridge (Mass.)

Altmann, J. (1983). Costs of reproduction in baboons (*Papio cynocephalus*). In: W.P. Aspey and S.I. Lustick (eds.) *The cost of survival in vertebrates,* pp. 67–88. Ohio State University Press, Columbus

Altmann, J. and Alberts, S. (1987). Body mass and growth rates in a wild primate population. *Oecologia* (in press)

Altmann, J., Altmann, S.A., Hausfater, G. and McCuskey, S.A. (1977). Lifehistory of yellow baboons: physical development, reproductive parameters and infant mortality. *Primates 18:* 315–30

Altmann, J., Altmann, S.A., Hausfater, G. (1986). Determinants of reproductive success in savannah baboons (*Papio cynocephalus*). In: T.H. Clutton-Brock (ed.) *Reproductive success,* Chicago University Press, Chicago

Altmann, S.A. (1962). A field study of the sociobiology of the rhesus monkey, *Macaca mulatta. Ann. NY Acad. Sci.* 102: 338-435

Altmann, S.A. (1967). The structure of primate social communication. In: S.A. Altmann (ed.) *Social communication among primates,* pp. 325–62. Chicago University Press, Chicago

Altmann, S.A. and Altmann, J. (1970). *Baboon ecology.* Chicago University Press, Chicago

Altmann, S.A. and Altmann, J. (1979) Demographic constraints on behaviour and social organisation. In: I. Bernstein and E.O. Smith (eds.) *Primate ecology and human origins,* pp. 47–64. Garland, New York

Alvarez, F. (1973). Periodic changes in the bare skin areas of *Theropithecus gelada. Primates 14:* 195–9

Andelman, S. (1986). Ecological and social determinants of cercopithecine mating patterns. In: D.I. Rubenstein and R.W. Wrangham (eds.) *Ecological aspects of social evolution,* pp. 201–16. Princeton University Press, Princeton, NJ

Andersen, D.C., Armitage, K.B. and Hoffman, R.S. (1976). Socioecology of marmots: female reproductive strategies. *Ecology 57:* 552–60

Anderson, C.M. (1981). Subtrooping in a chacma baboon (*Papio ursinus*) population. *Primates 23*: 445–58

Anderson, C.M. (1982). Baboons below the tropic of Capricorn. *J. human Evol. 11*: 205–17

Anderson, C.M. (1983). Levels of social organisation and male–female bonding in the genus *Papio*. *Am. J. phys. Anthropol. 60*: 15–22

Anderson, C.M. (1986). Female age: male preference and reproductive success in primates. *Int. J. Primatol. 7*: 305–26

Anderson, C.O. and Mason, W.A. (1974). Early experience and complexity of social organisation in groups of young rhesus monkeys (*Macaca mulatta*). *J. comp. Physiol. Psychol. 87*: 681–90

Anderson, C.O. and Mason, W.A. (1978). Competitive social strategies in groups of deprived and experienced rhesus monkeys. *Devel. Psychobiol. 11*: 289–99

Anderson, S.S. and Fedak, M.A. (1985). Grey seal males: energetic and behavioural links between size and sexual success. *Anim. Behav. 33*: 829–39

Angst, W. (1973). Pilot experiments to test group tolerance to a stranger in wild *Macaca fascicularis*. *Am. J. phys. Anthropol. 38*: 625–30

Angst, W. and Thommen, D. (1977). New data and a discussion of infant killing in Old World monkeys and apes. *Folia primatol. 27*: 198–229

Appleby, M.C. (1983a). The probability of linearity in hierarchies. *Anim. Behav. 31*: 600–8

Appleby, M.C. (1983b). Competition in a red deer stag social group: rank, age and relatedness of opponents. *Anim. Behav. 31*: 913–18

Arling, G.L. and Harlow, H.F. (1967). Effects of social deprivation on maternal behaviour of rhesus monkeys. *J. comp. Physiol. Psychol. 64*: 371–8

Ayala, F.J. and Campbell, C.A. (1974). Frequency-dependent selection. *Ann. Rev. Ecol. Syst. 5*: 115–37

Bachmann, C. and Kummer, H. (1980). Male assessment of female choice in hamadryas baboons. *Behav. Ecol. Sociobiol. 6*: 315–21

Badrian, A. and Badrian, N. (1984b). Social organisation of *Pan paniscus* in the Lomako Forest, Zaire. In: R.L. Sussman (ed.) *The pygmy chimpanzee*, pp. 325–46. Plenum Press, New York

Badrian, N. and Malenky, R.K. (1984a). Feeding ecology of *Pan paniscus* in the Lomako Forest, Zaire. In: R.L. Sussman (ed.) *The pygmy chimpanzee*, pp. 275–99. Plenum Press, New York

Baldwin, J.D. (1969). The ontogeny of social behaviour of squirrel monkeys (*Saimiri sciureus*) in a seminatural environment. *Folia primatol. 11*: 35–79

Baldwin, J.D. (1971). The social organisation of a semifree-ranging troop of squirrel monkeys (*Saimiri sciureus*). *Folia primatol. 14*: 23–50

Baldwin, J.D. and Baldwin J. (1972). The ecology and behaviour of squirrel monkeys (*Saimiri oerstedi*) in a natural forest in Western Panama. *Folia primatol. 18*: 161–84

Baldwin, P.J., McGrew, W.C. and Tutin, C.E.G. (1982). Wide-ranging chimpanzees at Mt. Assirik, Senegal. *Int. J. Primatol. 3*: 367–85

Balin, H. and Wan, L.S. (1968). The significance of circadian rhythms in the search for the moment of ovulation in primates. *Fertil. Steril. 19*: 228–43

Barash, D.P. (1982). *Sociobiology and behaviour*, 2nd edition. Elsevier, New York

Barnard, C. and Thompson, D.E.S. (1985). *Gulls and plovers: the ecology and behaviour of mixed-species feeding groups*. Croom Helm, London

Bateson, P.P.G. (1983). Optimal outbreeding. In: P. Bateson (ed.) *Mate choice*, pp. 257–77. Cambridge University Press, Cambridge

Bauchop, T. (1978). Digestion of leaves in vertebrate arboreal folivores. In: C.G. Montgomery (ed.) *The ecology of arboreal folivores*, pp. 193–204. Smithsonian Institution Press, Washington (DC)

Bauchop, T. and Martucci, R.W. (1968). Ruminant-like digestion of the langur monkey. *Science 161*: 698–700

Bazamo, E., Guillon, R., Forni, C., Fady, J.-C. and Bert, J. (1973) Modifications du comportement du babouin *Papio papio* dans son milieu naturel par l'apport d'alliments. *Folia primatol. 19*: 404–8

Bearder, S.K. and Doyle, G.A. (1974). Ecology of bushbabies, *Galago senegalensis* and *Galago crassicaudatus*, with some notes on the behaviour in the field. In: R.D. Martin, G.A. Doyle and A.C. Walker (eds.) *Prosimian biology*, pp. 109–30. Duckworth, London

Belonje, P.C. and van Niekerk, C.H. (1975). A review of the influence of nutrition upon the oestrous cycle and the early pregnancy in the mare. *J. Reprod. Fertil., Suppl. 23*: 167–9

Belovsky, G.E. and Jordan, P.A. (1978). The time-energy budget of a moose. *Theoret. pop. Biol. 14*: 76–104

Belzung, C. and Anderson, J.R. (1986). Social rank and response to feeding competition in rhesus monkeys. *Behav. Processes 12*: 307–16

Bercovitch, F.B. (1983). Time budgets and consortship in olive baboons (*Papio anubis*). *Folia primatol. 41*: 180–90

Berger, M.E. (1972). Population structure of olive baboons (*Papio anubis* (J.P. Fischer)) in the Laikipia District of Kenya. *E. Afr. Wildl. J. 10*: 159–64

Berman, C.M. (1980). Mother–infant relationships among free-ranging rhesus monkeys on Cayo Santiago: with a comparison with captive pairs. *Anim. Behav. 28*: 860–73

Berman, C.M. (1982). Ontogeny of social relationships with group companions among free-ranging infant rhesus monkeys: I. Social networks and differentiation. *Anim. Behav. 30*: 149–62

Berman, C.M. (1983). Effects of being orphaned: a detailed case study of an infant rhesus. In: R.A. Hinde (ed.) *Primate social relationships*, pp. 79–81. Blackwell Scientific Publications, Oxford

Bernstein, I.S. (1976). Dominance, aggression and reproduction in primate societies. *J. theoret. Biol. 60*: 459–72

Bernstein, I.S. (1981). Dominance: the baby and the bathwater. *Behav. Brain Sci. 3*: 419–58

Bernstein, I.S. and Ehardt, C. (1986). The influence of kinship and socialisation on aggressive behaviour in rhesus monkeys (*Macaca mulatta*). *Anim. Behav. 34*: 339–47

Berry, K.H. (1974). The ecology and social behaviour of the chuckwalla, *Sauromalus obesus obesus* Baird. *Univ. Calif. Publs Zool. 28*: 1–60

Bert, J., Ayats, H., Martino, A. and Colomb, H. (1967). Le sommeil nocturne chez le babouin *Papio papio* — Observations en milieu naturel et données éléctrophysiologiques. *Folia primatol. 6*: 28–43

Bertram, B. (1980). Vigilance and group size in ostriches. *Anim. Behav. 28*: 278–86

Bibeen, M., Symmes, D. and Masataka, N. (1986). Temporal and structural analysis of affiliative vocal exchanges in squirrel monkeys (*Saimiri sciureus*). *Behaviour 98*: 259–73

Bielert, C. (1982). Experimental examinations of baboon (*Papio ursinus*) sex stimuli. In: C. Snowdon, C. Brown and M. Petersen (eds) *Primate communication*, pp. 373–95. Cambridge University Press, Cambridge

Birdsell, J.B. (1979). Ecological influences on Australian aboriginal social organisation. In: I. Bernstein and E.O. Smith (eds.) *Primate ecology and human origins*, pp. 117–51. Garland, New York

Bishop, N. (1979). Himalayan langurs: temperate colobines. *J. human Evol. 8*: 251–81

Blackburn, M.W. and Calloway, D.H. (1976). Energy expenditure and consumption of mature, pregnant and lactating women. *J. Am. Diet. Assoc. 69*: 29–37

Boelkins, R.C. and Wilson, A.P. (1972). Intertroop social dynamics of the Cayo Santiago rhesus (*Macaca mulatta*) with special reference to changes in group membership by males. *Primates 13*: 125–40

Boesch, C. and Boesch, H. (1984). Possible causes of sex differences in the use of natural hammers by wild chimpanzees. *J. human Evol. 13*: 415–40

Boese, G.K. (1975). Social behaviour and ecological considerations of West African baboons (*Papio papio*). In: R.H. Tuttle (ed.) *Socioecology and psychology of primates*, pp. 205–30. Mouton, The Hague

Boggess, J. (1979). Troop male membership changes and infant killing in langurs (*Presbytis entellus*). *Folia primatol. 32*: 65–107

Borgia, G. (1981). Sexual competition in *Scatophagia stercoraria*: size- and density-related changes in male ability to capture females. *Behaviour 77*: 185–206

Bourlière, F., Bertrand, M. and Hunkeler, C. (1969). L'écologie de la mone de Lowe (*Cercopithecus campbelli lowei*) en Côte d'Ivoire. *Terre Vie 2*: 135–63

Bowden, D.M. and Williams, D.D. (1984). Aging. *Adv. Vet. Sci. Comp. Med. 28*: 305–41

Bowman, L.A., Dilley, S.R. and Keverne, E.B. (1978). Suppression of oestrogen-induced LH surges by social subordination in talapoin monkeys. *Nature, Lond. 275*: 56–8

Bramblett, C.A. (1967). Pathology of the Darajani baboon. *Am. J. phys. Anthropol. 26*: 331–40

Bramblett, C.A. (1970). Coalitions among gelada baboons. *Primates 11*: 327–34

Brattstrom, B.H. (1974). The evolution of reptilian social behaviour. *Am. Zool. 14*: 35–49

Brockelman, W.Y. and Srikosamatara, S. (1984). The maintenance and evolution of social structure in gibbons. In: H. Preuschoft, D. Chivers and W. Brockelman (eds.) *The lesser apes*, pp. 122–45. Edinburgh University Press, Edinburgh

Brody, S. (1945). *Bioenergetics and growth*. Rheinhold: New York

Buchanan, D. (1979). The genus *Pithecia*. In: R.A. Mittermeier and A.F. Coimbra-Filho (eds.) *Introduction to New World primatology* Smithsonian Institution Press, Washington (DC)

Budnitz, N. and Dainis, K. (1975). *Lemur catta*: ecology and behaviour. In: I. Tattersall and R.W. Sussman (eds.) *Lemur biology*, pp. 219–35. Plenum Press, New York

Buirski, P., Kellerman, H., Plutchik, R., Weininger, R. and Buirski, N. (1973). A field study of emotions, dominance and social behaviour in a group of baboons (*Papio anubis*). *Primates 14*: 67–78

Buirski, P., Plutchik, R. and Kellerman, H. (1978). Sex differences, dominance and personality in the chimpanzee. *Anim. Behav. 26*: 123–9

Buss, D.H. (1968). Gross composition and variation of the components of baboon milk during natural lactation. *J. Nutrition 96*: 421–6

Buss, D.H. and Reed, O.M. (1970). Lactation of baboons fed a low protein maintenance diet. *Lab. Anim. Care 26*: 709–12

Busse, C. (1977). Chimpanzee predation as a possible factor in the evolution of red colobus monkey social organisation. *Evolution 31*: 907–11

Busse, C. (1982). Social dominance and offspring mortality among female chacma baboons. *Int. J. Primatol. 3*: 267

Busse, C. and Hamilton, W.J. (1981). Infant carrying by male chacma baboons. *Science 212*: 1281–3

Bygott, D. (1974). *Agonistic behaviour in wild chimpanzees*. PhD thesis, University of Cambridge

Bygott, D., Bertram, B.C.R. and Hanby, J.P. (1979). Male lions in large coalitions gain reproductive advantages. *Nature, Lond. 282*: 839–41

Byrne, R.W. (1981). Distance vocalisations of Guinea baboons (*Papio papio*) in Senegal: an analysis of function. *Behaviour 78*: 283–312

Byrne, R.W. and Whiten, A. (1984). Tactical deception of familiar individuals in baboons (*Papio ursinus*). *Anim. Behav. 33*: 669–73

Byrne, R.W., Whiten, A. and Henzi, S.P. (in press). One-male groups and intertroop interactions of mountain baboons (*Papio ursinus*). *Int. J. Primatol.*

Caldecott, J.O. (1980). Habitat quality and populations of two sympatric gibbons (Hylobatidae) on a mountain in Malaya. *Folia primatol. 33*: 291–309

Caldecott, J.O. (1986a). *An ecological and behavioural study of the pig-tailed macaque.* Karger, Basel

Caldecott, J.O. (1986b). Mating patterns, societies and the ecogeography of macaques. *Anim. Behav. 34*: 208–20

Campanella, P.J. (1974). The evolution of mating systems in temperate zone dragonflies (Odonata: Anisoptera). II. *Libellula luctuosa* (Burmeister). *Behaviour 54*: 278–309

Campanella, P.J. and Wolf, L.L. (1974). Temporal leks as a mating system in a temperate zone dragonfly (Odonata: Anisoptera). I. *Plathemis lydia* (Drury). *Behaviour 51*: 49–87

Cant, J.H.G. (1980). What limits primates? *Primates 21*: 538–44

Carenza, L. and Zichella, L. (eds.). *Emotion and reproduction.* Academic Press, London

Caro, T.M. (1986). The functions of stotting: a review of the hypotheses. *Anim. Behav. 34*: 649–62

Castro, R. and Soini, P. (1977). Field studies on *Saguinus mystax* and other callitrichids in Amazonian Peru. In: D. Kleiman (ed.) *The biology and conservation of the Callitrichidae*, pp. 73–8. Smithsonian Institution Press, Washington (DC)

Caughley, G. (1976). The elephant problem — an alternative hypothesis. *E. Afr. Wildl. J. 14*: 265–83

Caughley, G. (1977) *Analysis of vertebrate populations.* Wiley, Chichester

Cavalli-Sforza, L.L. and Bodmer, W.F. (1971). *The genetics of human populations*, W.H. Freeman, San Francisco

Chalmers, A.F. (1978). *What is this thing called science?* Open University Press, Milton Keynes

Chamove, A.S., Eysenck, H.J. and Harlow, H.F. (1972). Personality in monkeys: factor analysis of rhesus social behaviour. *Q.J. exp. Psychol. 24*: 496–504

Chance, M.R.A. (1967). Attention structure as the basis of primate rank orders. *Man 2*: 503–18

Chapais, B. (1983a). Reproductive activity in relation to male dominance and the likelihood of ovulation in rhesus monkeys. *Behav. Ecol. Sociobiol. 12*: 215–28

Chapais, B. (1983b). Dominance, relatedness and the structure of female relationships in rhesus monkeys. In: R. Hinde (ed.) *Primate social relationships*, pp. 208–19. Blackwell Scientific Publications, Oxford

Chapais, B. and Schulman, S. (1980). An evolutionary model of female dominance relations in primates. *J. theoret. Biol. 82*: 48–89

Chapman, M. and Hausfater, G. (1979). The reproductive consequences of infanticide in langurs: a mathematical model. *Behav. Ecol. Sociobiol. 5*: 227–40

Charles-Dominique, P. (1974). Ecology and feeding behaviour of five sympatric lorisids in Gabon. In: R.D. Martin, G.A. Doyle and A.C. Walker (eds.) *Prosimian biology*, pp. 131–50, Duckworth, London

Charles-Dominique, P. (1977). *Ecology and behaviour of nocturnal primates.* Columbia University Press, New York

Charles-Dominique, P. and Hladik, C.M. (1971). Le lépilemur du sud de Madagascar: ecologie, alimentation et vie sociale. *Terre Vie 25*: 3–66

Charles-Dominique, P. and Martin, R.D. (1972). *Behaviour and ecology of nocturnal prosimians*. Parey, Hamburg

Cheney, D.L. (1977). The acquisition of rank and the development of reciprocal alliances among free-ranging immature baboons. *Behav. Ecol. Sociobiol. 2*: 303–18

Cheney, D.L. (1981). Intergroup encounters among free-ranging vervet monkeys. *Folia primatol. 35*: 124–46

Cheney, D.L. (1983). Extrafamilial alliances among vervet monkeys. In: R.A. Hinde (ed.) *Primate social relationships*, pp. 278–86. Blackwell Scientific Publications, Oxford

Cheney, D.L., and Seyfarth, R.M. (1977). Behaviour of adult and immature baboons during intergroup encounters. *Nature, Lond. 269*: 404–6

Cheney, D.L. and Seyfarth, R.M. (1980). Vocal recognition in free-ranging vervet monkeys. *Anim. Behav. 28*: 362–7

Cheney, D.L. and Seyfarth, R.M. (1982). How vervet monkeys perceive their grunts. *Anim. Behav. 30*: 739–51

Cheney, D.L. and Seyfarth, R.M. (1983) Nonrandom dispersal in free-ranging vervet monkeys: social and genetic consequences. *Am. Nat. 122*: 392–412

Cheney, D.L. and Seyfarth, R.M. (1985a). Social and non-social knowledge in vervet monkeys. In: L. Weiskrantz (ed.) *Animal intelligence*, pp. 187–202. Clarendon Press, Oxford

Cheney, D.L. and Seyfarth, R.M. (1985b). Vervet monkey alarm calls: manipulation through shared information? *Behaviour 94*: 150–66

Cheney, D.L. and Seyfarth, R.M. (1986). The recognition of social alliances by vervet monkeys. *Anim. Behav. 34*: 1722–31

Cheney, D. and Wrangham, R.W. (1987). Predation. In: B. Smuts, D. Cheney, R. Seyfarth, R. Wrangham and T. Struhsaker (eds.), *Primate societies*, pp. 227–39. Chicago University Press, Chicago

Cheney, D.L. Lee, P.C. and Seyfarth, R.M. (1981). Behavioural correlates of non-random mortality among free-ranging female vervet monkeys. *Behav. Ecol. Sociobiol. 9*: 153–61

Cheney, D.L., Seyfarth, R.M., Andelman, S.J. and Lee, P.C. (1986a). Factors affecting reproductive success in vervet monkeys. In: T.H. Clutton-Brock (ed.) *Reproductive success*. Chicago University Press, Chicago

Cheney, D.L., Seyfarth, R.M. and Smuts, B. (1986b). Social relationships and social cognition in nonhuman primates. *Science 234*: 1361–6

Chepko-Sade, B.D. and Olivier, T.J. (1979). Coefficient of genetic relationship and the probability of intra-genealogical fission in *Macaca mulatta*. *Behav. Ecol. Sociobiol. 5*: 263–78

Chepko-Sade, B.D. and Sade, D.S. (1979). Patterns of group-splitting within matrilineal kinship groups. *Behav. Ecol. Sociobiol. 5*: 67–86

Chiarelli, B. (1963). Sensitivity to P.T.C. in primates. *Folia primatol. 1*: 88–94

Chikazawa, D., Gordon, T.P., Bean, G.A. and Bernstein, I.S. (1979). Mother–daughter dominance reversals in rhesus monkeys (*Macaca mulatta*). *Primates 20*: 301–5

Chism, J. and Rowell, T.E. (1986). Mating and residence patterns of male patas monkeys. *Ethology 72*: 31–9

Chivers, D.J. (1969). On the daily behaviour and spacing of howling monkey groups. *Folia primatol. 10*: 48–102

Chivers, D.J. (1974). *The siamang in Malaya*. Karger, Basel

Chivers, D.J. and Raemakers, J.J. (1980). Long-term changes in behaviour. In: Chivers, D.J. (ed.) *Malayan forest primates*, pp. 209–60. Plenum Press, New York

van Citters, R., Smith, O., Franklin, D., Kemper, W. and Watson, N. (1967). Radio-telemetry of blood flow and blood pressure in feral baboons: a preliminary report.

In: D. Vagtborg (ed.) *The baboon in medical research*, Vol. 2, pp. 473–92. University of Texas Press, San Antonio (Tex.)

Clark, A.B. (1978). Sex ratio and local resource competition in a prosimian primate. *Science 201*: 163–5

Clarke, M.R. and Glander, K. (1984). Female reproductive success in a group of free ranging howler monkeys (*Alouatta palliata*) in Costa Rica. In: M.F. Small (ed.) *Female primates: studies by women primatologists* pp. 111–26. Liss, New York

Clegg, E.J. and Harrison, G.A. (1971). Reproduction in human high altitude populations. *Hormones 2*: 13–25

Clutton-Brock, T.H. (1972). *Feeding and ranging behaviour of the red colobus monkey.* PhD thesis, University of Cambridge

Clutton-Brock, T.H. (1973). Feeding levels and feeding sites of red colobus (*Colobus badius tephrosceles*) in the Gombe National Park. *Folia primatol. 19*: 368–79

Clutton-Brock, T.H. (1974). Primate ecology and social organisation. *Nature, Lond. 250*: 539–42

Clutton-Brock, T.H. (1975). Ranging behaviour of red colobus (*Colobus badius tephrosceles*) in the Gombe National Park. *Anim. Behav. 23*: 706–22

Clutton-Brock, T.H. (1982). Sons and daughters. *Nature, Lond. 298*: 11–13

Clutton-Brock, T.H. and Albon, S.D. (1982). Parental investment in male and female offspring in mammals. In: King's College Sociobiology Group (eds.) *Current problems in sociobiology*, pp. 223–47. Cambridge University Press, Cambridge

Clutton-Brock, T.H. and Harvey, P.H. (1976). Evolutionary rules and primate societies. In: P. Bateson and R. Hinde (eds.) *Growing points in ethology*, pp. 195–237. Cambridge University Press, Cambridge

Clutton-Brock, T.H. and Harvey, P.H. (1977). Primate ecology and social organisation. *J. Zool., Lond. 183*: 1–39

Clutton-Brock, T.H. and Harvey, P.H. (1979). Home range size, population density and phylogeny in primates. In: I. Bernstein and E.O. Smith (eds.) *Primate ecology and human origins*, pp. 201–14. Garland, New York

Cody, M.L. (1971). Finch flocks in the Mohave Desert. *Theoret. pop. Biol. 2*: 142–58

Coe, M.J., Cumming, D.H. and Phillipson, J. (1976). Biomass and production of large herbivores in relation to rainfall and primary production. *Oecologia 22*: 341–54

Coelho, A.M. (1974). Socio-bioenergetics and sexual dimorphism in primates. *Primates 15*: 263–9

Coelho, A.M. (1985). Baboon dimorphism: growth in weight, length and adiposity from birth to 8 years of age. In: E.S. Watts (ed.) *Non-human primate models for human growth and development*, pp. 125–59. Liss, New York

Coelho, A.M., Bramblett, C.A., Quick, L.A. and Bramblett, S.S. (1976). Resource availability and population density in primates: a socio-bioenergetic analysis of the energy budgets of Guatemalan howler and spider monkeys. *Primates 17*: 63–80

Cohen, J.E. (1971). *Casual groups of monkeys and men.* Harvard University Press, Cambridge (Mass.)

Cohen, J.E. (1975). The size and demographic composition of social groups of wild orang-utans. *Anim. Behav. 23*: 543–50

Collins, D.A. (1986). Interactions between adult male and infant yellow baboons (*Papio c. cynocephalus*) in Tanzania. *Anim. Behav. 34*: 430–43

Colvin, J. (1983a). Influences of the social situation on male emigration. In: R.A. Hinde (ed.) *Primate social relationships*, pp. 160–71. Blackwell Scientific Publications, Oxford

Colvin, J. (1983b). Familiarity, rank and the structure of rhesus male peer networks. In: R.A. Hinde (ed.) *Primate social relationships*, pp. 190–9. Blackwell Scientific Publications, Oxford

Cords, M. (1984). Mating patterns and social structure in redtail monkeys (*Cerco-*

pithecus ascanius). Z. Tierpsychol. 64: 313–29

Cords, M. (1986). Interspecific and intraspecific variation in diet of two forest guenons, *Cercopithecus ascanius* and *C. mitis. J. anim. Ecol. 55*: 811–28

Cords, M. and Rowell, T.E. (1986). Group fission in blue monkeys of the Kakamega Forest, Kenya. *Folia primatol. 46*: 70–82

Cords, M., Mitchell, B.J., Tsingalia, H.M. and Rowell, T.E. (1986). Promiscuous mating among blue monkeys in the Kakamega Forest, Kenya. *Ethology 72*: 214–26

Coulson, J. and Thomas, C.S. (1983). Mate choice in the kittiwake gull. In: P. Bateson (ed.) *Mate choice*, pp. 361–76. Cambridge University Press, Cambridge

van Couvering, J. (1980). Community evolution in East Africa during the late Cenozoic. In: A.K. Behrensmeyer and A.P. Hill (eds.) *Fossils in the making: vertebrate taphonomy and palaeoecology*, pp. 272–98. Chicago University Press, Chicago

Cox, C. and LeBoeuf, B.J. (1977). Female incitation of male competition: a mechanism in sexual selection. *Am. Nat. 111*: 317–35

Crook, J.H. (1966). Gelada baboon herd structure and movement: a comparative report. *Symp. zool. Soc. Lond. 18*: 237–58

Crook, J. and Gartlan, J.S. (1966). Evolution of primate societies. *Nature, Lond. 210*: 1200–3

Crook, J.H., Ellis, J.E. and Goss-Custard, J.D. (1976). Mammalian social systems: structure and function. *Anim. Behav. 24*: 261–74

Cubicciotti, D.D. and Mason, W.A. (1978). Comparative studies of social behaviour in *Callicebus* and *Saimiri*: heterosexual jealousy behaviour. *Behav. Ecol. Sociobiol. 3*: 311–22

Curtin, R. and Dolhinow, P.C. (1978). Primate social behaviour in a changing world. *Am. Sci. 66*: 468–75

Datta, S. (1981). *Dynamics of dominance among rhesus females.* PhD thesis, University of Cambridge

Datta, S. (1983a). Relative power and the acquisition of rank. In: R.A. Hinde (ed.) *Primate social relationships*, pp. 93–103. Cambridge University Press, Cambridge

Datta, S. (1983b). Relative power and the maintenance of dominance. In: R.A. Hinde (ed.) *Primate social relationships*, pp. 103–12. Cambridge University Press, Cambridge

Davidge, C. (1978). Ecology of baboons (*Papio ursinus*) at Cape Point. *Zool. Africana 13*: 329–50

Davies, N.B. (1983). Polyandry, cloaca-pecking and sperm competition in dunnocks, *Prunella modularis. Nature, Lond. 302*: 334–6

Davis, D.E. (1958). The role of density in aggressive behaviour in house mice. *Brit. J. anim. Behav. 2*: 207–10

Davis, D.E. (1959). Territorial rank in starlings. *Brit. J. anim. Behav. 7*: 214–21

Dawkins, R. (1976a). *The selfish gene.* Oxford University Press, Oxford

Dawkins, R. (1976b). Hierarchical organisation: a candidate principle for ethology. In: P. Bateson and R.A. Hinde (eds.) *Growing points in ethology*, pp. 7–54. Cambridge University Press, Cambridge

Dawson, G.A. (1979). The use of time and space by the Panamanian tamarin, *Saguinus oedipus. Folia primatol. 31*: 253–84

Deag, J.M. (1977). Aggression and submission in monkey societies. *Anim. Behav. 25*: 465–74

Deag, J.M. (1978). The adaptive significance of baboon and macaque social behaviour. In: P. Eblingaand and D.M. Stoddart (eds.) *Population control by social behaviour*, pp. 83–113. Institute of Biology, London

Deag, J.M., and Crook, J.H. (1971). Social behaviour and 'agonistic buffering' in the wild Barbary macaque *Macaca sylvana* L. *Folia primatol. 15*: 183–200

Delvoye, P., Badawi, M., Demaegd, M. and Robyn, C. (1978). Long-lasting lactation is associated with hyperprolactinemia and amenorrhea. In: C. Robyn and M. Harter (eds.) *Progress in prolactin physiology and pathology.* Elsevier-North Holland, New York

De Moor, P.P. and Steffens, F.E. (1972). The movements of vervet monkeys (*Cercopithecus aethiops*) within the ranges as revealed by radio-tracking, *J. anim. Ecol. 41*: 677–87

Denham, W.W. (1971). Energy relations and some basic properties of primate social organisation. *Am. Anthropol. 73*, 77–95

Depew, L.A. (1983). *Ecology and behaviour of baboons* (Papio anubis) *in the Shai Hills Game Production Reserve, Ghana.* MSc thesis, Cape Coast University, Ghana

DeVore, I. and Washburn, S.L. (1963). Baboon ecology and human evolution. In: F. Campbell and F. Bourlière (eds.) *African ecology and human evolution*, pp. 335–67. Aldine, New York

DeVore, I. and Hall, K.R.L. (1965). Baboon ecology. In: DeVore, I. (ed.) *Primate behaviour*, pp. 20–52. Holt, Rinehart & Winston, New York

Dickinson, A. (1980). *Contemporary animal learning theory.* Cambridge University Press, Cambridge

Dittus, W.P.J. (1975). Population dynamics of the toque monkeys, *Macaca sinica.* In: R.H. Tuttle (ed.) *Socioecology and psychology of primates*, pp. 125–51. Mouton, The Hague

Dittus, W.P.J. (1977). The social regulation of population density and age–sex distribution in the toque monkey. *Behaviour 63*: 281–322

Dittus, W.P.J. (1979). The evolution of behaviours regulating density and age-specific sex ratios in a primate population. *Behaviour 69*: 265–302

Dittus, W.P.J. (1980). The social regulation of primate populations: a synthesis. In: D.G. Lindburg (ed.) *The macaques*, pp. 263–86. Van Nostrand, New York

Dittus, W.P.J. (1984). Toque macaque food calls: semantic communication concerning food distribution in the environment. *Anim. Behav. 32*: 470–7

Dittus, W.P.J. (1986). Sex differences in fitness following a group take-over among toque macaques: testing models of social evolution. *Behav. Ecol. Sociobiol. 19*, 257–66

Dixson, A.F. (1983). Observations on the evolution and behavioural significance of 'sexual skin' in female primates. *Adv. Study Behav. 13*: 63–106

Dobbing, J. (1976). Vulnerable periods in brain growth and somatic growth. In: D.F. Roberts and A.M. Thompson (eds.) *The biology of human fetal growth*, pp. 137–47. Taylor & Francis, London

Dolhinow, P.C., McKenna, J. and von der Haar Laws, J. (1979). Rank and reproduction among female langur monkeys: age and improvement (they're not just getting older, they're getting better). *J. aggress. Behav. 5*: 19–30

Dominey, W.J. (1984). Alternative mating tactics and evolutionary stable strategies. *Am. Zool. 24*: L385–96

Doyle, G.A., Anderson, A. and Bearder, S.K. (1971). Reproduction in the lesser bushbaby (*Galago senegalensis moholi*) under semi-natural conditions. *Folia primatol. 14*: 15–22

Drickamer, L.C. (1974a). A ten-year summary of reproductive data for free-ranging *Macaca mulatta. Folia primatol. 21*: 61–80

Drickamer, L.C. (1974b). Social rank, observability and sexual behaviour of rhesus monkeys (*Macaca mulatta*). *J. Reprod. Fertil. 37*: 117–20

Dunbar, R.I.M. (1977). Feeding ecology of gelada baboons: a preliminary report. In: T.H. Clutton-Brock (ed.) *Primate ecology*, pp. 250–73, Academic Press, London

Dunbar, R.I.M. (1978). Sexual behaviour and social relationships among gelada baboons. *Anim. Behav. 26*: 167–78

Dunbar, R.I.M. (1979a). Structure of gelada baboon reproductive units. I. Stability of social relationships. *Behaviour* 69: 72–87

Dunbar, R.I.M. (1979b). Energetics, thermoregulation and the behavioural ecology of klipspringer. *Afr. J. Ecol. 17*: 217–30

Dunbar, R.I.M. (1979c). Population demography, social organisation and mating strategies. In: I. Bernstein and E.O. Smith (eds.) *Primate ecology and human origins,* pp. 65–88. Garland, New York

Dunbar, R.I.M. (1980a) Determinants and evolutionary consequences of dominance among female gelada baboons. *Behav. Ecol. Sociobiol. 7*: 253–65

Dunbar, R.I.M. (1980b). Demographic and life history variables of a population of gelada baboons (*Theropithecus gelada*). *J. anim. Ecol. 49*: 485–506

Dunbar, R.I.M. (1982a). Adaptation, fitness and the evolutionary tautology. In: King's College Sociobiology Group (eds.), *Current problems in sociobiology,* pp. 9–18. Cambridge University Press, Cambridge

Dunbar, R.I.M. (1982b). Intraspecific variations in mating stragegy. In: P. Klopfer and P. Bateson (eds.) *Perspectives in ethology,* Vol. 5, pp. 385–431. Plenum Press, New York

Dunbar, R.I.M. (1982c). Structure of social relationships in a captive group of gelada baboons: a test of some hypotheses derived from studies of a wild population. *Primates 23*: 89–94

Dunbar, R.I.M. (1983a). Lifehistory tactics and alternative strategies of reproduction. In: P. Bateson (ed.) *Mate choice,* pp. 423–33. Cambridge University Press, Cambridge

Dunbar, R.I.M. (1983b). Relationships and social structure in gelada and hamadryas baboons. In: R.A. Hinde (ed.) *Primate social relationships,* pp. 299–307. Blackwell Scientific Publications, Oxford

Dunbar, R.I.M. (1983c). Structure of gelada baboon reproductive units. II. Social relationships between reproductive females. *Anim. Behav. 31*: 556–64

Dunbar, R.I.M. (1983d). Structure of gelada baboon reproductive units. III. The male's relationships with his females. Anim. Behav. 31: 565–75

Dunbar, R.I.M. (1983e). Structure of gelada baboon reproductive units. IV. Integration at group level. *Z. Tierpsychol. 63*: 265–82

Dunbar, R.I.M. (1984a). *Reproductive decisions: an economic analysis of gelada baboon social strategies.* Princeton University Press, Princeton (NJ)

Dunbar, R.I.M. (1984b). Theropithecines and hominids: contrasting solutions to the same ecological problems. *J. human Evol. 12*: 647–58

Dunbar, R.I.M. (1984c). Use of infants by male gelada in agonistic contexts: agonistic buffering, progeny protection or soliciting support? *Primates 25*: 28–35

Dunbar, R.I.M. (1985). Monogamy on the rocks. *Nat. Hist. 94* (11): 40–7

Dunbar, R.I.M. (1986). The social ecology of gelada baboons. In: D. Rubenstein and R. Wrangham (eds.) *Ecological aspects of social evolution,* pp. 332–51, Princeton University Press: Princeton (NJ)

Dunbar, R.I.M. (1987). Demography and reproduction. In: B. Smuts, D. Cheney, R. Seyfarth, R. Wrangham and T. Struhsaker (eds.) *Primate societies,* pp. 240–9. Chicago University Press, Chicago

Dunbar, R.I.M. (in press). Habitat quality, population dynamics and group composition in colobus monkeys (*Colobus guereza*). *Int. J. Primatol*

Dunbar, R.I.M. and Dunbar, P. (1974a). On hybridisation between *Theropithecus gelada* and *Papio anubis* in the wild. *J. human Evol. 3*: 1187–92

Dunbar, R.I.M. and Dunbar, P. (1974b). Ecological relations and niche separation between sympatric terrestrial primates in Ethiopia. *Folia primatol. 21*: 36–60

Dunbar, R.I.M. and Dunbar, P. (1974c). Ecology and population dynamics of *Colobus guereza* in Ethiopia. *Folia primatol. 21*: 188–209

Dunbar, R.I.M. and Dunbar, P. (1975). *Social dynamics of gelada baboons*. Karger, Basel

Dunbar, R.I.M. and Dunbar, P. (1976). Contrasts in social structure among black-and-white colobus monkey groups. *Anim. Behav. 24*: 84–92

Dunbar, R.I.M. and Dunbar. P. (1980). The pairbond in klipspringer. *Anim. Behav. 28*: 219–29

Dunbar, R.I.M. and Dunbar, P. (1981). Grouping behaviour of walia ibex, with special reference to the rut. *Afr. J. Ecol. 19*: 251–63

Dunbar, R.I.M. and Dunbar, P. (in press). Maternal time budgets of gelada baboons. *Anim. Behav.*

Dunbar, R.I.M. and Nathan, M.F. (1972). Social organisation of the Guinea baboon, *Papio papio*. *Folia primatol. 17*: 321–34

Dunbar, R.I.M. and Sharman, M. (1983). Female competition for access to males affects birth rate in baboons. *Behav. Ecol. Sociobiol. 13*: 157–9

Dunbar, R.I.M. and Sharman, M. (1984). Is social grooming altruistic? *Z. Tierpsychol. 64*: 163–73

Duvall, S.W., Bernstein, I.S. and Gordon, T.P. (1976). Paternity and status in a rhesus monkey group. *J. Reprod. Fertil. 47*: 25–31

Eaton, G.G. (1974). Male dominance and aggression in Japanese macaque reproduction. In: W. Montague and W.A. Sadler (eds.) *Reproductive behaviour*, pp. 287–97. Plenum Press, New York

Ehardt, C.L. and Bernstein, I.S. (1986). Matrilineal overthrows in rhesus monkey groups. *Int. J. Primatol. 7*: 157–81

Eisenberg, J.F. Muckenhirn, N.A. and Rudran, R. (1972). The relation between ecology and social structure in primates. *Science 176*: 863–74

Eisenberg, J.F., Dittus, W.P.J., Fleming, T.H., Green, K., Struhsaker, T.T. and Thorington, R.W. (1981). *Techniques for the study of primate population ecology*. National Academy Press, Washington (DC)

Ellefson, J.O. (1968). Territorial behaviour in the common white-handed gibbon, *Hylobates lar* Linn. In: P. Jay (ed.) *Primates*, pp. 180–99. Holt, Rinehart & Winston: New York

Emlen, S.T. (1980). Ecological determinism and sociobiology. In: G. Barlow and J. Silverberg (eds.) *Sociobiology: beyond nature/nurture?* pp. 125–50. Westview Press, Boulder (CO)

Emlen, S.T. (1982). The evolution of helping. I. An ecological constraints model. *Am. Nat. 119*: 40–53

Emlen, S.T. (1984). Cooperative breeding in birds and mammals. In: J.R. Krebs and N.B. Davies (eds.) *Behavioural ecology*, 2nd edition, pp. 305–39. Blackwell Scientific Publications, Oxford

Emlen, S.T. and Oring, L. (1977). Ecology, sexual selection and the evolution of mating systems. *Science 197*: 215–23

Enomoto, T. (1974). The sexual behaviour of Japanese macques. *J. human Evol. 3*: 351–72

Enomoto, T., Seiki, K. and Haruki, Y. (1979). On the correlation between sexual behaviour and ovarian hormone level during the menstrual cycle in captive Japanese macaques. *Primates 20*: 563–70

Evans, S. (1983). The pair-bond of the common marmoset, *Callithrix jacchus jacchus*: an experimental investigation. *Anim. Behav. 31*: 651–8

Evans, S. and Poole, T.B. (1984). Long-term changes and maintenance of the pairbond in common marmosets, *Callithrix jacchus jacchus*. *Folia primatol. 42*: 33–41

Fa, J.E. (1986). *Use of time and resources by provisioned troops of monkeys*. Karger, Basel

Faiman, C., Reyes, F.I., Winter, J.S.D. and Hobson, W.C. (1981). Endocrinology of pregnancy in apes. In: C.E. Graham (ed.) *Reproductive biology of great apes*, pp. 45–68. Academic Press, New York

Fairbanks, L.A. (1980). Relationships among adult females in captive vervet monkeys: testing a model of rank-related attractiveness. *Anim. Behav. 28*: 853–9

Fairbanks, L.A. and Bird, J. (1978). Ecological correlates of interindividual distance in the St Kitts vervet (*Cercopithecus aethiops sabaeus*). *Primates 28*: 605–14

Fairbanks, L.A. and McGuire, M.T. (1984). Determinants of fecundity and reproductive success in captive vervets. *Am. J. Primatol. 7*: 27–38

Fairbanks, L.A. and McGuire, M.T. (1986). Age, reproductive value and dominance related behaviour in vervet monkey females: cross-generational influences on social relationships and reproduction. *Anim. Behav. 34*: 1710–21

Falconer, D.S. (1960). *An introduction to quantitative genetics*. Longman, London

Fedigan, L.M. (1983). Dominance and reproductive success. *Ybk phys. Anthropol. 26*: 91–129

Feer, F. (1979). Observations ecologiques sur le néotrague de Bates (*Neotragus batesi* de Winton 1903, Artiodactyle, Ruminant, Bovide) du nord-est du Gabon. *Terre Vie 33*: 159–239

Fernald, R.D. and Hirata, N.R. (1977). Field study of *Haplochromis burtoni*: quantitative behavioural observations. *Anim. Behav. 25*: 964–75

Ford, S.M. (1980). Callitrichids as phyletic dwarfs and the place of the Callitrichidae in Platyrrhini. *Primates 21*: 31–43

Fossey, D. and Harcourt, A.H. (1977). Feeding ecology of free-ranging mountain gorilla (*Gorilla gorilla berengei*). In: T.H. Clutton-Brock (ed.) *Primate ecology*, pp. 415–47. Academic Press, London

Fragaszy, D.M., Schwartz, S. and Shimosaka, D. (1982). Longitudinal observations of care and development of infant titi (*Callicebus moloch*). *Am. J. Primatol. 2*: 191–200

Frame, L.H. and Frame, G.A. (1977). Female African wild dogs emigrate. *Nature, Lond. 263*: 227–9

Freeland, W.J. (1979). Mangabey (*Cercocebus albigena*). Social organisation and population density in relation to food use and availability. *Folia primatol. 32*: 108–24

French, J.A., Abbott, D.H. and Snowdon, C.T. (1984). The effects of social environment on oestrogen excretion, scent marking and sociosexual behaviour in tamarins (*Saguinus oedipus*). *Am. J. Primatol. 6*: 155–67

Frisch, R. (1978). Nutrition, fatness and fertility: the effect of food intake on reproductive ability. In: W.H. Mosley (ed.) *Nutrition and human reproduction*, pp. 91–122. Plenum Press, New York

Frisch, R. (1982). Malnutrition and fertility. *Science 215*: 1272–3

Froehlich, J.W. and Thorington, R.W. (1982). The genetic structure and socioecology of howler monkeys (*Alouatta paliatta*) on Barro Colorado Island. In: E.G. Leigh, A.S. Rand and D.M. Windsor (eds.) *The ecology of tropical forests*, pp. 291–305. Smithsonian Institution Press, Washington (DC)

Furuishi, T. (1983). Interindividual distance and influence of dominance on feeding in a natural Japanese macaque troop. *Primates 24*: 445–55

Gadgil, M. (1972). Male dimorphism as a consequence of sexual selection. *Am. Nat. 106*: 574–80

Galat, G. and Galat-Luong, A. (1977). Démographie et régime alimentaire d'une troupe de *Cercopithecus aethiops sabaeus* en habitat marginal du nord Senegal. *Terre Vie 31*: 557–77

Galdikas, B.M.F. (1981). Orangutan reproduction in the wild. In: C. Graham (ed.)

Reproductive biology of great apes, pp. 281–300. Academic Press, New York

Galdikas, B.M.F. (1985). Adult male sociality and reproductive tactics among orang utans at Tranjung Puting. *Folia primatol. 45*: 9–24

Gale, G. (1979). *Theory of science.* McGraw-Hill, New York

Garber, P.A., Moya, L. and Malaga, C. (1984). A preliminary field study of the moustached tamarin monkey (*Saguinus mystax*) in northeastern Peru: questions concerned with the evolution of a communal breeding system. *Folia primatol. 42*: 17–33

Gartlan, J.S. (1968). Structure and function in primate society. *Folia primatol. 8*: 89–120

Gartlan, J.S. (1970). Preliminary notes on the ecology and behaviour of the drill, *Mandrillus leucophaeus* Ritgen, 1824. In: J.H. Napier and P. Napier (eds.) *Old World Monkeys*, pp. 445–80. Academic Press, London

Gartlan, J.S. and Brain, C.K. (1968). Ecology and social variability in *Cercopithecus aethiops* and *C. mitis.* In: P.C. Jay (ed.) *Primates*, pp. 253–92. Holt, Rinehart & Winston, New York

Gartlan, J.S. and Gartlan, S.C. (1973). Quelques observations sur les groupes exclusivement mâles chez *Erythrocebus patas. Ann. Fac. Sci. Cameroon 12*: 121–44

Gartlan, J.S. and Struhsaker, T.T. (1972). Polyspecific associations and niche separation of rain-forest anthropoids in Cameroon, West Africa. *J. Zool., Lond. 168*: 221–66

Gatinot, B.L. (1975). *Ecologie d'un colobe bai (Colobus badius temmincki Kuhn 1820) dans un mileau marginal au Senegal.* PhD thesis, Paris VI University

Gaulin, S.J.C. and Gaulin, C.K. (1982). Behavioural ecology of *Alouatta seniculus* in Andean cloud forest. *Int. J. Primatol. 3*: 1–32

Gautier-Hion, A. (1970). L'organisation sociale d'une bande de talapoins (*Miopithecus talapoin*) dans le nord-oest du Gabon. *Folia primatol. 12*: 116–41

Gautier-Hion, A. (1973). Social and ecological features of talapoin monkey — comparisons with sympatric cercopithecines. In: R.P. Michael and J.H. Crook (eds.) *Comparative ecology and behaviour of primates*, pp. 147–60. Academic Press, London

Gautier-Hion, A. (1980). Seasonal variations of diet related to species and sex in a community of *Cercopithecus* monkeys. *J. anim. Ecol. 49*: 237–69

Gautier-Hion, A. and Gautier, J.-P. (1976). Croissance, maturité sexuelle et sociale reproduction chez les cercopithécines forestiers africains. *Folia primatol. 26*: 165–84

Gautier-Hion, A. and Gautier, J.-P. (1978). Le singe de Brazza: une strategie originale. *Z. Tierpsychol. 48*: 84–104

Gautier-Hion, A. and Gautier, J.-P. (1985). Sexual dimorphism, social units and ecology among sympatric forest guenons. In: J. Ghesuiere, R.D. Martin and F. Newcombe (eds.) *Human sexual dimorphism*, pp. 61–77. Taylor & Francis, London

Geissmann, T. (1986). Mate change enhances duetting activity in the siamang (*Hylobates syndactylus*). *Behaviour 96*: 17–27

Geist, V. (1971). *Mountain sheep.* Chicago University Press, Chicago

Ghiglieri, M.P. ((1984). *The chimpanzees of Kibale Forest: a field study of ecology and social structure.* Columbia University Press, New York

Gibson, R.M. and Guinness, F.E. (1980). Differential reproduction among red deer (*Cervus elaphus*) stags on Rhum. *J. anim. Ecol. 49*: 199–208

Gillman, J. and Gilbert, C. (1956). The reproductive cycles of the chacma baboon with special reference to the problems of menstrual irregularities as assessed by the behaviour of the sex skin. *S. Afr. J. med. Sci. (Biol. Suppl.) 11*: 1–54

Ginsburg, B. and Allee, W.C. (1942). Some effects of conditioning on social dominance and subordination in inbred strains of mice. *Physiol. Zool. 15*: 485–506

Gittins, S.P. (1980). Territorial behaviour in the agile gibbon. *Int. J. Primatol. 1*: 381–99

Gittins, S.P. (1983). Use of forest canopy by the agile gibbon. *Folia primatol. 40*: 134–44

Glander, K. (1981). Feeding patterns in mantled howling monkeys. In: A. Kamil and T.D. Sargent (eds.) *Foraging behaviour: ecological, ethological and psychological approaches*, pp. 231–59. Garland, New York

Glassman, D.M., Coelho, A.M., Carey, K.D. and Bramblett, C.A. (1984). Weight growth in savannah baboons: a longitudinal study from birth to adulthood. *Growth 48*: 425–33

Goldfoot, D.A. (1982). Multiple channels of sexual communication in rhesus monkeys: role of olfactory cues. In: C.T. Snowdon, C.H. Brown and M.R. Petersen (eds.) *Primate communication*, pp. 413–28. Cambridge University Press, Cambridge

Goldizen, A.W. and Terborgh, J. (1986). Cooperative polyandry and helping behaviour in saddlebacked tamarins (*Saguinus fuscicollis*). In: J. Else and P. Lee (eds.) *Primate ecology and conservation*, pp. 191–8. Cambridge University Press, Cambridge

Goodall, A.G. (1977). Feeding and ranging behaviour of a mountain gorilla group (*Gorilla gorilla berengei*) in the Tshibinda-Kahuzi region (Zaire). In: T.H. Clutton-Brock (ed.) *Primate Ecology*, pp. 449–79. Academic Press, London

Goodall, J. (1965). Chimpanzees of the Gombe Stream Reserve. In: I. DeVore (ed.) *Primate behaviour*, pp. 425–73. Holt, Rinehart & Winston, New York

Goodall, J. (1967). Mother–offspring relations in free-ranging chimpanzees. In: D. Morris (ed.) *Primate ethology*, pp. 287–346. Weidenfeld & Nicolson, London

Goodall, J. (1968). The behaviour of free-living chimpanzees in the Gombe stream area. *Anim. Behav. Monogr. 1*: 161–311

Goodall, J. (1983). Population dynamics during a 15-year period in one community of free-living chimpanzees in the Gombe National Park. *Z. Tierpsychol. 61*: 1–60

Goosen, C. (1981). On the function of allogrooming in Old-World monkeys. In: A.B. Chiarelli and R.S. Corruccini (eds.) *Primate behaviour and sociobiology* pp. 110–20. Springer, Berlin

Gordon, I., Dunbar, R., Buckland, D. and Miller, D. (1987). Ponies, cattle and goats. In: T.H. Clutton-Brock and I. Ball (eds.) *Rhum: natural history of an island*, pp. 110–25. Edinburgh University Press, Edinburgh

Gosling, L.M. (1974). The social behaviour of Coke's hartebeest (*Alcephalus buselaphus cokei*). In: V. Geist and F.R. Walther (eds.) *The behaviour of ungulates and its relation to management*, pp. 488–571. IUCN Publications, Morges (Switzerland)

Gosling, L.M. (1986). Selective abortion of entire litters in the coypu: adaptive control of offspring production in relation to quality and sex. *Am. Nat. 127*: 772–95

Goss-Custard, J., Dunbar, R. and Aldrich-Blake, P. (1972). Survival, mating and rearing strategies in the evolution of primate social structure. *Folia primatol. 17*: 1–19

Gouzoules, H., Gouzoules, S. and Fedigan, L. (1982). Behavioural dominance and reproductive success in female Japanese monkeys (*M. fuscata*). *Anim. Behav. 30*: 1138–50

Gouzoules, S., Gouzoules, H. and Marler, P. (1984). Rhesus monkey (*Macaca mulatta*) screams: representational signalling in the recruitment of agonistic aid. *Anim. Behav. 32*: 182–93

Grafen, A. (1980). Opportunity cost, benefit and degree of relatedness. *Anim. Behav. 28*: 967–8

Grafen, A. (1982). How not to measure inclusive fitness. *Nature, Lond. 298*: 425–6

Grafen, A. (1984). Natural selection, kin selection and group selection. In: J.R. Krebs and N.B. Davies (eds.) *Behavioural ecology*, 2nd edition, pp. 62–84. Blackwell Scientific Publications, Oxford

Graham, G.G. (1968). The later growth of malnourished infants: effects of age, severity and subsequent diet. In: R.A. McCance and E.M. Widdowson (eds.) *Calorie deficiencies and protein deficiencies*, pp. 301–14, Churchill, London

Gray, J.P. (1985). *Primate sociobiology*. HRAF Press, New Haven (Conn.)

Greenberg, L.D. (1970). Nutritional requirements of macaque monkeys. In: R.S. Harris (ed.) *Feeding and nutrition of nonhuman primates*, pp. 117–57. Academic Press, New York

Grewal, B.S. (1980). Social relationships between adult central males and kinship groups of Japanese monkeys at Arashiyama with some aspects of troop organisation. *Primates 21*: 161–80

Griffin, D.R. (1981). *The question of animal awareness*, 2nd edition. Kaufmann, Los Altos (CA)

Griffin, D.R. (1982) (ed.) *Animal mind — human mind*. Springer, Berlin

Griffin, D.R. (1984). *Animal thinking*. Harvard University Press, Cambridge (Mass.)

Grossman, A., Moult, P., Gailand, R., Delitaza, G., Toff, W., Rees, L. and Besser, G. (1981). The opioid control of LH and FSH release: effects of met-enkephalin analogue and naxalone. *Clin. Endocrinol. 14*: 41–7

Grubb, P. (1973). Distribution, divergence and speciation of the drill and mandrill. *Folia primatol. 20*: 161–77

Guinness, F.E., Clutton-Brock, T.H. and Albon, S.D. (1978). Factors affecting calf mortality in red deer (*Cervus elaphus*). *J. anim. Ecol. 47*: 817–32

Guzman, M.A. (1968). Impaired physical growth and malnutrition in malnourished populations. In: N.S. Scrimshaw and J.E. Gordon (eds.) *Malnutrition, learning and behaviour*, pp. 42–54. MIT Press, Cambridge (Mass.)

Hadidian, J. and Bernstein, I.S. (1979). Female reproductive cycles and birth data from an Old World monkey colony. *Primates 20*: 429–42

Hall, K.R.L. (1963). Variations in the ecology of the chacma baboon, *Papio ursinus*. *Symp. zool. Soc. Lond. 10*: 1–28

Hall, K.R.L. (1965a). Behaviour and ecology of the wild patas monkey, *Erythrocebus patas*, in Uganda. *J. Zool., Lond. 148*: 15–87

Hall, K.R.L. (1965b). Behaviour and ecology of baboons, patas and vervet monkeys in Uganda. In: H. Vagtborg (ed.) *The baboon in medical research*, Vol. 2, pp. 43–61. University of Texas Press, San Antonio (Tex.)

Hall, K.R.L. (1967). Social interactions of the adult male and adult females of a patas monkey group. In: S.A. Altmann (ed.) *Social communication among Primates*, pp. 261–80. University of Chicago Press, Chicago

Hall, K.R.L. and DeVore, I. (1965). Baboon behaviour. In: I. DeVore (ed.) *Primate behaviour*, pp. 53–110. Holt, Rinehart & Winston, New York

Hamilton, W.D. (1964). The genetical evolution of social behaviour. I, II. *J. theoret. Biol. 7*: 1–52

Hamilton, W.D. (1971). Geometry for the selfish herd. *J. theoret. Biol. 7*: 295–311

Hamilton, W.J. (1986). Demographic consequences of a food and water shortage to desert chacma baboons, *Papio ursinus. Int. J. Primatol.* (in press)

Hamilton, W.J. and Arrowood, P. (1978). Copulatory vocalisations of chacma baboons (*Papio ursinus*), gibbons (*Hylobates hoolock*) and humans. *Science 200*: 1405–9

Hamilton, W.J., Buskirk, R.E.R. and Buskirk, W.H. (1975). Chacma baboon tactics during intertroop encounters. *J. Mammal. 56*: 857–70

Hamilton, W.J., Buskirk, R.E.R. and Buskirk, W.H. (1976). Defence of space and resources by chacma (*Papio ursinus*) baboon troops in an African desert and swamp. *Ecology 57*: 1264–72

Hamilton, W.J., Buskirk, R.E.R. and Buskirk, W.H. (1978). Omnivory and utilisation of food resources by chacma baboons, *Papio ursinus. Am. Nat. 112*: 911–24

Hamilton, W.J., Busse, C. and Smith, K.S. (1982). Adoption of infant orphan chacma baboons. *Anim. Behav. 30*: 29–34

Hansen, E.W. (1975). Selective responding by recently separated juvenile rhesus monkeys to the calls of their mothers. *Devel. Psychobiol. 9*: 83–8

Hanwell, A.S. and Peaker, M. (1977). Physiological effects of lactation on the mother. *Symp. zool. Soc. Lond. 41*: 297–311

Harcourt, A.H. (1978). Strategies of emigration and transfer by primates: with particular reference to gorillas. *Z. Tierpsychol. 48*: 401–20

Harcourt, A.H. (1981). Intermale competition and the reproductive behaviour of the great apes. In: C.E. Graham (ed.) *Reproductive biology of great apes*, pp. 301–18. Academic Press, New York

Harcourt, A.H. (1987). Dominance and fertility among female primates. *J. Zool., Lond.* (in press)

Harcourt, A.H. (in press). Alliances in contests and social intelligence. In: R. Byrne and A. Whiten (eds.) *Social expertise and evolution of intellect.* Cambridge University Press, Cambridge

Harcourt, A.H. and Stewart, K.J. (1981). Gorilla relationships: can differences during immaturity lead to contrasting reproductive tactics in adulthood? *Anim. Behav. 29*: 206–10

Harcourt, A.H. and Stewart, K.J. (1987). The influence of help in contests on dominance rank in primates: hints from gorillas. *Anim. Behav. 35*: 182–90

Harcourt, A.H., Stewart, K.J. and Fossey, D. (1976). Male emigration and female transfer in wild mountain gorillas. *Nature, Lond. 263*: 226–7

Harcourt, A.H., Stewart, K.J. and Fossey, D. (1981a). Gorilla reproduction in the wild. In: C.E. Graham (ed.) *Reproductive biology of great apes*, pp. 265–79. Academic Press, New York

Harcourt, A.H., Harvey, P.H., Larsen, S.G. and Short, R.V. (1981b). Testis weight, body weight and breeding system in primates. *Nature, Lond. 293*: 55–7

Harcourt, A.H., Fossey, D. and Sabater Pi, J. (1981c). Demography of *Gorilla gorilla. J. Zool., Lond. 195*: 215–23

Harcourt, C. and Nash, L. (1986). Social organisation of galagos in Kenyan coastal forests: I. *Galago zanzibaricus. Am. J. Primatol. 10*: 339–55

Harding, R.S.O. (1975). Meat-eating and hunting in baboons. In: R.H. Tuttle (ed.) *Socioecology and psychology of primates*, pp. 245–57. Mouton, The Hague

Harding, R.S.O. (1976). Ranging patterns of a troop of baboons (*Papio anubis*) in Kenya. *Folia primatol. 25*: 143–85

Harper, L.V. (1981). Offspring effects upon parents. In: D.J. Gubernick and P. Klopfer (eds.) *Parental care in mammals*, pp. 117–77. Plenum Press, New York

Hartman, C.G. (1932). Studies in the reproduction of the monkey *Macacus (Pithecus) rhesus*, with special reference to menstruation and pregnancy. *Contrib. Embryol. 23*: 1–161

Harvey, P.H. and Clutton-Brock, T.H. (1981). Primate home range size and metabolic needs. *Behav. Ecol. Sociobiol. 8*: 151–5

Harvey, P.H. and Mace, G.M. (1982). Comparisons between taxa and adaptive trends: problems of methodology. In: King's College Sociobiology Group (eds.) *Current problems in sociobiology*, pp. 343–62. Cambridge University Press, Cambridge

Harvey, P.H., Kavanagh, M. and Clutton-Brock, T.H. (1978). Sexual dimorphism in primate teeth. *J. Zool., Lond. 186*: 475–85

Hasegawa, T. and Hiraiwa, M. (1980). Social interactions of orphans observed in a free-ranging troop of Japanese monkeys. *Folia primatol. 33*: 129–58

Hasegawa, T. and Hiraiwa-Hasegawa, M. (1983). Opportunistic and restrictive matings among wild chimpanzees in the Mahali Mountains, Tanzania. *J. Ethol. 1*: 75–85

Hausfater, G. (1972). Intergroup behaviour of free-living rhesus monkeys (*Macaca mulatta*). *Folia primatol. 18*: 78–107

Hausfater, G. (1975). *Dominance and reproduction in baboons* (Papio cynocephalus). Karger, Basel

Hausfater, G. (1976). Predatory behaviour of yellow baboons. *Behaviour 56*: 44–68

Hausfater, G. and Bearce, W.H. (1976). *Acacia* tree exudates: their composition and use as a food source by baboons. *E. Afr. Wildl. J. 14*: 241–3

Hausfater, G. and Vogel, C. (1982). Infanticide in langur monkeys (genus *Presbytis*): recent research and a review of hypotheses. In: A.B. Chiarelli and R.S. Corruccini (eds.) *Advanced views in primate biology*, pp. 160-76. Springer, Berlin

Hausfater, G., Altmann, J. and Altmann, S.A. (1982a). Long-term consistency of dominance relations among female baboons (*Papio cynocephalus*). *Science 217*: 752–5

Hausfater, G., Aref, S. and Cairns, S.J. (1982b). Infanticide as an alternative male reproductive strategy in langurs: a mathematical model. *J. theoret. Biol. 94*: 391–412

Hegner, R.E., Emlen, S.T. and Demong, N.J. (1982). Spatial organisation of the white-fronted bee-eater. *Nature, Lond. 298*: 264–6

Hendy, H. (1986). Social interactions of free-ranging baboon infants. In: J. Else and P. Lee (eds) *Primate ontogeny, cognition and social behaviour*, pp. 267–80. Cambridge University Press, Cambridge

Henzi, S.P. (1981). Causes of testis-adduction in vervet monkeys (*Cercopithecus aethiops pygerythrus*). *Anim. Behav. 29*: 961–2

Henzi, S.P. and Lucas, J.W. (1980). Observations on the inter-troop movement of adult vervet monkeys (*Cercopithecus aethiops*). *Folia primatol. 33*: 220–35

Herbert, J. (1968). Sexual preference in the rhesus monkey (*Macaca mulatta*) in the laboratory. *Anim. Behav. 16*: 120–8

Hershkovitz, P. (1977). *Living New World Monkeys*. University of Chicago Press, Chicago

Hinde, R.A. (1969). Analysing the roles of the partners in a behavioural interaction — mother–infant relations in rhesus macaques. *Ann. NY Acad. Sci. 159*: 651–67

Hinde, R.A. (1976). Interactions, relationships and social structure. *Man 11*: 1–17

Hinde, R.A. (1978). Dominance and role — two concepts with dual meaning. *J. soc. Biol. Struct. 1*: 27–38

Hinde, R.A. and Atkinson, S. (1970). Assessing the roles of social partners in maintaining mutual proximity as exemplified by mother–infant relations in rhesus monkeys. *Anim. Behav. 18*: 169–76

Hinde, R.A. and Simpson, M.A. (1975). Qualities of mother–infant relationships in monkeys. In CIBA Foundation Symposium *Parent–infant interactions*, pp. 39–68

Hinde, R.A. and Stevenson-Hinde, J. (1976). Towards understanding relationships: dynamic stability. In: P.P.G. Bateson and R.A. Hinde (eds.) *Growing points in ethology*, pp. 451–79. Cambridge University Press, Cambridge

Hladik, C.M. (1977). A comparative study of the feeding strategies of two sympatric species of leaf monkeys: *Presbytis senex* and *Presbytis entellus*. In: T.H. Clutton-Brock (ed.) *Primate ecology*, pp. 324–53. Academic Press, London

Hladik, C.M. (1979). Diet and ecology of prosimians. In: G.A. Doyle and R.D. Martin (eds.) *The study of prosimian behaviour*, pp. 307–57. Academic Press, London

Hladik, C.M. and Charles-Dominique, P. (1974). The behaviour and ecology of the sportive lemur (*Lepilemur mustelinus*) in relation to its dietary peculiarities. In: R.D. Martin, G.A. Doyle and A.C. Walker (eds.) *Prosimian biology*, pp. 23–31. Duckworth, London

Hoage, R.J. (1977). Parental care in *Leontopithecus rosalia rosalia*: sex differences in carrying behaviour and the role of prior experience. In: D.G. Kleiman (ed.) *The biology and conservation of the Callitrichidae*, pp. 293–305. Smithsonian Institution Press, Washington (DC)

Hoogland, J. and Sherman, P. (1976). Advantages and disadvantages of bank swallow (*Riparia riparia*) coloniality. *Ecol. Monogr. 46*: 33–58

Hooley, J.M. and Simpson, M.A. (1981). A comparison of primiparous and multi-parous mother–infant dyads in *Macaca mulatta. Primates 22*: 379–92

Horr, D.A. (1972). The Bornean orang-utan: population structure and dynamics in relationship to ecology and reproductive strategy. In: L.A. Rosenblum (ed.) *Primate behaviour*, Vol. 4, pp. 25–80. Academic Press, New York

Horrocks, J. (1986). Life-history characteristics of a wild population of vervets (*Cercopithecus aethiops sabaeus*) in Barbados, West Indies. *Int. J. Primatol. 7*: 31–47

Horrocks, J. and Hunte, W. (1983a). Maternal rank and offspring rank in vervet monkeys: an appraisal of the mechanism of rank acquisition. *Anim. Behav. 31*: 772–81

Horrocks, J. and Hunte, W. (1983b). Rank relations in vervet sisters: a critique of the role of reproductive value. *Am. Nat. 122*: 417–21

Hoshino, J., Mori, A., Kudo, H. and Kawai, M. (1984). Preliminary report on the grouping of mandrills (*Mandrillus sphinx*) in Cameroons. *Primates 25*: 295–307

Howlett, T.A., Tomlin, S., Ngahfoong, L., Rees, L.H., Bullen, B.A., Skrinar, G.S. and MacArthur, J.W. (1984). Release of β-endorphin and met-enkephalin during exercise in normal women in response to training. *Brit. med. J. 288*: 1950–2

Hrdy, S.B. (1974). Male–male competition and infanticide among the langurs (*Presbytis entellus*) of Abu, Rajasthan. *Folia primatol. 22*: 19–58

Hrdy, S.B. (1976). Care and exploitation of nonhuman primate infants by conspecifics other than the mother. *Adv. Study Behav. 6*: 101–58

Hrdy, S.B. (1977). *The langurs of Abu.* Harvard University Press, Cambridge (Mass.)

Hrdy, S.B. (1979). Infanticide among animals: a review, classification and examination of the implications for the reproductive strategies of female. *Ethol. Sociobiol. 1*: 13–40

Hrdy, S.B. and Hrdy, D.B. (1976). Hierarchical relations among female hanuman langurs (Primate: Colobinae, *Presbytis entellus*). *Science 193*: 913–15

Hummer, R.L. (1970). Observations of the feeding of baboons. In: R.S. Harris (ed.) *Feeding and nutrition of nonhuman primates*, pp. 183–203. Academic Press, New York

Humphrey, N.K. (1976). The social function of intellect. In: P.P.G. Bateson and R.A. Hinde (eds.) *Growing points in ethology*, pp. 303–17. Cambridge University Press, Cambridge

Hutchins, M. and Barash, D.P. (1976). Grooming in primates: implications for its utilitarian function. *Primates 17*: 145–50

Huxley, J.S. (1934). A natural experiment on the territorial instinct. *British birds 27*: 270–7

Hytten, F.E. and Leitch, R.I. (1964). *The physiology of human pregnancy.* Blackwell, Oxford

Ikeda, H. (1982). Population changes and ranging behaviour of wild Japanese monkeys at Mt. Kawaradake in Kyushu, Japan. *Primates 23*: 338–47

Ingram, J.C. (1977). Interactions between parents and infants, and the development of independence in the common marmoset (*Callithrix jacchus*). *Anim. Behav. 25*: 811–27

Islam, M.A. and Husein, K.Z. (1982). A preliminary study on the ecology of the capped langur. *Folia primatol. 39*: 145–59

Iwamoto, T. (1979). Feeding ecology. In: M. Kawai (ed.) *Ecological and sociological studies of gelada baboons.*, pp. 280–330. Karger, Basel

Iwamoto, T. (1982). Food and nutritional condition of free ranging Japanese monkeys on Koshima Islet during winter. *Primates 23*: 153–70

Iwamoto, T. and Dunbar, R. (1983). Thermoregulation, habitat quality and the behavioural ecology of gelada baboons. *J. anim. Ecol. 52*: 357–66

Izard, M.K. and Simons, E.L. (1986). Management of reproduction in a breeding colony of bushbabies. In: J. Else and P. Lee (eds.) *Primate ecology and conservation*, pp. 315–24. Cambridge University Press, Cambridge.

Izawa, K. (1976). Group sizes and compositions of monkeys in the upper Amazon basin. *Primates 17*: 367–99

Izawa, K. (1980). Social behaviour of the wild black-capped capuchin (*Cebus apella*). *Primates 21*: 443–67

Izawa, K., Kimura, K. and Nieto, A.S. (1979). Grouping of the wild spider monkey. *Primates 20*: 503–12

Janis, C. (1976). The evolutionary strategy of the Equidae and the origins of rumen and caecal digestion. *Evolution 30*: 757–74

Janson, C. (1984). Female choice and mating system of the brown capuchin monkey *Cebus apella* (Primates: Cebidae). *Z. Tierpsychol. 65*: 177–200

Janson, C. (1985). Aggressive and individual food consumption in wild brown capuchin monkeys (*Cebus apella*). *Behav. Ecol. Sociobiol. 18*: 125–38

Janson, C. (1986). The mating system as a determinant of social evolution in capuchin monkeys. In: J. Else and P. Lee (eds.) *Primate ecology and conservation*, pp. 169–80. Cambridge University Press, Cambridge.

Jarman, M. and Jarman, P.J. (1973). Daily activity of impala. *E. Afr. Wildl. J. 11*: 75–92

Jarman, P.J. (1974). The social organisation of antelope in relation to their ecology. *Behaviour 48*: 215–67

Jolly, A. (1966a). *Lemur behaviour.* University of Chicago Press, Chicago

Jolly, A. (1966b). Lemur social behaviour and primate intelligence. *Science 153*: 501–6

Jolly, C.J. (1970). The large African monkeys as an adaptive array. In: J.H. Napier and P. Napier (eds.) *Old World Monkeys*, pp. 139–74. Academic Press, London

Jolly, C.J. (1972). The classification and natural history of *Theropithecus (Simopithecus)* (Andrews 1916), baboons of the African Plio-Pleistocene. *Bull. Brit. Mus. Nat. Hist. (Geol.) 22*: 1–123

Johnson, J.A. (1984). *Social relationships of juvenile olive baboons.* PhD thesis, University of Edinburgh

Johnson, J.A. (in press). Dominance rank in juvenile olive baboons (*Papio anubis*). *Anim. Behav.*

Johnson, R.L. and Southwick, C.H. (1984). Structural diversity and mother–infant relations among rhesus monkeys in India and Nepal. *Folia primatol. 43*: 189–248

Jones, C.B. (1980). The function of status in the mantled howler monkey, *Alouatta paliatta* Gray: intraspecific competition for group membership in a folivorous neotropical primate. *Primates 21*: 389–405

Jones, C.B. (1985). Reproductive patterns in mantled howler monkeys: estrus, mate choice and copulation. *Primates 26*: 130–42

Jouventin, P. (1975). Observations sur le socio-ecologie du mandrill. *Terre Vie 29*: 493–532

Jungers, W.L. and Susman, R.L. (1984). Body size and skeletal allometry in African apes. In: R.L. Susman (ed.) *The pygmy chimpanzee*, pp. 131–77. Plenum Press, New York

Kano, T. (1983). An ecological study of the pygmy chimpanzees (*Pan paniscus*) of Yalosidi, Republic of Zaire. *Int. J. Primatol. 4*: 1–31

Kano, T. and Mulavwa, M. (1984). Feeding ecology of the pygmy chimpanzees (*Pan paniscus*) of Wamba. In: R.L. Susman (ed.) *The pygmy chimpanzee*, pp. 233–74. Plenum Press, New York

Kaplan, J.R. (1978). Fight interference and altruism in rhesus monkeys. *Am. J. phys. Anthropol. 49*: 241–50

Kaufmann, J.H. (1965). A three-year study of mating behaviour in a freeranging band of rhesus monkeys. *Ecology 40*: 500–12

Kavanagh, M. (1978). The diet and feeding behaviour of *Cercopithecus aethiops tantalus*. *Folia primatol. 30*: 30–63

Kavanagh, M. (1981). Variable territoriality among tantalus monkeys in Cameroon. *Folia primatol. 36*: 76–98

Kavanagh, M. (1983). Birth seasonality in *Cercopithecus aethiops*. A social advantage from synchrony? In: P.K. Seth (ed.) *Perspectives in primate biology*, pp. 89–98. Today and Tomorrow, New Delhi

Kawai, M. (1958). On the rank system in a natural group of Japanese monkeys. I, II. *Primates 1*: 111–48

Kawai, M., Dunbar, R., Ohsawa, H. and Mori, U. (1983). Social organisation of gelada baboons: social units and definitions. *Primates 24*: 1–13

Kay, R.F., and Simons, E.L. (1980). The ecology of Oligocene African Anthropoidea. *Int. J. Primatol 1*: 21–37

Kay, R.N.B., Hoppe, P. and Maloiy, G.M.O. (1976). Fermentative digestion of food in the colobus monkey, *Colobus polykomos*. *Experientia 32*: 485–6

Keenleyside, M.H.A. (1972). Intraspecific intrusions into nests of spawning longear sunfish (Pisces: Centrarchidae). *Copeia 1972*, 272–8

Kerr, G.R. (1972). Nutritional requirements of subhuman primates. *Physiol. Rev. 52*: 415–67

Keverne, E.B. (1982). Olfaction and the reproductive behaviour of nonhuman primates. In: C.T. Snowdon, C.H. Brown and M.R. Petersen (eds.) *Primate communication*, pp. 396–412. Cambridge University Press, Cambridge

Keverne, E.B., Leonard, R.A., Scruton, D.M. and Young, S.K. (1978). Visual monitoring in social groups of talapoin monkeys (*Miopithecus talapoin*). *Anim. Behav. 26*: 933–44

King, J.A. (1955). Social behaviour, social organisation and population dynamics in a black-tailed prairiedog town in the Black Hills of South Dakota. *Contrib. Lab. Vert. Biol., Univ. Mich., Ann Arbor 67*: 1–123

Kinzey, W.G. (1974). Ceboid models for the evolution of hominoid dentition. *J. human Evol. 3*: 193–203

Kinzey, W.G. (1977). Diet and feeding behaviour of *Callicebus torquatus*. In: T.H. Clutton-Brock (ed.) *Primate ecology*, pp. 127–51. Academic Press, London

Kinzey, W.G. (1984). The dentition of the pygmy chimpanzee, *Pan paniscus*. In: R.L. Susman (ed.) *The pygmy chimpanzee*, pp. 65–88. Plenum Press, New York

Kleiber, M. (1961). *The fire of life*. Wiley, New York

Kleiman, D. (1977). Monogamy in mammals. *Q. Rev. Biol. 52*: 39–69

Kleiman, D. and Malcolm, J.R. (1981). The evolution of male parental investment in mammals, In: D.J. Gubernick and P.H. Klopfer (eds.) *Parental care in mammals*, pp. 347–87. Plenum Press, New York

Klein, L.L. and Klein, D.J. (1977). Feeding behaviour of the Columbian spider monkey. In: T.H. Clutton-Brock (ed.) *Primate ecology*, pp. 153–81. Academic Press, London

Knowlton, N. (1979). Reproductive synchrony, parental investment and the evolutionary dynamics of sexual selection. *Anim. Behav. 27*: 1022–33

Koyama, N. (1967). On dominance rank and kinship of a wild Japanese monkey troop in Arashiyama. *Primates 8*: 189–216

Koyama, N. (1970). Changes in dominance rank and division of a wild Japanese monkey troop at Arashiyama. *Primates 11*: 335–90

Koyama, N., Norikoshi, K. and Mano, T. (1975). Population dynamics of Japanese monkeys at Arashiyama. In: M. Kawai, S. Kondo and A. Ehara (eds.) *Contemporary primatology*, pp. 411–17. Karger, Basel

Koyama, T., Fujii, H. and Yonekawa, F. (1981). Comparative studies of gregariousness and social structure among seven feral *Macaca fuscata* groups. In: A.B. Chiarelli and R.S. Corruccini (eds.) *Primate behaviour and sociobiology*, pp. 52–63. Springer, Berlin

Krebs, J.R. and Dawkins, R. (1984). Animal signals, mind-reading and manipulation. In: J.R. Krebs and N.B. Davies (eds.) *Behavioural ecology*, 2nd edition, pp. 380–402. Blackwell Scientific Publications, Oxford

Krebs, J.R. and McCleery, R.H. (1984). Optimisation in behavioural ecology. In: J.H. Krebs and N.B. Davies (eds.) *Behavioural ecology*, pp. 91–121. Blackwell Scientific Publications, Oxford

Kruuk, H. (1972). *The spotted hyaena*. University of Chicago Press, Chicago

Kummer, H. (1967). Tripartite relations in hamadryas baboons. In: S.A. Altmann (ed.) *Social communication among primates*, pp. 63–72. University of Chicago Press, Chicago

Kummer, H. (1968). *Social organisation of hamadryas baboons*, Karger, Basel

Kummer, H. (1970). Cross-species modification of social behaviour in baboons. In: J.H. Napier and P. Napier (eds.) *Old World Monkeys*, pp. 353–63. Academic Press, London

Kummer, H. (1971). *Primate societies*. Aldine-Atherton, Chicago

Kummer, H. (1974). Distribution of interindividual distances in patas monkeys and gelada baboons: species and sex diferences. *Folia primatol. 21*: 153–60

Kummer, H. (1975). Rules of dyad and group formation among captive gelada baboons (*Theropithecus gelada*). In: S. Kondo, M. Kawai, A. Ehara and S. Kawamura (eds.) *Proceedings of the Fifth Congress of the International Primatological Society*, pp. 129–59. Japan Science Press, Tokyo

Kummer, H. (1978). On the value of social relationships to nonhuman primates: a heuristic scheme. *Soc. Sci. Inform. 17*: 687–705

Kummer, H. (1982). Social knowledge in free-ranging primates. In: D.R. Griffin (ed.) *Animal mind — human mind*, pp. 113–30. Springer, Berlin

Kummer, H., Götz, W. and Angst, W. (1974). Triadic differentiation: an inhibitory process protecting pair bonds in baboons. *Behaviour 49*: 62–87

Kummer, H., Banaja, A., Abo-Khatwa, A. and Ghandour, A. (1985). Differences in social behaviour between Ethiopian and Arabian hamadryas baboons. *Folia primatol. 45*: 1–8

Kurland, J.A. (1977). *Kin selection in the Japanese monkey*. Karger, Basel

Kuroda, S. (1979). Grouping of the pygmy chimpanzee. *Primates 20*: 161–83

Kuroda, S. (1984). Interactions over food among pygmy chimpanzees. In: R.L. Susman (ed.) *The pygmy chimpanzee*, pp. 301–24, Plenum Press, New York

Lancaster, J.B. (1971). Play-mothering: the relations between juvenile females and young infants among free-ranging vervet monkeys (*Cercopithecus aethiops*). *Folia primatol. 15*: 161–82

Lancaster, J.B. and Lee, R.B. (1965). The annual reproductive cycle in monkeys and apes. In: I. DeVore (ed.) *Primate behaviour*, pp. 486–513. Holt, Rinehart & Winston, New York

Landau, H.G. (1951). On dominance relations and the structure of animal societies. I. Effect of inherent characteristics. *Bull. Math. Biophys. 13*: 1–19

Lauer, C. (1980). Seasonal variability in spatial defence by free-ranging rhesus monkeys (*Macaca mulatta*). *Anim. Behav. 28*: 476–82

Lazarus, J. (1979). The early warning function of flocking in birds: an experimental study with captive *Quelea. Anim. Behav. 27*: 855–65

LeBoeuf, B.J. (1974). Male–male competition and reproductive success in elephant seals. *Am. Zool. 14*: 163–76

Lee, A.K. and Cockburn, A. (1985). *Evolutionary ecology of marsupials.* Cambridge University Press, Cambridge

Lee, P.C. (1983a). Ecological influences on relationships and social structure. In: R.A. Hinde (ed.) *Primate social relationships*, pp. 225–9. Blackwell Scientific Publications, Oxford

Lee, P.C. (1983b). Effects of the loss of the mother on social development. In: R.A. Hinde (ed.) *Primate social relationships*, pp. 73–9. Blackwell Scientific Publications, Oxford

Lee, P.C. (1983c). Context-specific unpredictability in dominance interactions. In: R.A. Hinde (ed.) *Primate social relationships*, pp. 35–44. Blackwell Scientific Publications, Oxford

Lee, P.C. (1984). Ecological constraints on the social development of vervet monkeys. *Behaviour 91*: 245–61

Lee, P.C. (1986). Environmental influences on development: play, weaning and social structure. In: J. Else and P.C. Lee (eds.) *Primate ontogeny, cognition and social behaviour*, pp. 227–37. Cambridge University Press, Cambridge

Lee, P.C. and Oliver, J.I. (1979). Competition, doominance and the acquisition of rank in juvenile yellow baboons (*Papio cynocephalus*). *Anim. Behav. 27*: 576–85

Leutenegger, W. (1973). Maternal–fetal weight relationships in primates. *Folia primatol. 20*: 280–93

Leutenegger, W. (1980). Monogamy in callitrichids: a consequence of phyletic dwarfism? *Int. J. Primatol. 1*: 95–8

Lindburg, D.G. (1971). The rhesus monkey in north India: an ecological and behavior study. In: L.A. Rosenblum (ed.) *Primate behaviour*, Vol. 1, pp. 1–106. Academic Press, New York

Loy, J. (1970). Behavioural responses of free-ranging rhesus monkeys to food shortage. *Am. J. phys. Anthropol. 33*: 263–72

Loy, J. (1971). Estrous behaviour in free-ranging rhesus monkeys (*Macaca mulatta*). *Primates 12*: 1–31

Lunn, S.F. and McNeilly, A.S. (1982). Failure of lactation to have a consistent effect on interbirth interval in the common marmoset, *Callithrix jacchus jacchus. Folia primatol. 37*: 99–105

McClintock, M.K. (1971). Menstrual synchrony and suppression. *Nature, Lond. 229*: 244–5

McClintock, M.K. (1978). Estrous synchrony and its mediation by airborne chemical communication (*Rattus norvegicus*). *Hormones Behav. 10*: 262–76

Macdonald, D.W. (1983). The ecology of carnivore social behaviour. *Nature, Lond. 301*: 379–84

Macdonald, D.W. and Moehlman, P.D. (1982). Cooperation, altruism and restraint in the reproduction of carnivores. In: P.P.G. Bateson and P.H. Klopfer (eds.) *Perspectives in ethology*, Vol. 5, pp. 433–67. Plenum Press, New York

Mace, G.M. and Harvey, P.H. (1983). Energetic constraints on home range size. *Am. Nat. 121*: 120–32

McFarland, D. and Houston, A. (1981). *Quantitative ethology: a state space approach.* Pitman, Oxford

McGrew, W.C. (1987). Helpers at the nest-box, or are cotton-top tamarins really Florida scrub jays? *Primate Report* (in press)

McGrew, W.C. and McLuckie, E.C. (1986). Philopatry and dispersion in the cotton-top tamarin, *Saguinus (o.) oedipus*: an attempted laboratory simulation. *Int. J. Primatol. 7*: 401–22

McGrew, W.C., Baldwin, P.J. and Tutin, C.E.G. (1981). Chimpanzees in a hot, dry and open habitat: Mt. Assirik, Senegal, West Africa. *J. human Evol.* 10: 227–44

Mack, D. (1979). Growth and development of infant red howling monkeys (*Alouatta seniculus*) in a free ranging population. In: J.F. Eisenberg (ed.) *Vertebrate ecology in the northern neotropics,* pp. 127–36. Smithsonian Institution Press, Washington (DC)

McKenna, J.J. (1979). The evolution of allomothering behaviour among colobine monkeys: function and opportunism in evolution. *Am. Anthropol.* 81: 818–40

McKenna, J.J. (1981). Primate infant caregiving behaviour: origins, consequences and variability with emphasis on the common Indian langur monkey. In: D.J. Gubernick and P.H. Klopfer (eds.) *Parental care in mammals,* pp. 389–416. Plenum Press, New York

McKey, D. and Waterman, P.G. (1982). Ranging behaviour of a group of black colobus (*Colobus satanas*) in the Douala-Edea Reserve, Cameroon. *Folia primatol.* 39: 264–304

Mackinnon, J.R. (1974). The behaviour and ecology of wild orangutans (*Pongo pygmaeus*). *Anim. Behav.* 22: 3–74

McNaughton, S.J. (1979). Grazing as an optimisation process: grass–ungulate relationships in the Serengeti. *Am. Nat.* 113: 691–703

McNeilly, A.S. (1979). Effects of lactation on fertility. *Brit. med. Bull.* 35: 151–4

Manzolillo, D.L. (1986). Factors affecting intertroop transfer by adult male *Papio anubis.* In: J. Else and P. Lee (eds.) *Primate ontogeny, cognition and social behaviour,* pp. 371–80. Cambridge University Press, Cambridge

McQueen, R.E. and van Soest, P.J. (1975). Fungal cellulose and hemicellulose prediction of forage digestibility. *J. Dairy Sci.* 58: 1482–91

Makwana, S.C. (1978). Field ecology and behaviour of the rhesus macaque (*Macaca mulatta*). *Primates* 19: 483–92

Marler, P. (1969). *Colobus guereza*: territoriality and group composition. *Science 163*: 93–5

Marler, P. (1972). Vocalisations of East African monkeys. II. Black and white colobus. *Behaviour 42*: 176–97

Marler, P. and Hobbett L. (1975). Individuality in a long-range vocalisation of wild chimpanzees. *Z. Tierpsychol.* 38: 97–109

Marsh, C.W. (1979a). Comparative aspects of social organisation in the Tana River red colobus, *Colobus badius rufomitratus. Z. Tierpsychol.* 51: 337–62

Marsh, C.W. (1979b). Female transfer and mate choice among Tana River red colobus. *Nature, Lond.* 281: 568–9

Marsh, C.W. (1981a). Ranging behaviour and its relation to diet selection in Tana River red colobus (*Colobus badius rufomitratus*). *J. Zool. Lond.* 195: 473–92

Marsh, C.W. (1981b). Time budgets of the Tana River red colobus. *Folia primatol.* 35: 30–50

Martensz, N.D., Vellucci, S.V., Keverene, E.B. and Herbert, J. (1986). β-endorphin levels in the cerebrospinal fluid of male talapoin monkeys in social groups related to dominance status and the leutinising hormone response to naxalone. *Neuroscience* 18: 651–8

Martin, R.D. (1973). A review of the behaviour and ecology of the lesser mouse lemur (*Microcebus murinus* J.F. Miller 1777). In: R.P. Michael and J.H. Crook (eds.) *Comparative ecology and behaviour of primates,* pp. 1–68. Academic Press, London

Maslow, A.H. (1936). The role of dominance in the social and sexual behaviour of infra-human primates. IV. The determination of hierarchy in pairs and in a group. *J. genet. Psychol.* 49: 161

Mason, W.A. (1965). Use of space by *Callicebus* groups. In: P.C. Jay (ed.) *Primates,* pp. 200–16. Holt, Rinehart & Winston, New York

Mason, W.A. (1971). Field and laboratory studies of social organisation in *Saimiri* and *Callicebus*. In: L.A. Rosenblum (ed.) *Primate Behaviour*, Vol. 2, pp. 107–37. Academic Press, New York

Massey, A. (1977). Agonistic aids and kinship in a group of pigtail macaques. *Behav. Ecol. Sociobiol. 2*: 31–40

Masui, K., Sugiyama, Y., Nishimura, A. and Ohsawa, H. (1975). The life table of Japanese monkeys at Takasakiyama. In: M. Kawai, S. Kondo and A. Ehara (eds.) *Contemporary primatology*, pp. 401–6. Karger, Basel

Matsumato, S., Igarashi, M. and Nagaoka, Y. (1968). Environmental anovulatory cycles. *Int. J. Fertil. 13*: 15–23

Maxim, P.E. and Buettner-Janusch, J. (1963). A field study of the Kenya baboon. *Am. J. phys. Anthropol. 21*: 165–89

Maynard Smith, J. (1976). Group selection. *Q. Rev. Biol. 51*: 277–83

Maynard Smith, J. (1982). *Evolution and the theory of games.* Cambridge University Press, Cambridge

Maynard Smith, J. and Parker, G.A. (1976). The logic of asymmetric contests. *Anim. Behav. 24*: 159–75

Maynard Smith, J. and Price, G.R. (1973). The logic of animal conflict. *Nature, Lond. 246*: 15–18

Meikle, D.B. and Vessey, S.H. (1981). Nepotism among rhesus monkey brothers. *Nature, Lond. 294*: 160–1

Meikle, D.B., Tilford, B.L. and Vessey, S.H. (1984). Dominance rank, secondary sex ratio and reproduction of offspring in polygamous primates. *Am. Nat. 124*: 173–87

Melnick, D.J. and Kidd, K.K. (1983). The genetic consequences of social group fission in a wild population of rhesus monkeys (*Macaca mulatta*). *Behav. Ecol. Sociobiol. 12*: 229–36

Ménard, N., Vallet, D. and Gautier-Hion, A. (1985). Démographie et reproduction de *Macaca sylvanus* dans differents habitats en Algérie. *Folia primatol. 44*: 65–81

Mendoza, S.P. and Mason, W.A. (1986). Parental division of labour and differentiation of attachments in a monogamous primate (*Callicebus moloch*). *Anim. Behav. 34*: 1336–47

Menzel, E.W. (1978). Cognitive mapping in chimpanzees. In: S.H. Hulse, H. Fowler and W.K. Honig (eds.) *Cognitive processes in animal behaviour*, pp. 375–422. Lawrence Erlbaum, Hillsdale

Menzel, E.W. and Juno, C. (1985). Social foraging in marmoset monkeys and the question of intelligence. In: L. Weiskrantz (ed.) *Animal intelligence* pp. 145–57. Clarendon Press, Oxford

Meyer, P.L. (1970). *Introductory probability and statistical applications.* 2nd edition, Addison-Wesley, Reading (Mass.)

Michael, R.P. and Keverne, E.B. (1968). Pheromones: their role in the communication of sexual status in primates. *Nature, Lond. 218*: 746–9

Michael, R.P. and Saayman, G.S. (1967). Individual differences in sexual behaviour of male rhesus monkeys (*Macaca mulatta*) under laboratory conditions. *Anim. Behav. 15*: 460–6

Michael, R.P. and Zumpe, D. (1970). Rhythmic changes in the copulatory frequency of rhesus monkeys (*Macaca mulatta*) in relation to the menstrual cycle and a comparison with the human cycle. *J. Reprod. Fertil. 21*: 199–201

Michael, R.P., Bonsall, R.W. and Zumpe, D. (1978). Consort bonding and operant behaviour by female rhesus monkeys. *J. comp. Physiol. Psychol. 92*: 837–45

Milton, K. (1980). *The foraging strategy of howler monkeys.* Columbia University Press, New York

Milton, K. (1982). Dietary quality and demographic regulation in a howler monkey population. In: E.G. Leigh, A.S. Rand and D.M. Windsor (eds.) *The ecology of a tropical forest*, pp. 273–89. Smithsonian Institution Press, Washington (DC)

Milton, K. (1985). Mating patterns of woolly spider monkeys, *Brachyteles arachnoides:* implications for female choice. *Behav. Ecol. Sociobiol. 17:* 53–9

Milton, K. and McBee, R. (1982). Structural carbohydrate digestion in a New World primate, *Alouatta palliata* Gray. *Comp. Biochem. Physiol. 74:* 29–31

Milton, K. and May, M.L. (1976). Body weight, diet and home range area in primates. *Nature, Lond. 259:* 459–62

Milton, K., van Soest, P.J. and Robertson, J.B. (1980). Digestive efficiencies of wild howler monkeys. *Physiol. Zool. 53:* 402–9

Missakian, E.A. (1972). Genealogical and cross-genealogical dominance relations in a group of free-ranging rhesus monkeys (*Macaca mulatta*) on Cayo Santiago. *Primates 13:* 169–80

Missakian, E.A. (1973). The timing of fission among free-ranging rhesus monkeys. *Am. J. phys. Anthropol. 38:* 621–4

Mitani, J.C. (1984). The behavioural regulation of monogamy in gibbons (*Hylobates muelleri*). *Behav. Ecol. Sociobiol. 15:* 225–9

Mitani, J.C. and Rodman, P.S. (1979). Territoriality: the relation of ranging pattern and home range size to defendability, with an analysis of territoriality among primate species. *Behav. Ecol. Sociobiol. 5:* 241–51

Mitchell, B., McCowan, D. and Nicholson, I.A. (1976). Annual cycles of body weight and condition in Scottish red deer, *Cervus elaphus. J. Zool. Lond. 180:* 107–27

Mitchell, G. and Brandt, E.M. (1972). Paternal behaviour in primates. In: F.E. Poirier (ed.) *Primate socialisation*, pp. 173–206. Random House, New York

Moehlman, P.D. (1979). Jackal helpers and pup survival. *Nature, Lond. 277:* 382–3

Mohnot, S.M. (1971). Some aspects of social changes and infant-killing in the hanuman langur, *Presbytis entellus* (Primates: Cercopithecidae), in Western India. *Mammalia 35:* 175–98

Mohnot, S.M., Gadgil, M. and Makwana, S.C. (1981). On the dynamics of the hanuman langur populations of Jodhpur (Rajasthan, India). *Primates 22:* 182–91

Moore, J. (1984). Female transfer in primates. *Int. J. Primatol. 5:* 537–89

Moore, J. and Ali, R. (1984). Are dispersal and inbreeding avoidance related? *Anim. Behav. 32:* 94–112

Mori, A. (1975). Signals found in the grooming interactions of wild Japanese monkeys of the Koshima troop. *Primates 16:* 107–40

Mori, A. (1977a). The social organisation of the provisioned Japanese monkey troops which have extraordinarily large population sizes. *J. Anthropol. Soc. Nippon 85:* 325–45

Mori, A. (1977b). Intra-troop spacing mechanism of the wild Japanese monkeys of the Koshima troop. *Primates 18:* 331–57

Mori, A. (1979a). Analysis of population changes by measurement of body weight in the Koshima troop of Japanese monkeys. *Primates 20:* 371–97

Mori, A. (1979b). An experiment on the relation between the feeding speed and the caloric intake through leaf eating in Japanese monkeys. *Primates 20:* 185–95

Mori, U. (1979a). Reproductive behaviour. In: M. Kawai (ed.) *Ecological and sociological studies of gelada baboons*, pp. 183–97, Karger, Basel

Mori, U. (1979b). Unit formation and the emergence of a new leader. In: M. Kawai (ed.) *Ecological and sociological studies of gelada baboons*, pp. 156–81. Karger, Basel

Mori, U. and Dunbar, R.I.M. (1985). Changes in the reproductive condition of female gelada baboons following the takeover of one-male units. *Z. Tierpsychol. 67:* 215–24

Moss, C.J. and Poole, J.H. (1983). Relationships and social structure of African elephants. In: R.A. Hinde (ed.) *Primate social relationships*, pp. 314–25. Blackwell Scientific Publications, Oxford

Mount, L.E. (1979). *Adaptation to thermal environment.* Arnold, London

Müller, H. (1980). *Variations of social behaviour in a baboon hybrid zone* (Papio anubis × Papio hamadryas) *in Ethiopia*. PhD thesis, University of Zurich

Müller-Schwarze, D., Stagge, B. and Müller-Schwarze, C. (1982). Play behaviour; persistence, decrease and energetic compensation during food shortage in deer fawns. *Science, NY 215*: 85–7

Nagel, U. (1971). Social organisation in a baboon hybrid zone. *Proc. 3rd Int. Congr. Primatol. 3*: 48–57. Karger, Basel

Nagel, U. (1973). A comparison of anubis baboons, hamadryas baboons and their hybrids at a species border in Ethiopia. *Folia primatol. 19*: 104–65

Nagel, U. (1979). On describing primate groups as systems: the concept of ecosocial behaviour. In: I. Bernstein and E.O. Smith (eds.) *Primate ecology and human origins*, pp. 313–40. Garland, New York

Nagy, K.A. and Milton, K. (1979). Energy metabolism and food consumption by wild howler monkeys. *Ecology 60*: 475–80

Naismith, D.J. and Ritchie, C.D. (1975). The effect of breast-feeding and artificial feeding on body weights, skinfold measurements and food intakes of 42 primiparous women. *Proc. Nutr. Soc. 34*: 116A–17A

Nash, L. (1976). Troop fission in free-ranging baboons in the Gombe Stream National Park, Tanzania. *Am. J. phys. Anthropol. 44*: 63–78

Nash, L. (1978). The development of the mother–infant relationship in wild baboons (*Papio anubis*). *Anim. Behav. 26*: 746–59

Nash, L. (1986). Social organisation of two sympatric galagos at Gedi, Kenya. In: J. Else and P. Lee (eds.) *Primate ecology and conservation*, pp. 125–32. Cambridge University Press, Cambridge

Nash, V.J. and Chamove, A.S. (1981). Personality and dominance behaviour in stump-tailed macaques. In: A.B. Chiarelli and R.S. Corruccini (eds.) *Primate behaviour and sociobiology*, pp. 88–92, Springer, Berlin

Netto, W.J. and van Hooff, J. (1986). Conflict interference and the development of dominance relationships in immature *Macaca fascicularis*. In: J. Else and P. Lee (eds.) *Primate ontogeny, cognition and social behaviour*, pp. 291–300. Cambridge University Press, Cambridge

Neville, M.K. (1972). The population structure of red howler monkeys (*Alouatta seniculus*) in Trinidad and Venezuela. *Folia primatol. 17*: 56–86

Newton, P.N. (1986). Infanticide in an undisturbed forest population of hanuman langurs, *Presbytis entellus*. *Anim. Behav. 34*: 785–9

Neyman, P. (1977). Aspects of the ecology of free-living cotton-top tamarins (*Saguinus o. oedipus*). In: D.G. Kleiman (ed.) *The biology and conservation of the Callitrichidae*, pp. 39–72. Smithsonian Institution Press, Washington (DC)

Neyman, P. (1980). *Ecology and social organisation of the cotton-top tamarin (Saguinus oedipus)*. PhD thesis, University of California, Berkeley

Nicholson, S.E. (1981). The historical climatology of Africa. In: T.M.L. Wigley, M.J. Ingram and C. Farmer (eds.) *Climate and history*, pp. 249–70. Cambridge University Press, Cambridge

Nishida, T. (1968). The social group of wild chimpanzees in the Mahali Mountains. *Primates 19*: 167–224

Nishida, T. (1979). Social structure among wild chimpanzees of the Mahali Mountains. In: D.A. Hamburg and E.R. McCown (eds.) *The great apes*, pp. 72–121. Benjamin/Cummings, Menlo Park (CA)

Nishida, T. (1983). Alpha status and agonistic alliance in wild chimpanzees (*Pan troglodytes schweinfurthii*). *Primates 24*: 318–36

Noë, R. (1986). Lasting alliances among adult male savannah baboons. In: J. Else and P. Lee (eds.) *Primate ontogeny, cognition and social behaviour*, pp. 381–92. Cambridge University Press, Cambridge van Noordwijk, M.A. (1985). Sexual behaviour of Sumatran long-tailed macaques. *Z. Tierpsychol.* 70: 277–96

van Noordwijk, M.A. and van Schaik, C.P. (1985). Male migration and rank acquisition in wild long-tailed macaques (*Macaca fascicularis*). *Anim. Behav.* 33: 849–61

van Noordwijk, M.A. and van Schaik, C.P. (in press). Competition among female long-tailed macaques (*Macaca fascicularis*). *Anim. Behav.*

Norton, G.W. (1986). Leadership: decision processes of group movement in yellow baboons. In: J. Else and P. Lee (eds.) *Primate ecology and conservation*, pp. 145–56. Cambridge University Press, Cambridge

Oates, J.F. (1977a). The guereza and its food. In: T.H. Clutton-Brock (ed.) *Primate ecology*, pp. 275–321. Academic Press, London

Oates, J.F. (1977b). The social life of a black-and-white colobus monkey, *Colobus guereza. Z. Tierpsychol.* 45: 1–60

Oates, J.F. (1978). Water-plant and soil consumption by guereza monkeys (*Colobus guereza*): a relationship with minerals and toxins in the diet. *Biotropica 10*: 241–53

Ohsawa, H. and Dunbar, R.I.M. (1984). Variations in the demographic structure and dynamics of gelada baboon populations. *Behav. Ecol. Sociobiol. 15*: 231–40

O'Keefe, J. (1985). Is consciousness the gateway to the hippocampal cognitive map? A speculative essay on the neural basis of mind. In: D.A. Oakley (ed.) *Brain and mind*, pp. 59–98. Methuen, London

Oppenheimer, J.R. (1968). *Behaviour and ecology of the white-faced monkey*, Cebus capuchinus, on Barro Colorado Island, C.Z. PhD thesis, University of Illinois, Urbana

Orians, G.H. (1969). On the evolution of mating systems in birds and mammals. *Am. Nat. 103*: 589–603

Owen-Smith, N. (1977). On territoriality in ungulates and an evolutionary model. *Q. Rev. Biol. 52*: 1–38

Oyama, S. (1986). *The ontogeny of information.* Cambridge University Press, Cambridge

Packer, C. (1977). Reciprocal altruism in *Papio anubis. Nature, Lond. 265*: 441–3

Packer, C. (1979a). Inter-troop transfer and inbreeding avoidance in *Papio anubis. Anim. Behav. 27*: 1–36

Packer, C. (1979b). Male dominance and reproductive activity in *Papio anubis. Anim. Behav. 27*: 37–45

Packer, C. (1980). Male care and exploitation of infants in *Papio anubis. Anim. Behav. 28*: 512–20

Packer, C. and Pusey, A.E. (1979). Female aggression and male membership in troops of Japanese macaques and olive baboons. *Folia primatol. 31*: 212–18

Packer, C. and Pusey, A.E. (1982). Cooperation and competition within coalitions of male lions: kin selection or game theory? *Nature, Lond. 296*: 740–2

Parke, R.D. and Suomi, S.T. (1981). Adult male–infant relationships: human and non-human primate evidence. In: K. Immelman, G. Barlow, L. Petrinovich and M. Main (eds.) *Behavioural development*, pp. 700–25. Cambridge University Press, Cambridge

Parker, G.A. (1970). Sperm competition and its evolutionary consequences in the insects. *Biol. Rev. 45*: 525–67

Parker, G.A. (1974a). Assessment strategy and the evolution of animal conflicts. *J. theoret. Biol. 47*: 223–43

Parker, G.A. (1974b). Courtship persistence and female-guarding as male time investment strategies. *Behaviour 48*: 157–84

Parker, G.A. (1984). Evolutionarily stable strategies. In: J.R. Krebs and N.B. Davies (eds.) *Behavioural ecology*, 2nd edition, pp. 30–61. Blackwell Scientific Publications, Oxford

Patterson, J.D. (1973). Ecologically differentiated patterns of aggressive and sexual behaviour in two troops of Ugandan baboons. *Papio anubis. Am. J. phys. Anthropol.* *38*: 641–8

Patterson, J.D. (1976). *Variations in ecology and adaptation of Ugandan baboons* Papio cynocephalus anubis. PhD thesis, University of Calgary

Paul, A. and Thommen, D. (1984). Timing of birth, female reproductive success and infant sex ratio in semi-freeranging Barbary macaques (*Macaca sylvana*). *Folia primatol.* *42*: 2–16

Pearson, R. (1978). *Climate and evolution.* Academic Press, London

Pélaez, F. (1982). Greeting movements among adult males in a colony of baboons: *Papio hamadryas, P. cynocephalus* and their hybrids. *Primates 23*: 233–44

Perrill, S.A., Gerhardt, H.C. and Daniel, R. (1978). Sexual parasitism in the green tree frog (*Hyla cinerea*). *Science 200*: 1179–80

Petter, J.J. and Hladik, C.M. (1970). Observations sur le domaine vital et la densité de population de *Loris tardigradus* dans les forêts de Ceylon. *Mammalia 34*: 394–409

Platt, J.R. (1964). Strong influence. *Science 146*: 347–53

Pollard, J.H. (1977). *Handbook of numerical and statistical techniques*, Cambridge University Press, Cambridge

Pollock, J.I. (1975). Field observations on *Indri indri*: a preliminary report. In: I. Tattersal and R.W. Sussman (eds.) *Lemur biology*, pp. 287–311. Plenum Press, New York

Pollock, J.I. (1977). The ecology and sociology of feeding in *Indri indri*. In: T.H. Clutton-Brock (ed.) *Primate ecology*, pp. 37–68. Academic Press, London

Popp, J.L. (1978). *Male baboons and evolutionary principles*. PhD thesis, Harvard University

Popp, J.L. and DeVore, I. (1979). Aggressive competition and social dominance theory: synopsis. In: D.A. Hamburg and E.K. McCown (eds.) *The great apes*, pp. 57–95. Benjamin Cummings, Menlo Park (CA)

Portman, O.W. (1970). Nutritional requirements (NRC) of nonhuman primates. In: R.S. Harris (ed.) *Feeding and nutrition of nonhuman primates*, pp. 87–115. Academic Press, New York

Post, D. (1978). *Feeding and Ranging Behaviour of the Yellow Baboon* (Papio cynocephalus). PhD thesis, Yale University

Post, D., Hausfater, G. and McCuskey, S.A. (1980). Feeding behaviour of yellow baboons (*Papio cynocephalus*): relationship to age, gender and dominance rank. *Folia primatol. 34*: 170–95

Prins, H.H.T., Ydenberg, R.C. and Drent, R.H. (1980). The interaction of Brent geese *Branta bernicla* and sea plantain *Plantago maritima* during spring staging: field observations and experiments. *Acta Bot. Neerl. 29*: 585–96

Pulliam, H.R. (1973). On the advantages of flocking. *J. theoret. Biol. 38*: 419–22

Pusey, A.E. (1980). Inbreeding avoidance in chimpanzees. *Anim. Behav. 28*: 543–52

Pyke, G.H., Pulliam, H.R. and Charnov, E.L. (1977). Optimal foraging: a selective review of theory and tests. *Ann. Rev. Biol. 52*: 137–54

Quris, R. (1975). Ecologie et organisation sociale de *Cercocebus galeritus agilis* dans le nord-est du Gabon. *Terre Vie 29*: 337–98

Raemakers, J. (1979). Ecology of sympatric gibbons. *Folia primatol. 31*: 227–45

Raemakers, J. (1980). Causes of variation between months in the distance travelled daily by gibbons. *Folia primatol. 34*: 46–60

Ralls, K. (1977). Sexual dimorphism in mammals: avian models and unwarranted unanswered questions. *Am. Nat. 111*: 917–38

Ransom, T.W. (1981). *Beach troop of the Gombe*. Bucknell University Press, Lewisberg (PA)

Ransom, T.W. and Ransom, B.S. (1971). Adult male–infant relations among baboons

(*Papio anubis*). *Folia primatol. 16*: 179–95

Ransom, T.W. and Rowell, T.E. (1972). Early social development of feral baboons. In: F.E. Poirier (ed.) *Primate socialisation*, pp. 105–44. Random House, New York

Rasmussen, D.R. (1978). *Environmental and behavioural correlates of changes in range use in a troop of yellow* (Papio cynocephalus) *and a troop of olive* (P. anubis) *baboons*. PhD thesis, University of California, Riverside

Rasmussen, D.R. (1979). Correlates of patterns of range use of a troop of yellow baboons (*Papio cynocephalus*). I. Sleeping sites, impregnable females, births and male emigrations and immigrations. *Anim. Behav. 27*: 1098–112

Rasmussen, D.R. (1981a). Communities of baboon troops (*Papio cynocephalus*) in Mikumi National Park, Tanzania. *Folia primatol. 36*: 232–42

Rasmussen, D.R. (1981b). Pairbond strength and stability and reproductive success. *Psychol. Rev. 88*: 274–90

Rasmussen, D.R. (1983). Correlates of patterns of range use of a troop of yellow baboons (*Papio cynocephalus*). II. Spatial structure, cover density, food gathering and individual behaviour patterns. *Anim. Behav. 31*: 834–56

Rasmussen, K.R.L. (1983). Age-related variation in the interactions of adult females with adult males in yellow baboons. In: R.A. Hinde (ed.) *Primate social relationships*, pp. 47–53. Blackwell Scientific Publications, Oxford

Rasmussen, K.R.L. (1985). Changes in the activity budgets of yellow baboons (*Papio cynocephalus*) during sexual consortships. *Behav. Ecol. Sociobiol. 17*: 161–70

Rasmussen, K.R.L. (1986). Spatial patterns and peripheralisation of yellow baboons (*Papio cynocephalus*) during sexual consortships. *Behaviour 97*: 161–80

Reiss, M.J. (1984). Kin selection, social grooming and the removal of ectoparasites: a theoretical investigation. *Primates 25*: 185–91

Reynolds, V. and Reynolds, F. (1965). Chimpanzees of the Budongo Forest. In: I. DeVore (ed.) *Primate behaviour*, pp. 368–424. Holt, Rinehart & Winston, New York

Rhine, R.J., Forthman, D.L., Stillwell-Barnes, R., Westlund, B.J. and Westlund, H.D. (1979). Movement patterns of yellow baboons (*Papio cynocephalus*): the location of subadult males. *Folia primatol. 32*: 241–51

Richard, A. (1970). A comparative study of the activity patterns and behaviour of *Alouatta villosa* and *Ateles geoffroyi*. *Folia primatol. 12*: 241–63

Richard, A. (1978). *Behavioural variation: case study of a Malagasy lemur*. Bucknell University Press, Lewisburg (PA)

Richard, A. (1985). *Primates in nature*. W.H. Freeman, San Francisco

Richards, S.M. (1974). The concept of dominance and methods of assessment. *Anim. Behav. 22*: 914–30

Ridley, M. (1986). The number of males in a primate troop. *Anim. Behav. 34*: 1848–58

Riopelle, A.J., Hale, P.A. and Watts, E.S. (1976). Protein deprivation in primates. VII. Determinants of size and skeletal maturity at birth in rhesus monkeys. *Human Biol. 48*: 203–22

Ripley, S. (1970). Leaves and leaf-eaters: the social organisation of foraging gray langurs *Presbytis entellus thersites*. In: J.H. Napier and P. Napier (eds.) *Old World Monkeys*, pp. 481–509. Academic Press, London

Ripley, S. (1980). Infanticide in langurs and man, adaptive advantage or social pathology. In: M.N. Cohen, R.S. Malpas and H.G. Klein (eds.) *Biosocial mechanisms of population regulation*, pp. 349–90. Yale University Press, New Haven (CT)

Riss, D. and Goodall, J. (1977). The recent rise to the alpha-rank in a population of free-living chimpanzees. *Folio primatol. 27*: 134–51

Roberts, P. (1971). Social interactions of *Galago crassicaudatus*. *Folia primatol. 14*: 171–81

Rodman, P.S. (1973). Population composition and adaptive organisation among orang-utans of the Kutai Reserve. In: R.P. Michael and J.H. Crook (eds.) *Comparative ecology and behaviour of primates*, pp. 171–209. Academic Press, London

Rodman, P.S. (1977). Feeding behaviour of orang-utans of the Kutai Reserve, East Kalimantan. In: T.H. Clutton-Brock (ed.) *Primate Ecology*, pp. 383–413. Academic Press, London

Rosenblum, L.A. and Youngstein, K.P. (1974). Developmental changes in compensatory dyadic response in mother and infant monkeys. In: M. Lewis and L.A. Rosenblum (eds.) *The effect of the infant on its caregiver*, pp. 141–61. Wiley, New York

Rosenzweig, M.L. (1968). Net primary productivity of terrestrial communities: prediction from climatological data. *Am. Nat. 102*: 67–74

Rowell, T.E. (1966a). Forest-living baboons in Uganda. *J. Zool., Lond. 147*: 344–54

Rowell, T.E. (1966b). Hierarchy in the organisation of a captive baboon group. *Anim. Behav. 14*: 430–43

Rowell, T.E. (1967). A quantitative comparison of the behaviour of a wild and a caged baboon group. *Anim. Behav. 15*: 499–509

Rowell, T.E. (1969). Longterm changes in a population of Ugandan baboons. *Folia primatol. 11*: 241–54

Rowell, T.E. (1970). Baboon menstrual cycles affected by social environment. *J. Reprod. Fertil. 21*: 133–41

Rowell, T.E. (1972a). *Social behaviour of monkeys*. Penguin, Harmondsworth

Rowell, T.E. (1972b). Female reproduction cycles and social behaviour in primates. *Adv. Study Behav. 4*: 69–105

Rowell, T.E. (1974). The concept of social dominance. *Behav. Biol. 11*: 131–54

Rowell, T.E. (1978). How female reproduction cycles affect interaction patterns in groups of patas monkeys. In: D.J. Chivers and J. Herbert (eds.) *Recent advances in primatology, Vol. 1, Behaviour*, pp. 489–90. Academic Press, London

Rowell, T.E. and Chism, J. (1986a). Sexual dimorphism and mating systems: jumping to conclusions. *Human Evol. 1*: 215–19

Rowell, T.E. and Chism, J. (1986b). The ontogeny of sex differences in the behaviour of patas monkeys. *Int. J. Primatol. 7*: 83–109

Rowell, T.E. and Hartwell, K. (1978). The interaction of behaviour and reproductive cycles in patas monkeys. *Behav. Biol. 24*: 141–67

Rowell, T.E. and Richards, S.M. (1979). Reproductive strategies of some African monkeys. *J. Mammal. 60*: 58–69

Rowley, I., Russell, E. and Brocker, M. (1986). Inbreeding: benefits may outweigh costs. *Anim. Behav. 34*: 939–41

Rubenstein, D.I. (1980). On the evolution of alternative mating strategies. In: J.E.R. Staddon (ed.) *Limits to action*, pp. 65–100. Academic Press, London

Rubenstein, D.I. and Wrangham, R.W. (1980). Why is altruism towards kin so rare? *Z. Tierpsychol. 54*: 381–7

Rudran, R. (1973). The reproductive cycle of two subspecies of purple-faced langurs (*Presbytis senex*) with relation to environmental factors. *Folia primatol. 19*: 41–60

Rudran, R. (1979). The demography and social mobility of a red howler (*Alouatta seniculus*) population in Venezuela. In: J.F. Eisenberg (ed.) *Vertebrate ecology in the northern neotropics*, pp. 107–26. Smithsonian Institution Press, Washington (DC)

de Ruiter, J.R. (1986). The influence of group size on predator scanning and foraging behaviour of wedgecapped capuchin monkeys (*Cebus olivaceus*). *Behaviour 98*: 240–58

Ruiz de Elvira, M.-C. and Herndon, J.G. (1986). Disruption of sexual behaviour by high ranking rhesus monkeys (*Macaca mulatta*). *Behaviour 96*: 227–40

Rutberg, A.T. (1983). The evolution of monogamy in primates. *J. theoret. Biol.* *104*: 93–112

Saayman, G.S. (1970). The menstrual cycle and sexual behaviour in a troop of free ranging chacma baboons (*Papio ursinus*). *Folia primatol.* *12*: 81–110

Sabater Pi, J. (1979). Feeding behaviour and diet of chimpanzees (*Pan troglodytes troglodytes*) in the Okorobiko Mountains of Rio Muni (West Africa). *Z. Tierpsychol.* *50*: 265–81

Sackett, G.P. (1968). The persistence of abnormal behaviour in monkeys following isolation rearing. In: R. Rorter (ed.) *The role of learning in psychotherapy*, pp. 112–23. Churchill, London

Sackett, G.P. (1982). Can single processes explain effects of postnatal influences on primate development? In: R.N. Emde and R.J. Harman (eds.) *The development of attachment and affiliative systems*, Plenum Press, New York

Sackett, G.P., Holm, R.A., Davis, A.E. and Farhenbruck, E.E. (1975). Prematurity and low birth weight in pigtail macaques: incidence, prediction and effects on infant development. In: S. Kondo, M. Kawai, A. Ehara and S. Kawamura (eds.) *Proceedings of the Fifth Congress of the International Primatological Society*. Japan Science Press, Tokyo

Sade, D.S. (1967). Determinants of dominance in a group of free-ranging rhesus monkeys. In: S.A Altmann (ed.) *Social communication among primates*, pp. 99–114. University of Chicago Press, Chicago

Sade, D.S. (1972). Sociometrics of *Macaca mulatta*. I. Linkages and cliques in grooming matrices. *Folia primatol.* *18*: 196–223

Sade, D.S., Cushing, K., Cushing, P., Dunaid, J., Figueroa, A., Kaplan, J., Lauer, C., Rhodes, D. and Schneider, J. (1976). Population dynamics in relation to social structure on Cayo Santiago. *Ybk phys. Anthropol.* *20*: 253–62

Sadleier, R.M.S. (1969). *The ecology of reproduction in wild and domestic mammals.* Methuen, London

Sailer, L.D., Gaulin, S.J.C., Boster, J.S. and Kurland, J.A. (1985). Measuring the relationship between dietary quality and body size in primates. *Primates 26*: 14–27

Samuels, A. and Henrickson, R.V. (1983). Outbreak of severe aggression in captive *Macaca mulatta*. *Am. J. Primatol.* *5*: 277–81

Samuels, A., Silk, J.B. and Altmann, J. (1987). Continuity and change in dominance relations among female baboons. *Anim. Behav.* (in press)

Saunders, C. and Hausfater, G. (1978). Sexual selection in baboons (*Papio cynocephalus*): a computer simulation of differential reproduction with respect to dominance rank in males. In: D.J. Chivers and J. Herbert (eds.) *Recent advances in primatology, Vol. 1, Behaviour*, pp. 567–71. Academic Press, London

van Schaik, C.P. (1983). Why are diurnal primates living in groups? *Behaviour 87*: 120–44

van Schaik, C.P. and van Hooff, J. (1983). On the ultimate causes of primate social systems. *Behaviour 85*: 91–117

van Schaik, C.P. and van Noordwijk, M.A. (1985a). Interannual variability in fruit abundance and the reproductive seasonality of Sumatran long-tailed macaques. *J. Zool., Lond. 206*, 533–49

van Schaik, C.P. and van Noordwijk, M.A. (1985b). The evolutionary effect of the absence of felids on the social organisation of the Simeulue monkey (*Macaca fascicularis fusca*, Miller 1903). *Int. J. Primatol. 6*: 180–200

van Schaik, C.P., van Noordwijk, M.A., Wasone, M.A. and Sitriono, E. (1983a). Party size and early detection in Sumatran forest primates. *Primates 24*: 211–21

van Schaik, C.P., de Boer, R.J. and den Tonkelaar, I. (1983b). The effects of group size on time budgets and social behaviour in wild long-tailed macaques (*Macaca fascicularis*). *Behav. Ecol. Sociobiol. 13*: 173–81

Schaller, G.B. (1967). *The deer and the tiger.* University of Chicago Press, Chicago

Schaller, G.B. (1972). *The Serengeti lion.* University of Chicago Press, Chicago

Schenkel, R. and Schenkel-Hulliger, L. (1967). On the sociology of free-ranging colobus (*Colobus guereza caudatus* Thomas 1885). In: D. Stark, R. Schneider and H.-J. Kuhn (eds.) *Progress in primatology*, pp. 185–94. Karger, Basel

Schilling, A., Perret, M. and Predine, J. (1984). Sexual inhibition in a prosimian primate: a pheromone-like effect. *J. Endocrinol. 102*: 143–51

Schmidt-Nielsen, K. (1984). *Scaling: why is animal size so important?* Cambridge University Press, Cambridge

Schoener, T.W. (1971). Theory of feeding strategies. *Ann. Rev. Ecol. Syst. 2*: 369–404

Schubert, G. (1982). Infanticide by usurper hanuman langur males: a sociobiological myth. *Soc. Sci. Inform. 21*: 199–244

Schulman, S.R. and Chapais, B. (1980). Reproductive value and rank relations among macaque sisters. *Am. Nat. 115*: 580–93

Schürmann, C.L. and van Hooff, J. (1986). Reproductive strategies of the orang-utan: new data and a reconsideration of existing sociosexual models. *Int. J. Primatol. 7*: 265–87

Seay, B.M., Alexander, B.K. and Harlow, H.F. (1964). Maternal behaviour of socially deprived rhesus monkeys. *J. abnorm. soc. Psychol. 69*: 345–54

Sekulic, R. (1983). Male relationships and infant death in red howler monkeys (*Alouatta seniculus*). *Z. Tierpsychol 61*: 185–202

Seth, P.K. and Seth, S. (1986). Ecology and behaviour of rhesus monkeys in India. In: J. Else and P. Lee (eds.) *Primate ecology and conservation*, pp. 89–104. Cambridge University Press, Cambridge

Seyfarth, R.M. (1976). Social relationships among adult female baboons. *Anim. Behav. 24*: 917–38

Seyfarth, R.M. (1977). A model of social grooming among adult female monkeys. *J. theoret. Biol. 65*: 671–98

Seyfarth, R.M. (1978a). Social relationships among adult male and female baboons. I. Behaviour during sexual consortships. *Behaviour 64*: 204–26

Seyfarth, R.M. (1978b). Social relationships among adult male and female baboons. II. Behaviour throughout the female reproductive cycle. *Behaviour 64*: 227–47

Seyfarth, R.M. (1980). The distribution of grooming and related behaviours among adult female vervet monkeys. *Anim. Behav. 28*: 798–813

Seyfarth, R.M. (1983). Grooming and social competition in primates. In: R.A. Hinde (ed.) *Primate social relationships*, pp. 182–90. Blackwell Scientific Publications, Oxford

Seyfarth, R.M. and Cheney, D. (1984). Grooming, alliances and reciprocal altruism in vervet monkeys. *Nature, Lond. 308*: 541–3

Seyfarth, R.M., Cheney, D. and Marler, P. (1980). Vervet monkey alarm calls. *Anim. Behav. 28*: 1070–94

Sharman, M. (1981). *Feeding, ranging and social organisation of the Guinea baboon.* PhD thesis, University of St Andrews

Sharman, M. and Dunbar, R.I.M. (1982). Observer bias in selection of study group in baboon field studies. *Primates 23*: 567–73

Shotake, T. (1980). Genetic variability within and between herds of gelada baboons in central Ethiopian highlands. *Anthropol. Contemp. 3*: 270

Shotake, T. (1981). Population genetical study of natural hybridisation between *Papio anubis* and *P. hamadryas*. *Primates 22*: 285–308

Sibly, R.M. (1981). Strategies of digestion and defecation. In: C.R. Townsend and P. Calow (eds.) *Physiological ecology*, pp. 109–39. Blackwell Scientific Publications, Oxford

Siebel, M. and Taylor, M.F. (1982). Emotional aspects of infertility. *Fertil. Steril.* 37: 137

Siegfried, W.R. and Underhill, L.G. (1975). Flocking as an anti-predator strategy in doves. *Anim. Behav.* 23: 504–8

Sigg, H. (1980). Differentiation of female positions in hamadryas one-male-units. *Z. Tierpsychol.* 53: 265–302

Sigg, H. and Stolba, A. (1981). Home range and daily march in a hamadryas baboon troop. *Folia primatol.* 36: 40–75

Sigg, H., Stolba, A., Abegglen, J.-J. and Dasser, V. (1982). Life history of hamadryas baboons: physical development, infant mortality, reproductive parameters and family relationships. *Primates* 23: 473–87

Silk, J.B. (1982). Altruism among female *Macaca radiata*: explanations and analysis of patterns of grooming and coalition formation. *Behaviour* 79: 162–88

Silk, J.B. (1983). Local resource competition and facultative adjustment of sex ratio in relation to competitive abilities. *Am. Nat.* 121: 56–64

Silk, J.B. and Boyd, R. (1983). Cooperation, competition and mate choice in matrilineal macaque groups. In: S.K. Wasser (ed.) *Social behaviour of female vertebrates*, pp. 315–47. Academic Press, New York

Silk, J.B., Clark-Wheatly, C.B., Rodman, P.S. and Samuels, A. (1981a). Differential reproductive success and facultative adjustment of sex ratios among captive female bonnet macaques. *Anim. Behav.* 29: 1106–20

Silk, J.B., Samuels, A. and Rodman, P.S. (1981b). The influence of kinship, rank and sex on affiliation and aggression between adult female and immature bonnet macaques (*Macaca radiata*). *Behaviour* 78: 111–37

Silk, J.B., Samuels, A. and Rodman, P.S. (1981c). Hierarchical organisation of female *Macaca radiata*. *Primates* 22: 84–5

Simonds, P.E. (1965). The bonnet macaque of south India. In: I. DeVore (ed.) *Primate behaviour*, pp. 175–96. Holt, Rinehart & Winston, New York

Simpson, M.J. (1973). The social grooming of male chimpanzees. In: R.P. Michael and J.H. Crook (eds.) *Comparative ecology and behaviour of primates*, pp. 411–506. Academic Press, London

Simpson, M.J. and Howe, S. (1986). Group and matriline differences in the behaviour of rhesus monkey infants. *Anim. Behav.* 34: 444–59

Simpson, M.J. and Simpson, A. (1982). Birth sex ratios and social rank in rhesus monkey mothers. *Nature, Lond.* 300: 440–1

Simpson, M.J. and Simpson, A. (1985). Short-term consequences of different breeding histories for captive rhesus macaque mothers and young. *Behav. Ecol. Sociobiol.* 18: 83–9

Simpson, M.J., Simpson, A., Hooley, J. and Zunz, M. (1981). Infant-related influences on birth intervals in rhesus monkeys. *Nature, Lond.* 290: 49–51

Simpson, M.J., Simpson, A. and Howe, S. (1986). Changes in the rhesus mother–infant relationship through the first four months of life. *Anim. Behav.* 34: 1528–39

Sinclair, A.R.E. (1977). *The African buffalo*. Chicago University Press, Chicago

Skinner, J.D. and Skinner, C.P. (1974). Predation on the cattle egret (*Bulbulcus ibis*) and masked weaver (*Ploceus velatus*) by the vervet monkey (*Cercopithecus aethiops*). *S. Afr. J. Sci.* 70: 157–8

Small, M.F. (1981). Body fat, rank and nutritional status in a captive group of rhesus macaques. *Int. J. Primatol.* 2: 91–5

Small, M.F. and Smith, D.G. (1984). Sex differences in maternal investment by *Macaca mulatta*. *Behav. Ecol. Sociobiol.* 14: 313–14

Small, M.F. and Smith, D.G. (1986). The influence of birth timing upon infant growth and survival in captive rhesus macaques (*Macaca mulatta*). *Int. J. Primatol.* 7: 289–304

Smith, R.H. and Lecount, A. (1979). Some factors affecting survival of desert mule deer fawns. *J. Wildl. Mgmt 43*: 657–65

Smith, R.H., Butler, T.N. and Pace, N. (1975). Weight growth of colony raised chimpanzees. *Folia primatol. 24*: 29–59

Smith, C.C. (1977). Feeding behaviour and social organisation in howler monkeys. In: T.H. Clutton-Brock (ed.) *Primate ecology*, pp. 97–126. Academic Press, New York

Smith, D.G. (1980). Paternity exclusion in six captive groups of rhesus monkeys (*Macaca mulatta*). *Am. J. phys. Anthropol. 53*: 243–9

Smith, D.G. (1981). The association between rank and reproductive success of male rhesus monkeys. *Am. J. Primatol. 1*: 83–90

Smith, D.G. (1982). A comparison of the demographic structure and growth of free-ranging and captive groups of rhesus monkeys (*Macaca mulatta*). *Primates 23*: 24–30

Smith, S.M. (1978). The 'underworld' in a territorial sparrow: adaptive strategy for floaters. *Am. Nat. 112*: 571–82

Smuts, B.B. (1985). *Sex and friendship in baboons*. Aldine, New York

Snowdon, C.T. and Cleveland, J. (1980). Individual recognition of contact calls by pygmy marmosets. *Anim. Behav. 28*: 717–27

van Soest, P.J. (1977). Plant fibre and its role in herbivore nutrition. *Cornell Vet. 67*: 307–26

van Soest, P.J. (1980). The limitations of ruminants. In: *Proceedings of the Cornell Nutrition Conference for Feed Manufacturers*, pp. 78–90

Soini, P. (1982). Ecology and population dynamics of the pygmy marmoset, *Cebuella pygmaea*. *Folia primatol. 39*: 1–21

Sokal, R.R. and Rolf, F.J. (1969). *Biometry*. W.H. Freeman, San Francisco

Sourd, C. and Gautier-Hion, A. (1986). Fruit selection by a forest guenon. *J. anim. Ecol. 55*: 235–44

Southwick, C.H. (1967). An experimental study of intra-group agonistic behaviour in rhesus monkeys. *Behaviour 28*: 182–209

Southwick, C.H. and Siddiqi, M.F. (1977). Population dynamics of rhesus monkeys in northern India. In: Rainier of Monaco and G.H. Bourne (eds.) *Primate Conservation*, pp. 339–62. Academic Press, New York

Southwick, C.H., Beg, M.H. and Siddiqi, M.R. (1965). Rhesus monkeys in north India. In: I. DeVore (ed.) *Primate behaviour*, pp. 111–59. Holt, Rinehart and Winston, New York

Sparks, J. (1967). Allogrooming in primates: a review. In: D. Morris (ed.) *Primate Ethology*, pp. 148–75. Weidenfeld & Nicolson, London

Spencer-Booth, Y. and Hinde, R.A. (1971a). Effects of brief separation from mothers during infancy on behaviour of rhesus monkeys 6–24 months later. *J. Child Psychol. Psychiat. 12*: 157–72

Spencer-Booth, Y. and Hinde, R.A. (1971b). Effects of 6 days separation from mother on 18- to 32-week-old rhesus monkeys. *Anim. Behav. 19*: 174–91

Spies, H. and Chappel, S.C. (1984). Mammals: nonhuman primates. In: G.E. Lemming (ed.) *Marshall's physiology of reproduction*, 4th edition, pp. 659–712. Churchill Livingstone, London

Stacey, P.B. (1986). Group size and foraging efficiency in yellow baboons. *Behav. Ecol. Sociobiol. 18*: 175–87

Stammbach, E. (1979). On social differentiation in groups of captive female hamadryas baboons. *Behaviour 67*: 322–38

Stammbach, E. and Kummer, H. (1982). Individual contributions to a dyadic interaction: an analysis of baboon grooming. *Anim. Behav. 30*: 964–71

Stein, D.M. and Stacey, P.B. (1981). A comparison of infant–adult male relation in a one-male group with those in a multi-male group for yellow baboons (*Papio cyno-*

cephalus). *Folia primatol. 36*: 264–76

Stern, B.R. and Smith, D.G. (1984). Sexual behaviour and paternity in three captive groups of rhesus monkeys (*Macaca mulatta*). *Anim. Behav. 32*: 23–32

Stevenson, M.F. (1978). The behaviour and ecology of the common marmoset. In: H. Rothe (ed.) *The biology and conservation of the Callitrichidae* Vol. 2. Smithsonian Institution Press, Washington (DC)

Stevenson, M.F. (1986). Captive breeding of callitrichids: a comparison of reproduction and propagation in different species. In: J. Else and P. Lee (eds.) *Primate ecology and conservation*, pp. 301–14. Cambridge University Press, Cambridge

Stevenson-Hinde, J. (1983). Consistency over time. In: R.A. Hinde (ed.) *Primate social relationships*, pp. 30–3. Blackwell Scientific Publications, Oxford

Stevenson-Hinde, J. and Zunz, M. (1978). Subjective assessment of individual rhesus monkeys. *Primates 19*: 473–82

Stevenson-Hinde, J., Stillwell-Barnes, R. and Zunz, M. (1980). Subjective assessment of rhesus monkeys over four successive years. *Primates 21*: 66–82

Stolba, A. (1979). *Entscheidungsfindung in Verbänden von* Papio hamadryas. PhD thesis, University of Zurich

Stoltz, L. (1977). *The population dynamics of baboons in the Transvaal.* PhD thesis, University of Pretoria

Stoltz, L. and Saayman, G.S. (1970). Ecology and behaviour of baboons in the Transvaal. *Ann. Transvaal Mus. 26*: 99–143

Struhsaker, T.T. (1967a). Ecology of vervet monkeys (*Cercopithecus aethiops*) in the Masai-Amboseli game reserve, Kenya. *Ecology 48*: 891–904

Struhsaker, T.T. (1967b). Auditory communication among vervet monkeys (*Cercopithecus aethiops*). In S.A. Altmann (ed.) *Social communication among primates*, pp. 281–324. University of Chicago Press, Chicago

Struhsaker, T.T. (1967c). Social structure among vervet monkeys (*Cercopithecus aethiops*). *Behaviour 29*: 83–121

Struhsaker, T.T. (1968). Correlates of ecology and social organisation among African Cercopithecines. *Folia primatol. 11*: 80–118

Struhsaker, T.T. (1971). Social behaviour of mother and infant vervet monkeys (*Cercopithecus aethiops*). *Anim. Behav. 18*: 233–50

Struhsaker, T.T. (1973). A recensus of vervet monkeys in the Masai-Amboseli Game Reserve, Kenya. *Ecology 54*: 930–2

Struhsaker, T.T. (1974). Correlates of ranging behaviour in a group of red colobus monkeys (*Colobus badius tephrosceles*). *Am. Zool. 14*: 177–84

Struhsaker, T.T. (1975). *The red colobus monkey.* University of Chicago Press, Chicago

Struhsaker, T.T. (1977). Infanticide and social organisation in the redtail monkey (*Cercopithecus ascanius schmidti*) in the Kibale forest, Uganda. *Z. Tierpsychol. 45*: 75–84

Struhsaker, T.T. and Gartlan, J.S. (1970). Observations on the behaviour and ecology of the patas monkey (*Erythrocebus patas*) in the Waza Reserve, Cameroon. *J. Zool., Lond. 161*: 49–63

Struhsaker, T.T. and Leland, L. (1985). Infanticide in a patrilineal society of red colobus monkeys. *Z. Tierpsychol. 69*: 89–132

Strum, S.C. (1981). Processes and products of change: baboon predatory behaviour at Gilgil, Kenya. In: R.S. Harding and G. Teleki (eds.) *Omnivorous primates: gathering and hunting in human evolution*, pp. 255–302. Columbia University Press, New York

Strum, S.C. (1982). Agonistic dominance in male baboons: an alternative view. *Int. J. Primatol. 3*: 175–202

Strum, S.C. (1983). Use of females by male olive baboons (*Papio anubis*). *Am. J. Primatol. 5*: 93–109

Strum, S.C. (1984). Why females use infants. In: D.M. Taub (ed.) *Primate paternalism*, pp. 146–85. Van Nostrand Reinhold, New York

Strum, S.C. and Western, D. (1982). Variations in fecundity with age and environment in olive baboons (*Papio anubis*). *Am. J. Primatol. 3*: 61–76

Suarez, B. and Ackerman, D.R. (1971). Social dominance and reproductive behaviour in male rhesus monkeys. *Amer. J. phys. Anthropol. 35*: 219–22

Sugiyama, Y. (1965). On the social change of hanuman langurs (*Presbytis entellus*) in their natural condition. *Primates 6*: 381–418

Sugiyama, Y. (1967). Social organisation of hanuman langurs. In: S.A. Altmann (ed.) *Social communication among primates*, pp. 221–36. University of Chicago Press, Chicago

Sugiyama, Y. (1968). Social organisation of chimpanzees in the Budongo forest, Uganda. *Primates 9*: 225–58

Sugiyama, Y. (1973). The social structure of wild chimpanzees: a review of field studies. In: R.P. Michael and J.H. Crook (eds.) *Comparative ecology and behaviour of primates*, pp. 375–410. Academic Press, London

Sugiyama, Y. (1976). Life history of male Japanese macaques. *Adv. Study Behav. 7*: 255–84

Sugiyama, Y. and Koman, J. (1979). Social structure and dynamics of wild chimpanzees at Bossou, Guinea. *Primates 20*: 323–39

Sugiyama, Y. and Ohsawa, H. (1982a). Population dynamics of Japanese monkeys with special reference to the effect of artificial feeding. *Folia primatol. 39*: 238–63

Sugiyama, Y. and Ohsawa, H. (1982b). Populatin dynamics of Japanese monkeys at Ryozenyama III. Female desertion of the troop. *Primates 23*: 31–44

Sullivan, J.T. (1966). Studies of the hemicelluloses of forage plants. *J. anim. Science 25*: 83–6

Suomi, S.J. (1981). Genetic, maternal and environmental influences on social development in rhesus monkeys. In: A.B. Chiarelli and R.S. Corruccini (eds.) *Primate behaviour and sociobiology*, pp. 81–7. Springer, Berlin

Sussman, R.W. (1977). Feeding behaviour of *Lemur catta* and *Lemur fulvus*. In: T.H. Clutton-Brock (ed.) *Primate ecology*, pp. 1–36. Academic Press, London

Suzuki, A. (1969). An ecological study of chimpanzees in a savanna woodland. *Primates 10*: 103–48

Suzuki, A. (1979). The variation and adaptation of social groups of chimpanzees and black and white colobus monkeys. In: I. Bernstein and E.O. Smith (eds.) *Primate ecology and human origins*, pp. 153–73. Garland, New York

Swartz, K.B. and Rosenblum, L.A. (1981). The social context of parental behaviour: a perspective on primate socialisation. In: D.J. Gubernick and P.H. Klopfer (eds.) *Parental care in mammals*, pp. 417–54. Plenum Press, New York

Szalay, F.S. and Delson, E. (1979). *Evolutionary history of the primates*. Academic Press, New York

Takahata, Y. (1980). The reproductive biology of a free-ranging troop of Japanese monkeys. *Primates 21*: 303–29

Takasaki, H. (1981). Troop size, habitat quality and home range area in Japanese macaques. *Behav. Ecol. Sociobiol. 9*: 277–81

Tanner, N.M. (1981). *On becoming human*. Cambridge University Press, Cambridge

Taub, D.M. (1980). Female choice and mating strategies among wild Barbary macaques (*Macaca sylvanus* L.), In: D.G. Lindburg (ed.) *The macaques*, pp. 287–344. Van Nostrand Reinhold, New York

Taub, D.M. (1984). Male caretaking behaviour among wild Barbary macaques (*Macaca sylvanus*). In: D.M. Taub (ed.) *Primate paternalism*, pp. 20–55. Van Nostrand Reinhold, New York

Taylor, C.R., Heglund, N.C., McMahon, T.A. and Looney, T.R. (1980). Energetic cost of generating muscular force during running: a comparison of large and small animals. *J. exp. Biol. 86*: 9–18

Taylor, C.R., Heglund, N.C. and Maloiy, G.M.O. (1982). Energetics and mechanics of terrestial locomotion. I. Metabolic energy consumption as a function of speed and body size in birds and mammals. *J. exp. Biol. 97*: 11–21

Teas, J., Richie, T., Taylor, H. and Southwick, C. (1980). Population patterns and behavioural ecology of rhesus monkeys (*Macaca mulatta*) in Nepal. In: D.G. Lindburg (ed.) *The macaques*, pp. 247–62. Van Nostrand Reinhold, New York

Teas, J., Richie, T., Taylor, H., Siddiqi, M.F. and Southwick, C.H. (1981). Natural regulation of rhesus monkey populations in Kathmandu, Nepal. *Folia primatol. 35*, 117–23

Teleki, G. (1973). *The predatory behaviour of wild chimpanzees*. Bucknell University Press, Lewisburg (PA)

Teleki, G., Hunt, E.E. and Pfifferling, J.H. (1976). Demographic observations (1963–1973) on the chimpanzees of Gombe National Park, Tanzania. *J. human Evol. 5*: 559–98

Tenaza, R. (1975). Territory and monogamy among Kloss' gibbons (*Hylobates klossi*) in Siberut Island, Indonesia. *Folia primatol. 24*: 60–80

Terborgh, J. (1983). *Five New World primates*. Princeton University Press, Princeton (NJ)

Terborgh, J. (1986). The social systems of New World primates: an adaptionist view. In: J.G. Else and P.C. Lee (eds.) *Primate ecology and conservation*, pp. 199–211. Cambridge University Press, Cambridge

Terborgh, J. and Goldizen, A.W. (1985). On the mating system of the cooperatively breeding saddle-backed tamarin (*Saguinus fuscicollis*). *Behav. Ecol. Sociobiol. 16*: 293–300

Thierry, B. and Anderson, J.R. (1986). Adoption in anthropoid primates. *Int. J. Primatol. 7*: 191–216

Thompson-Handler, N., Malenky, R.K. and Badrian, N. (1984). Sexual behaviour of *Pan paniscus* under natural conditions in the Lomako forest, Equateur, Zaire. In: R.L. Sussman (ed.) *The pygmy chimpanzee*, pp. 347–68. Plenum Press, New York

Thorne, E.T., Dean, R.E. and Hepworth, W.G. (1976). Nutrition during gestation in relation to successful reproduction in elk. *J. Wildl. Mgmt 40*: 330–5

Tilson, R.L. (1977). Social organisation of Simakou monkeys (*Nasalis concolor*) in Siberut island, Indonesia. *J. Mammal. 58*: 202–12

Tilson, R.L. (1981). Family formation strategies of kloss's gibbons. *Folia primatol. 35*: 259–87

Tilson, R.L. and Tenaza, R.R. (1976). Monogamy and duetting in an Old World monkey. *Nature, Lond. 263*: 230–1

Tinbergen, N. (1963). On the aims and methods of ethology. *Z. Tierpsychol. 20*: 410–33

Tokura, H., Hara, F., Okada, M., Mekata, F. and Ohsawa, W. (1975). Thermoregulatory responses at various ambient temperatures in some primates. In: S. Kondo, M. Kawai and A. Ehara (eds.) *Contemporary primatology*, pp. 171–6. Karger, Basel

Torres de Assumpçao, C. and Deag, J.M. (1979). Attention structure in monkeys: a search for a common trend. *Folia primatol. 31*: 285–300

Treisman, M. (1975a). Predation and the evolution of gregariousness. I. Models for concealment and evasion. *Anim. Behav. 23*: 779–800

Treisman, M. (1975b). Predation and the evolution of gregariousness. II. An economic model for predator–prey interaction. *Anim. Behav. 23*: 801–25

Trivers, R.L. (1971). The evolution of reciprocal altruism. *Q. Rev. Biol. 46*: 35–57

Trivers, R.L. (1972). Parental investment and sexual selection. In: B. Campbell (ed.) *Sexual selection and the descent of man*, pp. 136–79. Aldine, Chicago

Trivers, R.L. (1974). Parent–offspring conflict. *Am. Zool. 14*: 249–64

Trivers, R.L. and Willard, D.E. (1973). Natural selection of parental ability to vary the sex ratio. *Science 179*: 90–2

Troisi, A. and Schino, G. (1986). Diurnal and climatic influences on allogrooming behaviour in a captive group of Java monkeys. *Anim. Behav. 34*: 1420–6

Tsingalia, H.M. and Rowell, T.E. (1984). The behaviour of adult male blue monkeys. *Z. Tierpsychol. 64*: 253–68

Tutin, C.E.G. (1979). Mating patterns and reproductive strategies in a community of wild chimpanzees. *Behav. Ecol. Sociobiol. 6*: 29–38

Tutin, C.E.G. and McGinnis, P.R. (1981). Chimpanzee reproduction in the wild. In: C.E. Graham (ed.) *Reproductive biology of great apes*, pp. 239–64. Academic Press, New York

Tutin, C.E.G., McGrew, M.C. and Baldwin, P.J. (1983). Social organisation of savanna-dwelling chimpanzees, *Pan troglodytes verus*, at Mt. Assirik, Senegal. *Primates 24*: 154–73

Vaitl, E. (1978). Nature and implications of the complexly organised social system in nonhuman primates. In: D.J. Chivers and J. Herbert (eds.) *Recent advances in primatology, Vol. 1, Behaviour*, pp. 17–30. Academic Press, London

Vandermark, N.L. and Free, M.J. (1970). Temperature effects. In: A.D. Johnson, W.R. Gomes and N.L. Vandermark (eds.) *The testis*, pp. 233–312. Academic Press, New York

Vehrencamp, S.L. (1983). A model for the evolution of despotic versus egalitarian societies. *Anim. Behav. 31*: 667–82

Vessey, S.H. and Meikle, D.B. (1984). Free-living rhesus monkeys: adult male interactions with infants and juveniles. In: D.M. Taub (ed.) *Primate paternalism*, pp. 113–26. Van Nostrand Reinhold, New York

Vine, I. (1971). Risk of visual detection and pursuit by a predator and the selective advantage of flocking behaviour. *J. theoret. Biol. 30*: 406–22

de Waal, F. (1977). The organisation of agonistic relations within two captive groups of Java-monkeys (*Macaca fascicularis*). *Z. Tierpsychol. 44*: 225–82

de Waal, F. (1982). *Chimpanzee politics*. Allen & Unwin: London

de Waal, F. (1984a). Coping with social tension: sex differences in the effect of food provision to small rhesus monkey groups. *Anim. Behav. 32*: 765–73

de Waal, F. (1984b). Sex differences in the formation of coalitions among chimpanzees. *Ethol. Sociobiol. 5*: 239–55

de Waal, F. (in press). Aggression and aggression control: the basis for primate social organisation. In: V. Standen and R. Foley (eds.) *Comparative socio-ecology of mammals and man*, Blackwells Scientific Publications, Oxford

de Waal, F. and Luttrell, L.M. (1986). The similarity principle underlying social bonding among female rhesus monkeys. *Folia primatol. 46*: 215–34

de Waal, F., and van Roosmalen, A. (1979). Reconciliation and consolation among chimpanzees. *Behav. Ecol. Sociobiol. 5*: 55–66

Wada, K. and Ichiki, Y. (1980). Seasonal home range use by Japanese monkeys in the snowy Shiga Heights. *Primates 21*: 468–83

Wade, T.D. (1978). Status and hierarchy in nonhuman primate societies. In: P. Klopfer and P. Bateson (eds.) *Perspectives in ethology*, Vol. 3, pp. 109–34. Plenum Press, New York

van Wagenen, G. (1936). The coagulating function of the cranial lobe of the prostrate gland in the monkey. *Anat. Rec. 66*: 411

van Wagenen, G. (1945) Optimal mating time for pregnancy in the monkey. *Endocrinology 37*: 307–12

van Wagenen, G. and Catchpole, H.R. (1956). Physical growth of the rhesus monkey

(*Macaca mulatta*). *Am. J. phys. Anthropol. 14*: 245–73

Walters, J. (1980). Interactions and the development of dominance relationships in female baboons. *Folia primatol. 34*: 61–89

Ward, P. (1965). Feeding ecology of the black-faced dioch *Quelea quelea* in Nigeria. *Ibis 107*: 173–214

Ward, P. and Zahavi, A. (1973). The importance of certain assemblages of birds as 'information centres' for food findings. *Ibis 115*: 517–34

Waser, P.M. (1976). *Cercocebus albigena*: site attachment, avoidance and intergroup spacing. *Am. Nat. 110*: 911–35

Waser, P.M. (1977a). Feeding, ranging and group size in the mangabey, *Cercocebus albigena*. In: T.H. Clutton-Brock (ed.) *Primate ecology*, pp. 182–222. Academic Press, London

Waser, P.M. (1977b). Individual recognition, intragroup cohesion and intergroup spacing: evidence from sound playback to forest monkeys. *Behaviour 60*: 28–74

Waser, P.M. (1982). Primate polyspecific associations: do they occur by chance? *Anim. Behav. 30*: 1–8

Waser, P.M., Austad, S. and Keane, B. (1986). When should animals tolerate inbreeding? *Am. Nat. 128*: 529–37

Wasser, S.K. (1983). Reproductive competition and cooperation among female yellow baboons. In: S.K. Wasser (ed.) *Social behaviour of female vertebrates*, pp. 349–90. Academic Press, New York

Wasser, S.K. and Barash, D.P. (1983). Reproductive suppression among female mammals: implications for biomedicine and sexual selection theory. *Q. Rev. Biol. 58*: 513–38

Waterman, P.G. and Choo, G.M. (1981). The effects of digestibility-reducing compounds in leaves on food selection by some Colobinae. *Malays. appl. Biol. 10*: 147–62

Watts, D.P. (1985). Relations between group size and composition and feeding competition in mountain gorilla groups. *Anim. Behav. 32*: 72–85

Wells, K.D. (1978). Territoriality in the green frog (*Rana clamitans*): vocalisations and agonistic behaviour. *Anim. Behav. 26*: 1051–63

Western, D. and van Praet, C. (1973). Cyclical changes in the habitat and climate of an East African ecosystem. *Nature, Lond. 241*: 104–6

Whiten, A., Byrne, R.W. and Henzi, P. (in press). The behavioural ecology of mountain baboons. *Int. J. Primatol.*

Whitten, A.L. (1983). Diet and dominance among female vervet monkeys (*Cercopithecus aethiops*). *Am. J. Primatol. 5*: 139–59

Wilson, A.P. and Boelkins, R.C. (1970). Evidence for seasonal variation in aggressive behaviour by *Macaca mulatta*. *Anim. Behav. 18*: 719–24

Wilson, E.O. (1975). *Sociobiology: the new synthesis*. Belknap Press, Cambridge (Mass.)

Wilson, E.O. and Bossert, W.H. (1971). *A primer of population biology*. Sinauer, Sunderland (Mass.)

Wilson, M.E. (1981). Social dominance and female reproductive behaviour in rhesus monkeys (*Macaca mulatta*). *Anim. Behav. 29*: 472–82

Wilson, M.E., Gordon, T.P. and Bernstein, I.S. (1978). Timing of births and reproductive success in rhesus monkey social groups. *J. med. Primatol. 7*: 202–12

Winkler, P., Loch, H. and Vogel, C. (1984). Life history of hanuman langurs (*Presbytis entellus*): reproductive parameters, infant mortality and troop development. *Folia primatol. 43*: 1–23

Winkless, N. and Browning, I. (1975). *Climate and the affairs of men*. Peter Davies, London

Wirtz, P. (1981). Territorial defence and territory take-over by satellite males in the waterbuck *Kobus ellipsiprymnus* (Bovidae). *Behav. Ecol. Sociobiol. 3*: 397–427

Wirtz, P. and Wawra, M. (1986). Vigilance and group size in *Homo sapiens*. *Ethology 71*: 283–6

Witt, R., Schmidt, G. and Schmidt, J. (1981). Social rank and Darwinian fitness in a multimale group of Barbary macaques (*Macaca sylvana* L.). *Folia primatol. 36*: 201–11

Wittenberger, J.F. (1980). Group size and polygamy in social mammals. *Am. Nat. 115*: 197–222

Wittenberger, J.F. and Tilson, R.L. (1980). The evolution of monogamy: hypotheses and evidence. *Ann. Rev. Ecol. Syst. 11*: 197–232

Wolf, K.E. and Fleagle, J.G. (1977). Adult male replacement in a group of silvered leaf-monkeys (*Presbytis cristata*) at Kuala Selangor, Malaysia. *Primates 18*: 949–55

Wolfe, L.D. (1979). Sexual maturation among members of a transported troop of Japanese macaques (*Macaca fuscata*). *Primates 20*: 411–18

Wolfe, L.D. (1984). Female rank and reproductive success among Arashiyama B Japanese macaques (*Macaca fuscata*). *Int. J. Primatol. 5*: 133–43

Wolfenden, G.E. and Fitzpatrick, J.W. (1984). *The Florida scrub jay: demography of a cooperative-breeding bird*. Princeton University Press, Princeton (NJ)

Wolfheim, J.H. (1977). A quantitative analysis of the organisation of a group of captive talapoin monkeys (*Miopithecus talapoin*). *Folia primatol. 27*: 1–27

Wood, C.A. and Lovett, R.R. (1974). Rainfall, drought and the solar cycle. *Nature, Lond. 251*: 594–6

Woodruff, G. and Premack, D. (1979). Intentional communication in the chimpanzee: the development of deception. *Cognition 7*: 313–62

Wrangham, R.W. (1974). Artificial feeding of chimpanzees and baboons in their natural habitat. *Anim. Behav. 22*: 83–93

Wrangham, R.W. (1977). Behaviour of feeding chimpanzees in the Gombe National Park, Tanzania. In: T.H. Clutton-Brock (ed.) *Primate ecology*, pp. 503–38. Academic Press, London

Wrangham, R.W. (1979). On the evolution of ape social systems. *Soc. Sci. Inform. 18*: 335–68

Wrangham, R.W. (1980). An ecological model of female-bonded primate groups. *Behaviour 75*: 262–300

Wrangham, R.W. (1981) Drinking competition in vervet monkeys. *Anim. Behav. 29*: 904–10

Wrangham, R.W. (1983). Ultimate factors determining social structure. In: R.A. Hinde (ed.) *Primate social relationships*, pp. 255–62. Blackwell Scientific Publications, Oxford

Wrangham, R.W. (1986). Ecology and social relationships in two species of chimpanzees. In: D.I. Rubenstein and R.W. Wrangham (eds.) *Ecological aspects of social evolution*, pp. 352–78. Princeton University Press, Princeton (NJ)

Wrangham, R.W. and Smuts, B.B. (1980). Sex differences in the behavioural ecology of chimpanzees in the Gombe National Park, Tanzania. *J. Reprod. Fertil. Suppl. 28*: 13–31

Wrangham, R.W. and Waterman, P.G. (1981). Feeding behaviour of vervet monkeys on *Acacia tortilis* and *Acacia xanthophloea*: with special reference to reproductive strategies and tannin production. *J. anim. Ecol. 50*: 715–31

Wright, P.C. (1978). Home range, activity pattern and agonistic encounters of a group of night monkeys (*Aotus trivirgatus*) in Peru. *Folia primatol. 29*: 43–55

Wright, P.C. (1986). Ecological correlates of monogamy in *Aotus* and *Callicebus*. In: J. Else and P. Lee (eds.) *Primate ecology and conservation*, pp. 159–68. Cambridge University Press, Cambridge

Yamagiwa, J. (1985). Socio-sexual factors of troop fission in wild Japanese monkeys

(*Macaca fuscata yakui*) on Yakushima Island, Japan. *Primates 26*: 105–20

Yen, S.S.C. and Lein, A. (1984). Mammals: man. In: G.E. Lemming (ed.) *Marshall's physiology of reproduction*, 4th edition, pp. 713–88. Churchill-Livingstone, London

Yoshiba, K. (1968). Local and intertroop variability in ecology and social behaviour of common Indian langurs. In: P.C. Jay (ed.) *Primates*, pp. 217–42. Holt, Reinhart & Winston, New York

Young, G.H. and Hawkins, R.J. (1979). Infant behaviour in mother-reared and harem-reared baboons (*Papio cynocephalus*). *Primates 20*: 87–94

Zacur, H., Chapanis, N., Lake, C., Ziegler, M. and Tyson, J. (1976). Galactorrhea-amenorrhea: psychological interaction with neuroendocrine function. *Am. J. Obstet. Gynecol. 125*: 859–62

Zihlman, A.L. (1984). Body build and tissue composition in *Pan paniscus* and *Pan troglodytes*, with comparisons to other hominoids. In: R.L. Susman (ed.) *The pygmy chimpanzee*, pp. 179–200. Plenum Press, New York

Zimen, E. (1976). On the regulation of pack size in wolves. *Z. Tierpsychol. 40*: 300–41

Index